One of the founding fathers of organic chemistry and also a great teacher, the German scientist Justus von Liebig transformed scientific education, medical practice, and agriculture in Great Britain. William H. Brock's fresh interpretation of Liebig's stormy career shows how he moved chemistry into the sociopolitical marketplace, demonstrating its significance for society in food production, nutrition, and public health. Through his controversial ideas on artificial fertilizers and recycling, his theory of disease, and his stimulating suggestions concerning food and nutrition, he warned the world of the dangers of failing to recycle sewage or to replace soil nutrients. Liebig also played the role of an elder statesman of European science by commenting, via popular lectures and expansions of his readable *Chemical Letters,* on such issues as scientific methodology and materialism.

This is the first English-language biography of Justus von Liebig (1803–73) since 1895.

Justus von Liebig

Justus von Liebig

The Chemical Gatekeeper

WILLIAM H. BROCK
University of Leicester

PUBLISHED BY THE PRESS SYNDICATE OF THE UNIVERSITY OF CAMBRIDGE
The Pitt Building, Trumpington Street, Cambridge, United Kingdom

CAMBRIDGE UNIVERSITY PRESS
The Edinburgh Building, Cambridge CB2 2RU, UK
40 West 20th Street, New York NY 10011–4211, USA
477 Williamstown Road, Port Melbourne, VIC 3207, Australia
Ruiz de Alarcón 13, 28014 Madrid, Spain
Dock House, The Waterfront, Cape Town 8001, South Africa

http://www.cambridge.org

First published 1997
First paperback edition 2002

Typeface Sabon.

A catalogue record for this book is available from the British Library

Library of Congress Cataloguing in Publication data
Brock, W. H. (William Hodson)
Justus von Liebig: the chemical gatekeeper / William H. Brock.
p. cm.
Includes bibliographical references (p. –).
ISBN 0 521 56224 4
1. Liebig, Justus, Freiherr von, 1803–1873. 2. Chemists – Germany –
Biography. I. Title.
QD22.L7B76 1997
540′.92–dc20
[B] 96-25468 CIP

ISBN 0 521 56224 4 hardback
ISBN 0 521 52473 3 paperback

Contents

Preface

Liebig has always been revered in Germany as the father of organic chemistry, as an agricultural chemist, and for the Extract of Meat, which is still sold there in its original form. Elsewhere, Liebig's name is less familiar, though chemistry teachers continue to associate him with a condenser used in organic preparations (which he did not invent) and with the early history of organic chemistry. In fact, Liebig had an international reputation during his lifetime, when he enjoyed a particularly close relationship with Great Britain, whose pattern of scientific education, agriculture, and medical practice he helped to change.

Liebig visited Great Britain six times between 1837 and 1855, rubbing shoulders with English and Scottish landowners, politicians, industrialists, men of science, and members of the Royal Family. Through British friends in high places he was able to obtain a post in the British Army Medical Service in India for his eldest (and most favoured) son Georg. He arranged to have all his books published in English through the London publishing house of Taylor & Walton (in Gower Street), as well as translations or abstracts of all his chemical papers in such journals as *The Lancet*, the *Philosophical Magazine*, the *Chemical Gazette*, and from 1843 the *Quarterly Journal of the Chemical Society*. A number of his letters were even published in *The Times* during the 1860s. Liebig strongly influenced the development of chemistry teaching in Britain, he transformed English farming and the practice of agricultural chemistry, and stimulated medical and engineering ideas concerning public health, sanitation, and the nature of disease. He also dared to criticise Victorian reverence for the seventeenth-century philosopher and statesman Francis Bacon. In 1843, several of his English friends, including Wheatstone, tried to persuade him to apply for the vacant Chair of Chemistry at King's College, London, but the college's affiliation with the Church of England precluded any serious offer being made to him, and the position went instead to his erstwhile pupil William Allen Miller. Two years later, when the Royal College of Chemistry was about to be established, it was again hoped that Liebig might personally come to London to superintend it and to re-create his

Giessen research school in the laboratory in Oxford Street; in the event, the post of director was given on Liebig's personal recommendation to his greatest pupil, Wilhelm August Hofmann, who spent twenty years in London. Together with Liebig's fifty or so other British pupils who held industrial or academic posts throughout the British Isles, Hofmann ensured that Liebig's name and work were kept prominently before the public eye. And after 1865, when sadly (and according to contemporaries, disastrously for Britain), Hofmann left London for Berlin, Liebig's name was kept before the public through his commercial activities with Liebig's Extract of Meat and Liebig's Concentrated Milk Baby Food.

It would not be farfetched to claim, therefore, that Liebig, though born and bred in Darmstadt and fulfilling his academic destiny in Giessen and Munich, was very much an honorary Englishman. To be sure, most of his books and writings, notably the *Chemische Briefe*, appeared in French and Italian and other, more exotic European languages, he had numerous European pupils and he travelled much within Europe. Even so, his connections with Britain were undoubtedly special. Liebig's relationship with Britain and with the British was, as much recent research has suggested, a symbiosis. Liebig looked to Britain for a respectful and sympathetic audience for his ideas on and discoveries in chemical science. Here was an overseas audience that, though frequently provoked into controversy by his work, was often more sympathetic towards it than those in the German nations. Equally, young professionalising groups of British chemists, doctors, engineers, and educators, found "LIEBIG" an effective publicity slogan, agent, and figurehead in their campaigns to reverse the individualistic and utilitarian tendencies in British culture and society. Liebig himself captured the mood for this need in his famous critical observation, which he made to Michael Faraday in 1844: "What struck me most in England was the perception that only those works of a practical tendency awake attention and command respect; while the purely scientific, which possess far greater merit, are almost unknown."

Liebig could be irascible, pigheaded, quarrelsome, opinionated, and sometimes devious, but unlike his younger contemporary Hermann Kolbe, whom he much admired, he never became quirky, obstreperous, and an embarrassment. Quarrels were quickly patched up, and by deliberately sidelining himself from the theoretical problems of organic chemistry after 1840, he avoided the frequently painful controversies over atomic weights and the issue of chemical structure, leaving what Alan Rocke has called "the quiet revolution" to his pupils and younger colleagues such as Hofmann, Frankland, Williamson, Kekulé, and Kolbe. Instead, Liebig moved chemistry into the market place, into a sociopolitical context, by arguing and demonstrating its significance for the benefit of society in food production, nutrition, and public health. Like Linus Pauling in the twentieth century, his stance as a public scientist was often controversial, and controversy was exacerbated by his stubbornness in admitting that the models and systems of agricultural and animal chemistry that he had articulated were lacking in precision. But he is to be admired for the courage with which he ventured into such questions and honoured for the stimulus that he gave to the development of what the twentieth century recognises as biochemistry. Released from teaching by his position at Munich in 1852, he was able to relax from the punishing schedule of work he had kept up since joining Kastner as a student at Bonn and Erlangen. At Munich, through popular lectures and expansions of his readable *Chemische Briefe,* he was able to play the role of an elder statesman of science and to comment on broader issues such as scientific methodology, to oppose rank materialism, or, like a Hebrew prophet, warn the nations of the earth of the dangers of failing to recycle sewage or replace soil nutrients that were harvested as animal and human food. Above all, after signal failures in his prime to make money from chemistry, he was able to demonstrate, through the exploitation of the cadavers of South American cattle, how chemistry or, rather, chemical physiology could be exploited commercially. His success here was surely a model for the industrialisation of the food processing industry upon which twentieth-century society is rooted.

Although this biography, being written in English, emphasizes Liebig's British connections, I have tried nevertheless to adopt an international perspective by portraying Liebig as a gatekeeper and, as Edmund Muspratt admiringly said, "Man of the century." In adopting the gatekeeper image, I have in mind the ways in which Liebig acted as an entrepreneur and propagandist for the extension of chemistry's boundaries. For Liebig, chemistry was a mature field of knowledge loosely separated or bounded by other disciplines. His essential message was that these adjacent fields of endeavour populated by workers trained in very different traditions of theory and practice would benefit from some chemical fertilizer. Once done, these boundary sciences would not only make more sense and prove more fruitful but would enable important new interdisciplines to emerge.

German historians of pharmacy, as well as a number of Liebig specialists, have drawn attention to the ways pharmacy and mining became catalysts for the creation of the modern chemistry teaching and research laboratory. At the same time, attention has been drawn to the fact that apothecaries were the founders of several chemical firms and industries. The movement to "modernise" the training of pharmacists and to improve their social status was well under way in the late-eighteenth-century German states, as was the cameralism that stimulated interest in mineral resources, their identification, analysis, and metallurgical exploitation; but the completion of these processes owed much to the efforts of Geiger and Liebig in the reconstructed *Annalen der Pharmacie* in the 1830s. In an important editorial of 1835, "On the study of pharmacy in its relation to medicine and in the training of pharmacists," Geiger and Liebig agreed that, with the development of purer and active-ingredient-only drugs and a greater understanding of their *chemical* constitution, doctors and medical schools needed to take pharmacy much more seriously. And, of course, pharmacy was no longer *materia medica* and the memorisation of a few prescriptions but a discipline that had been transformed by organic chemistry. In another editorial of 1838, soon after returning from a trip to Britain where something similar was afoot through the activities of Jacob Bell, Liebig emphasised the close relationship between pharmacy and industry.

These statements of Liebig's concerning the borders between pharmacy and medicine, pharmacy and industry, and pharmacy and chemistry can be seen as a broader effort on his part to establish chemistry as the most significant and fundamental science for the modern age. These efforts, which began as early as 1826 in his correspondence with Hessen government officials over a salt works near Giessen, reached their climax in the well-known and notorious 1840 review of the state of chemistry teaching in Prussia. This polemic was penned at the same moment that Liebig was completing the *Agricultural Chemistry,* which was also published in French, German, and English in 1840. As Regina Zott and others have demonstrated, Liebig saw his *Agricultural Chemistry* as an example of the power of chemistry to transform an empirical activity like farming or agriculture into a science. In practice, he saw the book as revolutionising plant physiology and as demonstrating to "botanists" that their discipline could become properly scientific only if its students were trained in experimental chemistry.

Two years later Liebig continued his argument with physiologists in *Animal Chemistry,* where in its "Preface" he argued that animal physiology and pathology were inextricably "mixed up together." It was at these borderlands that chemists could make their greatest contributions to society. As gatekeeper to these borderlands, Liebig said of himself, as Macaulay had said of Francis Bacon, that he was the voice that was in a position to draw attention to the inexhaustible supplies of neglected wealth that the commercial and industrialised nations of the world could mine by following the chemical road.

Liebig's central message was extremely successful, so that now it actually takes some effort to understand that despite the glamourous propaganda of twentieth-century cosmologists and physicists or molecular biologists chemistry really is the most fundamentally central and useful of all the sciences. Liebig's aphorism that soap is a measure of civilization is really a metaphor of the fact that our modern complex societies cannot function without chemists and the understanding of chemistry. Liebig's other message as gatekeeper concerned the vital importance of good husbandry and of the significance of chemical cycles in nature. The very success of his first message, however, with its support for High Farming using mineral and ammoniated fertilizers, or the belief that disease could be eradicated only by an understanding of the chemistry of living processes, tended to suppress his other essentially "ecological" perspective during his own lifetime. It is only fairly recently that the relentless growth of world population, increasingly frequent famines caused by overintensive farming, and Malthusian warning of limits to growth have led to a reappraisal of Liebig's "law of the minimum," his recognition that the nutrient that is present in the least amount relative to the amount required determines the yield of a plant or the health of an organism. And only relatively recently has come recognition of the significance of his denunciations and warnings concerning *Raubbau,* our prevalent current exploitive system of civilisation that pillages Nature and breaks her own chemical laws of recycling. The fixation of nitrogen by the Haber–Bosch process solved the late-Victorian need to recycle, but soon we may be faced by a shortage of the phosphorus that is also an essential ingredient of living systems. These themes of ecological conscience have certainly led to a revival of interest in Liebig's ideas in Germany, and in Britain they have even led the poet Tony Harrison to introduce Liebig into his thought-provoking verse drama *Square Rounds.*

Liebig's career, therefore, deserves attention not only because he helped to transform the teaching of chemistry and created the modern research school, but because he transformed the ways doctors, pharmacists, and physiologists saw chemistry and how chemists themselves saw their roles intellectually and as possessing "civic worth" in a modern society. For Liebig, chemistry was the fundamental, or central, science. Other sciences were diminished and distorted if their exponents were ignorant of it. At the borderlands between different sciences, there were new disciplines such as chemical physics, pharmaceutical chemistry, agricultural chemistry, physiological chemistry (or biochemistry), and industrial chemistry calling out for research and for able recruits. It was at these borderlands that chemists could make their greatest contribution to society.

There has been no English-language biography of Liebig since the 1895 publication of a brief life by a public school chemistry teacher and researcher, William Shenstone. This drew heavily on the brilliant account of his former teacher given by A. W. Hofmann in a Faraday Lecture to the

London Chemical Society in 1875, as well as on the Cantor Lectures of 1875–76 delivered at the Royal Society of Arts by another Anglo-German pupil, Ludwig Thudichum. German readers, appropriately for someone who was one of the giants of nineteenth-century chemistry, have been better served: an unofficial biography in 1904 by a leading historian Adolph Kohut, who did not, however, have access to most of the family papers; and the massive official biography by Liebig's pupil and family friend Jakob Volhard that was published in 1909. Although he was not a professional historian, Volhard was enormously conscientious and his biography is an impressive achievement. Time after time in writing the present biography, when convinced that I was exploiting a document for the first time, I found that Volhard had been there before and that he had referred to the same evidence. Apart from not being translated into English, the principal drawback of Volhard's biography is its lack of an index. My new life of Liebig should not be regarded as replacing Volhard but as complementing his pioneering achievement.

Although several German popular biographies, encyclopaedia entries, and commemorative essays, all based upon Volhard, appeared in the 1920s and 1930s, notably a nationalistic one by Richard Blunck in 1938 (reprinted 1946), it was not until 1973 that Irene Strube penned a 100-page portrait for readers in the then German Democratic Republic. Not surprisingly, she laid stress on the social implications of Liebig's work by concentrating on his agricultural chemistry, giving short shrift to his equally important research in animal chemistry or the popularisation of chemistry in the mid-Victorian period. By then, the 1970s, the professionalisation of history of science in Britain and America was in full flood. In the United States in 1964, Frederic L. Holmes produced a rich and scholarly essay to accompany an English-language reprint of Liebig's *Animal Chemistry;* he capped this a decade later with a stimulating, archive-based sketch of Liebig' s career and work for that great cooperative enterprise of the 1970s, the multivolumed *Dictionary of Scientific Biography.* There, in contrast to Strube, Holmes stressed the role of Liebig's intellectual and practical activities as one of the new breed of organic chemists who emerged in the 1830s. In 1971, Holmes's Yale student Margaret Rossiter examined the impact that Liebig had had on American agriculture between 1840 and 1880. Her stimulating study of Liebig's influence on the Americans was subsequently published in 1975. In Britain, complementing Holmes and Rossiter, Jack Morrell's seminal *Ambix* essay of 1972 emphasized the significance of the chemical school that Liebig had established and made world-famous at Giessen, and he portrayed Liebig as a breeder of chemists and of chemical research. In so doing, he raised influential questions concerning the role of patronage in science and the social conditions that had allowed this thriving academy to emerge.

My own interest in Liebig began in the 1960s when I contemplated research on Edward Frankland – now happily and better executed by my friend Colin Russell. Frankland led me in 1973 via a Deutsche Akademie Austauschdienst scholarship to Germany, where I was able to explore the rich and scarcely tapped Liebig archives at the Liebig Museum in Giessen and at the Bavarian State Library in Munich. Such archives had just begun to be exploited by a young American doctoral student Bernard Gustin, whose Chicago thesis of 1975 questioned the traditional portrait of Liebig as the founder of practical chemistry teaching. It was Gustin who first drew English-speaking historians to the hitherto unaddressed significance of pharmacy, both in Liebig's teaching career and in the industrialisation of Germany. Gustin's work has been ably continued by Pat Munday, whose stimulating Cornell thesis of 1990, again based upon the critical use of archives, offered a reappraisal of Liebig's early training in Germany and Paris, the creation of his Giessen laboratory, and his decision in the late 1830s to abandon pure organic chemistry and move into what would now be recognized as biochemistry and its applications to agriculture, medicine, and nutrition. I gladly acknowledge the way my own study has benefitted from the research of Holmes, Rossiter, Morrell, Gustin, and Munday, as well as from the excellent studies of Liebig's agricultural work made by Vance Hall and Mark R. Finlay, and the illuminating analysis of Liebig's many pupils made by Professor Joseph S. Fruton. All Liebig scholars are also permanently indebted to the bibliographical work of Carlo Paoloni.

In a review of Liebig studies in 1981, I drew attention to the richness of the archival remains of Liebig in the hope that it would stimulate further work. I like to think that it has. Apart from my own edition of the correspondence between Liebig and Hofmann (1984), a good deal of the regular correspondence between Liebig and his pupils and friends has now been carefully transcribed and published through the dedicated scholarship of Dr. Emil Heuser (Leverkusen), with the backing and magnificent support of Dr. Heinrich Propfe. Sadly, neither of them has lived to read this biography. With the forthcoming edition of the complete and unexpurgated correspondence of Liebig and Wöhler edited by Professor Christoph Meinel, there will perhaps remain only the daunting task of publishing the extensive correspondence of Liebig and Kopp concerning the editing and publishing of *Annalen der Chemie*. Much else has been done in recent years to reestablish the significance of Liebig in German and world scientific culture by the activities of the revamped Liebig

Gesellschaft and its associated wonderful Liebig Museum at Giessen. Here the quiet authority of Professor Konrad Mengel, Dr. Siegfried Heilenz, and their team of helpers, and especially the dedicated enthusiasm of Wilhelm Lewicki, a descendant of Liebig's daughter Nanny Thiersch, have very considerably aided the professional activities of historians such as myself.

In addition to the historians and Liebig lovers already mentioned and to whom I am indebted, it is a pleasure to acknowledge the award of an Edelstein International Fellowship in the History of the Chemical Sciences, which enabled me to spend ten months in Philadelphia at the Chemical Heritage Foundation in 1990–91 and two months in Jerusalem at the Edelstein Center for the History of Science, Technology, and Medicine in 1992. Although these periods in America and Israel were primarily dedicated to the completion of my *Fontana History of Chemistry* (London, 1992), which was published in the United States as *The Norton History of Chemistry* (New York, 1993), they also gave me time to consolidate my thoughts on Liebig. A further period of leave from the University of Leicester in the autumn of 1993, again spent in Philadelphia, saw the completion of over half of the text. The remainder was finished at Leicester betwixt teaching and administration. I am enormously grateful to Professor A. Thackray and Dr. O. T. Benfey and the other staff at Philadelphia, Dr. A. S. Travis at Jersusalem, as well as the University of Leicester for leave of absence. Finally, I owe D. M. Knight, J. B. Morrell, C. A. Russell, and W. A. Smeaton my warm thanks for their support and encouragement over more than thirty years of our mutually sustaining interests in the history of chemistry.

He was not the maker of that road; he was not the discoverer of that road; he was not the person who first surveyed and mapped that road. But he was the person who first called public attention to an inexhaustible mine of wealth, which had been utterly neglected, and which was accessible by that road alone. By doing so he caused that road, which had previously been trodden only by peasants and higglers, to be frequented by a higher order of travellers.

Macaulay on Francis Bacon, quoted by Liebig of his own role in agricultural chemistry, *J. Roy. Agri. Soc. of England*, 17(1856), 326

Liebig himself seems to have occupied the role of a gate.

Thomas Pynchon, *Gravity's Rainbow* (1973; Penguin edition 1987), p. 411

1

From Pharmacy to Chemistry

> Chemical laboratories, in which instruction in chemical analysis was imparted, existed nowhere at that time. What passed by that name were more like kitchens. . . . No one really understood how to teach it.[1]

In 1890, when sifting through his father's papers, Georg von Liebig came across an autobiographical sketch written by Justus von Liebig on or soon after his sixtieth birthday in 1863. Published by Georg in the *Berichte der Deutschen chemischen Gesellschaft* in 1890, the sketch was rapidly translated into English by the Liverpool chemist and historian of chemistry, J. Campbell Brown. These reminiscences became the basis of most later accounts of Liebig's early life and career. Like most reconstructions of a person's life written half a century after childhood, Liebig's autobiography has to be treated with circumspection. His treatment of his family's circumstances may well have been distorted by his parents' reminiscences as much as by his own inclination as a baron of Hessen-Darmstadt, to "improve on" his birthright. These factors, together with his desire to illustrate certain educational precepts through personal anecdotes, all tended to produce an account that does not entirely dovetail with what the historian can discover from other sources.

Liebig tells us nothing about his birth and childhood before he reached the age of fourteen, except to say that his father, Johann Georg Liebig (1775–1850), was a drysalter and hardware merchant (*Materialist*) in the ducal city of Darmstadt, who dealt in and made up varnishes, paints, boot and grate polishes and pigments for householders and various local trades and businesses.[2] For this purpose Johann had fitted up a laboratory-cum-

1. Georg von Liebig, "Liebig: Eigenhändige biographische Aufzeichnungen," *Berichte* 23 (1890), Referat pp. 785–816, and *Deutsche Rundschau* (17 January 1891), 30–39; English translation by J. Campbell Brown, *Chemical News* 63 (1891), 265–66, 276–77; *Annual Reports Smithsonian Institution* (Washington, D.C., 1893), pp. 257–68, from which citations are made.
2. Liebig sent Hofmann a recipe for the boot polish by which his father had earned a tidy sum each year. See Liebig to Hofmann, 28 January 1852, in W. H. Brock, ed., *Justus von*

workshop in an allotment garden not far from his home, where the young Liebig was inducted into the ancient mysteries of "bucket" chemistry. All this can be verified, but as Pat Munday has demonstrated, nothing in Liebig's early life is quite straightforward.[3] Even the date of Liebig's birth is controversial. Different sources give 3 May 1803 (his mother), 8 May (a Munich gravestone), and 14 May (church records); following A. W. Hofmann, Liebig's greatest student, historians have opted for 10 May 1803.[4] Interestingly, the church recorder wrote Liebig's father's name as it was pronounced: "Liebich."

Liebig's birthplace survived until it became unsafe in 1920, when it was carefully dismantled and rebuilt as the "Liebighaus," which opened in 1928. This museum was completely destroyed in the Allied bombing of Darmstadt in 1944, though the museum's publications provide a good picture of what it was like.[5] Situated at 30 Grosse Kaplaneigasse, Liebig's terraced home opened directly onto a narrow hallway that led to a paved courtyard at the rear. There were three small rooms to the right of the hall, one of which served as the shop. A narrow staircase led to three more small rooms on the first floor and a further three bedrooms in the attics. From one of these attic rooms there was a clear view of the royal palace and Court Library. Although it was "pokey" by modern standards, there must have been ample space and accommodation for the Liebigs and their five surviving children.

Justus Liebig was Johann's second child and son. The older brother, Johann Ludwig "Louis" (1801–30), seems to have been a disappointment to his parents. It was obviously intended that he should inherit the family business and extend its range and status by making it a pharmacy. In 1816 Louis was apprenticed to a manufacturer in Frankfurt but was sent back the following year to be reapprenticed to a pharmacist in St. Goar, following which he worked in many German towns before his early death, having made it clear to his father in 1822 that he did not want to return to Darmstadt. Another younger brother died at the age of five, and four sisters died in infancy. By the time Justus Liebig had become famous in 1830, therefore, his only siblings were two younger brothers Johann Georg (1811–43) and Karl (1818–70) and a sister Elizabeth (1820–90),

Liebig und August Wilhelm Hofmann in ihrer Briefen (Weinheim, 1984), pp. 123–24.

3. P. Munday, *Sturm und Dung. Justus von Liebig and the Chemistry of Agriculture* (PhD thesis, Cornell University, 1990); and "Sturm and Dung," *Ambix* 37(1990), 1–19.
4. A. W. Hofmann, Faraday Lecture for 1875, *The Life Work of Liebig* (London, 1876), frontispiece and p. 50. Reprinted in *Faraday Lectures of the Chemical Society* (London, 1928).
5. H. Hohmann, *Justus von Liebigs Geburtshaus* (Darmstadt, c. 1928); Ernst Berl, "The Liebig House and the Kekulé Room at Darmstadt," *J .Chem. Educ.* 6 (1929), 1869–81.

Photograph of Liebig's birthplace in Darmstadt taken in the 1920s before the house was transformed into a museum. (Courtesy W. Lewicki)

who later married Liebig's Giessen colleague Friedrich Knapp. In place of Louis's refusal to take over the family business and because of Justus's academic aspirations, the family hopes were placed on Karl, who studied pharmacy at the University of Giessen under his brother in 1842 before taking over the family's materials business the following year. The shop's name "Georg Liebig & Sohn" was retained after Karl's death in 1870, and the business was still in the hands of Karl's son Georg Liebig, when Volhard (1909) published the standard German biography of Justus Liebig. By then the family's business also included a factory in Hannover.

Liebig's mother, Maria Caroline Möser (1781–1855), who ran the hardware shop, was the illegitimate child of an itinerant Swabian tailor Christoph Einselin and the daughter of a Darmstadt farmer named Fuchs. Caroline, as she was familiarly known, had been adopted by the Mösers, whose name she had taken.[6] That Liebig's mother, who was noted for her quick wit, had Jewish ancestry seems likely; there were anxious attempts by Aryan local historians in the 1930s to extinguish the possibility that Liebig had descended from anything but pure Hessian stock. Liebig's prominent peasant's nose (*Bauernase*) and impressive eyes were undoubtedly inherited from his mother. She also had a calming influence on her husband, whose impetuosity and occasional vehemence were to be unfortunate characteristics of their son Justus.

Hessen is one of the *Länder* (administrative regions) of modern Germany. Situated on the Rhine, it was formerly divided into numerous independent principalities, but toward the end of the eighteenth century it had been partitioned into the two dominions of Hessen-Kassel ruled by an electoral prince, with its university situated in Marburg; and Hessen-Darmstadt, a grand duchy devoted to agriculture and vinoculture with its Lutheran University at Giessen. Although Hessen-Kassel was absorbed by Prussia in 1866, Hessen-Darmstadt retained its independence in return for a loose alliance with France. With the outbreak of the Franco-Prussian war in 1870, Hessen-Darmstadt joined the German Federation and thus became part of the United German Empire. The original grand duke of Hessen-Darmstadt, Ludewig (or Ludwig) I (1753–1848), was pro-French; indeed, he owed his duchy's independence to Napoleon.[7] Liebig was to benefit from this connection and allegiance. Although Ludwig freed the peasantry from serfdom in 1812 and gave his subjects a written constitution in 1820, these reforms did little to reduce the traditional feudal powers of the local Hessen nobility, and despite some desultory attempts to introduce machinery and industry to the state, much of the peasantry continued to live in poverty little better than serfdom. Hessen-Darmstadt was said to have more nobility with extensive estates than any other part of Germany. While studying with Liebig at Giessen in 1846, the American student Eben Horsford was horrified by the poverty he witnessed in the countryside around Giessen.[8]

During the reign of Ludwig II, who succeeded as Landgrave after his father's abdication in 1830, the country's political and economic stagnation worsened, leading to a series of disturbances that culminated in the 1848 revolution. Having inherited his father's considerable debts, Ludwig II raised taxes and quashed a parliamentary rebellion by calling an election, which he rigged to ensure a loyal parliamentary following. Famines, and the fact that the countryside no longer seemed able to support the

6. See E. Berl, "Justus von Liebig," *J. Chem. Educ.* 15 (1938), 553–62.
7. Ludwig's portrait appears in *J. Chem. Educ.* 13 (1936), 315.
8. S. Rezneck, "Horsford," *Technology and Culture* 11 (1970), 376.

peasantry that had returned from Napoleonic battlefields, had also increased the pressure to emigrate. Emigration was officially opposed in 1826, though rising prices, cholera, and general disorder led to a wave of emigration to America in the 1830s.[9] Such actions radicalised a considerable number of students at the University of Giessen in the 1830s, though Liebig himself appears to have been largely oblivious to these events. Not until the 1850s did he begin to view them in Malthusian terms of overpopulation. A factor here was the death of Ludwig II in 1848; he was succeeded by another Ludwig who survived until 1877. Ludwig III was not to Liebig's liking, and this was one of the several factors that was to prompt Liebig's move from the University of Giessen to the University of Munich in 1852.

Johann Liebig was by all accounts an inventive man. The son of a shoemaker, whose Hessian Oberwald ancestors can be traced back to 1577, he belonged to the shopkeepers' guild in Darmstadt, taught himself sufficient chemistry to manufacture useful chemicals on a small scale and to acquire a certain fame locally during the 1820s for illuminating his home (or was it only the shop?) with acetylene gas prepared from heated bones.[10] He was ambitious that his two elder sons should rise socially from the family's lower-middle-class ranks (*Kleinbürger*) to that of the upper middle class (*Bildungsbürger*). In Liebig's youth, Darmstadt's population was about 15,000 people. Times were undoubtedly hard for Liebig's family during the Napoleonic wars that stretched through Liebig's childhood until he was twelve in 1815, and during the recriminations that followed in their wake. The tiny duchy, allied to France, had to help Napoleon with supplies and army recruits. Cut off from raw materials by Britain's effective blockade of the Continent, Johann Liebig's business must have suffered, although supplies of sulphur astutely bought by Frau Liebig before the blockade began had proved profitable.

Any family hardship did not, however, prevent Johann Liebig from sending the young Justus Liebig to Darmstadt's Ludwig-Georgs-Gymnasium, a fine traditional grammar school run on strictly classical lines by a scholar-headmaster, Johann Zimmermann (1754–1829).[11] This was apparently after some earlier schooling with a Dr. Graul. Both Justus and his older brother Louis entered the Gymnasium at Easter 1811. Liebig was only eight, and since the average age in the first class was ten, he must

9. H. Lindenberger, *Georg Büchner* (Carbondale, 1964), pp. 5–9; A. H. J. Knight, *Georg Büchner* (Oxford, 1951), p. 30; Mark Walker, *Germany and the Emigration 1816–1885* (Cambridge, Mass., 1964).
10. Liebig to his father 25 February 1821, in E. Berl, *Briefe von Justus von Liebig nach neuen Funden* (Giessen, 1928), pp. 17–18.
11. *Hessische Biographien*, vol. 3 (Darmstadt, 1934; reprint Wiesbaden, 1973), p. 100.

have been precocious to have coped. Surviving records examined by Volhard show that at the end of the fourth form (*Tertia*) in 1813, Liebig was placed twenty-third out of twenty-eight, and when he was in the fifth form (*Sekunda*) at Easter 1815 he was seventeenth out of twenty-seven. This does not suggest the complete indifference to the curriculum that he expressed mockingly in his autobiography, especially if we bear in mind that Liebig was always almost two years younger than the average age of his form throughout his schooling.

> With this [observational and empirical] bent of mind it is easy to understand that my position at school was very deplorable; I had no ear memory, and retained nothing or very little of what is learned through this sense. I found myself in the most uncomfortable position in which a boy could possibly be; languages and everything that is acquired by their means, that gains praise and honour in the school were out of my reach; and when the venerable rector of the Gymnasium, on one occasion of his examination of my class, came to me and made a most cutting remonstrance with me for my want of diligence, how I was the plague of my teachers and the sorrow of my parents, and what did I think was to become of me, and when I answered that I would be a chemist, the whole school and the good old man himself broke into an uncontrollable fit of laughter, for no one at that time had any idea that chemistry was a thing that could be studied.[12]

The anecdote is plausible enough – especially the headmaster's raillery – though whether Liebig himself really understood the difference between pharmacy and chemistry at this time is doubtful. In another anecdote collected by Kohut, the assistant headmaster Johann Storck accused Liebig of being a sheep's head (*Schafskopf*), and this became Liebig's playground nickname.[13]

In fact the Darmstadt school was a good one; a notable earlier pupil had been Emmanuel Merck (1794–1855), who took over his father's long-established Darmstadt pharmacy in 1816 after studying with Trommsdorff at Erfurt.[14] Liebig and Merck were to become collaborators in the 1820s. Three others of Liebig's immediate classroom contemporaries were Georg Gervinus, Jacob Kaup, and Ludwig Reuling. All three became distinguished men: Gervinus as a historian, Kaup as a naturalist and

12. Liebig, "Autobiography," p. 260.
13. A. Kohut, *Justus von Liebig* (Giessen, 1904), p. 9. Kohut, who unlike Volhard was a professional historian, was not given access to Liebig's papers. Nevertheless, his biography is an interesting complement to Volhard's.
14. Fritz Ebner, *Merck und Darmstadt* (Darmstadt, undated).

Darwinian,[15] and Reuling as a conductor of the Imperial Opera House in Vienna.[16]

The implication of Liebig's autobiography is that he dropped out of the Darmstadt Gymnasium in 1817 when he was fourteen because the curriculum so clearly did not suit him.[17] A more likely reason is that his father could no longer afford the fees, for following Napoleon's defeat in 1815, tough measures were taken by Metternich at the Vienna Congress against Hessen-Darmstadt for having supported the French. Rich agricultural areas of the duchy were ceded to other states, taxes were increased, and appalling weather in 1816 and 1817 caused crop failures and famine. For the next forty years Hessen-Darmstadt became the poorhouse of Germany, which fed political radicalism and stimulated peasant emigration on a wide scale.

Having left school at fourteen without taking his *Abitur,* or leaving certificate, Liebig's chances of preferment were grim; entry to a university would be difficult (though, as we shall see, not impossible), and teaching and the civil service would be beyond the pale. Given Liebig's evident interest in his father's chemical preparations and the fact that the profession of apothecary had, since the eighteenth century, become a position of esteem, Johann decided to appprentice his second son to the apothecary Gottfried Pirsch (1792–1870), who kept a business at Heppenheim in Baden-Württenburg to the south of Darmstadt and a few miles to the north of Heidelberg.[18]

Liebig's reminiscences again tell only half the story:

> Since the ordinary career of a gymnasium student was not open to me, my father took me to an apothecary at Heppenheim, in the Hessian Bergstrasse; but at the end of ten months [in fact only six months] he was so tired of me that he sent me home again to my father. I wished to be a chemist, but not a druggist. The ten months [sic] sufficed to make me completely acquainted alike with the use and manifold

15. For Gervinus and Kaup, see *Allgemeine Deutsche Biographie* (Munich, 1875–1912), vols. 9 and 15.
16. Reuling (1802–79) composed seventeen operas. See *Baker's Biographical Dictionary of Musicians,* 7th ed. (Oxford, 1984), p. 1883.
17. J. Volhard, *Justus von Liebig* , 2 vols. (Leipzig, 1909), hereafter "Volhard, *Liebig,*" noted a memorandum of 1852 in Liebig's handwriting which claimed he left school when he was seventeen; the school's records did not confirm this date. See Volhard, *Liebig,* vol. I, p. 15.
18. Munday, *Sturm und Dung,* p. 20. This handsome timbered pharmacy, subsequently renamed the Liebig-Apotheker on the basis of Liebig's brief connection, survives in the Bergstrasse as a restaurant. See W. H. Hein, *Die Deutsche Apotheke* (Stuttgart, 1960), p. 176; and Kohut, *Justus von Liebig,* p. 3.

applications of the thousand and one different things which are found in a druggist's shop.[19]

In fact, as surviving correspondence between Pirsch and Liebig's father shows, times were too hard for Johann Liebig to pay the indenture fee. Despite gifts of tobacco, with no fee and no apprenticeship agreement ever made, Pirsch simply returned Liebig to his father, just as Louis had been earlier. The successful adult Baron Liebig obviously found his boyhood poverty embarrassing and so he invented stories of causing explosions in the pharmacy that had earned him dismissal as an unruly apprentice. This episode formed one of the scenes from Liebig's life pictured in the *Liebigbilder* advertising cards that were circulated by the Liebig Meat Company in the 1880s and that was taken over and embroidered by Volhard in his standard biography.[20]

For the next two years, from 1817 to 1819, and until he was seventeen, Liebig remained at home, helping his father and reading chemistry books that he was able to borrow from Duke Ludwig's Court Library. This privilege seems to have developed from the fact that his father, who presumably had the equivalent of a British "by Royal Appointment" with Duke Ludwig, was allowed to borrow technical books from the Court Library, and young Justus was often sent to return them.

> The lively interest that I took in my father's labours naturally led me to read the books which guided him in his experiments, and such a passion for these books was gradually developed in me that I became indifferent to every other thing that ordinarily attracts children. Since I did not fail to fetch the books from the Court Library myself, I became acquainted with the librarian, Hess,[21] who occupied himself successfully with botany, and as he took a fancy to the little fellow [*sic,* Liebig was probably a teenager]. I got, through him, all the books I could desire for my own use. Of course the reading of books went on without any system. I read the books just as they stood upon the shelves, whether from below upwards or from right to left was all the same to me; my 14-year-old head was like an ostrich stomach for their contents, and amongst them I found side by side upon the shelves the thirty-two volumes of Macquer's "Chemical Dictionary,"[22] Basil Valentine's "Triumphal Chariot of Antimony," Stahl's "Phlogistic

19. Liebig, "Autobiography," p. 260.
20. Volhard, *Liebig,* vol. I, p. 17.
21. Johannes Hess (1786–1837), architect and botanist, as well as Court Librarian. See *Hessische Biographien,* vol. 3, p. 121.
22. P. J. Macquer, *Dictionnaire de chymie,* 2 vols. (Paris, 1766); 2nd ed., 4 vols. (Paris, 1778); *Chemisches Wörterbuch,* 6 vols. (Leipzig, 1781–83; 2nd ed., 7 vols., 1788–90). Liebig may have been using a mixed edition that looked like 12 volumes.

> Chemistry," thousands of essays and treatises in Göttling's periodicals,[23] the works of Kirwan, Cavendish, etc.[24]

In the light of our earlier scepticism towards parts of Liebig's autobiography, we might well be suspicious as to whether the son of a local hardwaresman would be allowed to borrow books from an aristocratic library. However, Liebig undoubtedly speaks the truth here, though there is no reason to think that he was as privileged a borrower as he implies. Ludwig's library seems to have functioned more as a state or public library than as a private one, and there are independent sources that confirm that other Darmstadters were allowed to use it. On the other hand, if Liebig's memory of his browsing is accurate, then his reading was out of date; apart from a picture of more recent chemistry that he could have obtained from Göttling's *Almanach oder Taschenbuch für Scheidekünstler und Apotheker* (1780–1802), his reading material was pre-Lavoisian. Nevertheless, although the reading was useless as far as current chemistry was concerned, in retrospect Liebig believed that he benefited closely from it.

> It developed in me the faculty, which is peculiar to chemists more than to other natural philosophers, of thinking in terms of phenomena; it is not very easy to give a clear idea of phenomena to anyone who cannot recall in his imagination a mental picture of what he sees and hears, – like the poet and artist, for example. Most closely akin is the peculiar power of the musician, who while composing thinks in tones which are as much connected by laws as the logically arranged conceptions in a conclusion or series of conclusions. There is in the chemist a form of thought by which all ideas become visible in the mind as the strains of an imagined piece of music. This form of thought is developed in Faraday in the highest degree, whence it arises that to one who is not acquainted with this method of thinking, his scientific works seem barren and dry, and merely a series of researches strung together, while his oral discourse when he teaches or explains is intellectual, elegant, and of wonderful clearness.[25]

To achieve this faculty, however, it was necessary to combine reading with practice:

23. B. Valentine (pseud.), *Triumph Wagen Antimoni* (Leipzig, 1604) and later editions; E. Stahl, *Fundamenta Chymiae Dogmatico-Rationalis et Experimentalis* (Nürnberg, 1723); Göttling (1755–1809), Professor of Chemistry at Jena, published a large number of practical pharmacy texts. J. R. Partington, *A History of Chemistry*, vol. iii (London, 1962), pp. 595–96. R. Kirwan (1733–1812) and H. Cavendish (1731–1810) both published on pneumatic chemistry.
24. Liebig, "Autobiography," pp. 257–58.
25. *Ibid.*, p. 258.

> The faculty of thinking in phenomena can only be cultivated if the mind is constantly trained, and this was effected in my case by my endeavouring to perform, so far as my means would allow me, all the experiments whose description I read in the books. These means were very limited, and hence it arose that, in order to satisfy my inclinations, I repeated such experiments as I was able to make, a countless number of times, until I ceased to see anything new in the process, or till I knew thoroughly every aspect of the phenomenon which presented itself. The natural consequence of this was the development of a memory of the sense, that is to say of the sight, a clear perception of the resemblance or differences of things or of phenomena, which afterwards stood me in good stead.[26]

All chemists will recognize this as a valid description of the eye and ear memory that is so essential in the successful practitioner and that today is needfully reinforced by a good three-dimensional sense of the shapes of molecules.

By the time he was seventeen, trade had improved and Liebig decided that he would be a pharmaceutical chemist and possibly an industrialist. One of Liebig's chemical notebooks from this period, from 1819 to 1821, has survived and shows him taking a lively interest in recipes for dye stuffs, pigments, and varnishes, though not in the fulminates that his reminiscences imply he had been fascinated by since a schoolboy. As Munday has commented: "At this point in [Liebig's] life, chemistry was significant only for its immediate application to the apothecary trade or to a manufactury like his father's."[27] Liebig was still doing bucket chemistry, as his work on fulminates illustrates. He had witnessed a pedlar in the Darmstadt market prepare a mercury fulminate for the toy torpedoes or firecrackers that he was selling. Recognising the ingredients as mercury, nitric acid, and alcohol, it had been a simple matter for Liebig to make up his own crackers as an exciting sales item for his parents' shop.[28]

The term *fulmination* had been used by the alchemists to refer to explosions, and gold and silver fulminates may well have been accidentally prepared during the seventeenth century. However, the powerful detonating silver fulminate was only formally discovered by Berthollet in France in 1789 when he added ammonia to silver oxide in an attempt to discover the composition of ammonia. Subsequently, in 1800, an English aristocratic inventor Charles Howard (1774–1816), as well as the Italian Luigi Brugnatelli (1761–1818), had prepared mercury and silver fulminates by a variety of methods. There was much confused discussion in the litera-

26. *Ibid.*, p. 258.
27. Munday, *Sturm und Dung*, p. 22.
28. Volhard, *Liebig*, vol. i, p. 11.

ture of the period, which Liebig picked on, as to whether these fulminates contained ammonia.[29] Here was an appropriate subject for elucidation for anyone foolhardy enough to investigate. This rather dangerous work, which had cost other investigators eyes and fingers and which must have been done in his father's workshop, led to Liebig's first paper, published in J. A. Buchner's *Repertorium für die Pharmacie* in 1822. This Bavarian journal, founded by the reforming pharmacist A. F. Gehlen (1775–1815) in the year of his death, was much concerned with raising the professional status of pharmacists, a campaign that Liebig was soon to join.

Liebig's first paper, which was little more than a demonstration of the identity of composition of the various silver fulminate preparations described in formularies, together with his own recommendation of a foolproof method for preparing silver fulminate, was actually submitted to Buchner by the professor of chemistry at the University of Erlangen, K. W. G. Kastner (1783–1857), who included a curious, fulsome introductory note:

> These first proofs of the experimental diligence of a young chemist are commended to the reader's indulgence. The author has already dedicated himself to chemistry at Bonn with enthusiasm and intends to go about it in the same spirit here in Erlangen.[30]

Kastner writes like an indulgent father; indeed, he was Liebig's *in loco parentis* from 1821.

The University of Bonn

The University of Berlin had been established in the Prussian capital city of Berlin in 1809 to offset the closure of the prestigious University of Halle by Napoleon. As Friedrich Schleiermacher wrote at the time, it was founded to demonstrate "that Prussia, instead of surrendering the function it had so long practised, that of striving above all else for a higher intellectual culture as the source of power, proposed to begin anew; that Prussia . . . would not allow itself to be isolated, but desired, rather, in this respect also, to remain in living union with the whole of natural Germany."[31] Underwritten by the educational philosophy of freedom of

29. Partington, *History of Chemistry*, vol. iv (London, 1964), pp. 257–58.

30. J. Liebig, "Einige Bermerkungen über die Bereitung und Zusammensetzung des Brugnatellischen und Howardischen Knallsilbers," *Büchners Repertorium der Pharmacie* 12 (1822), 412–26. Kastner's remark prefaces the article.

31. F. Schleiermacher, *Gelegenheit Gedanken über Universitäten* (Berlin, 1808), cited in F. Paulsen, *German Universities* (London, 1906), p. 50. See Charles E. McClelland, *State, Society, and University in Germany 1700–1914* (Cambridge, 1980).

learning and teaching, and of learning for its own sake rather than for a career (*Brotstudien*), Berlin became the model for other Prussian universities, including Bonn. A Catholic University had been created at Bonn in 1786, only to be suppressed by the French. However, after Bonn was ceded to Prussia, following the Congress of Vienna in 1815, King Friedrich Wilhelm III seized the opportunity of re-creating the university in the former Electoral Palace and at the neighbouring Poppelsdorff Schloss. His adviser, the Prussian minister of culture Karl Altenstein, was determined that the new university should cultivate the sciences as well as the humanities and theology (there were both a Protestant and a Catholic faculty of theology). To this end he seems to have deliberately sought to hire members of Germany's oldest and most prestigious learned society, the Leopoldina, which had been founded in 1672. This brought Kastner from Halle and several other Professors from the University of Erlangen, including the chemical geologist Karl Bischof (1792–1870), who was to succeed Kastner in 1822, and the great botanist Christian Nees von Esenbeck (1776–1858). The policy was so successful that the Leopoldina library and natural science and history collections were also brought from Erlangen to Bonn.[32] The rebirth of the Leopoldina at Bonn was also to lead Nees's fellow naturalist Lorenz Oken to the idea of annual gatherings of doctors and scientists in different German university towns. Although Liebig was only a cursory attender at these meetings of the Gesellschaft Deutsche Naturforscher und Aerzte, which began at Leipzig in 1822, they marked the blossoming of German science and formed a model for the meetings of the British Association for the Advancement of Science (BAAS) that Liebig did attend.

Although now virtually unknown, in the 1820s Kastner must have been regarded in many quarters along with Wolfgang Döbereiner (1780–1849) at Jena and Friedrich Stromeyer (1776–1835) at Göttingen, as a leading chemist among all the German states. The author of three or four texts, he was successively professor of chemistry at the University of Heidelberg (1805–12), the university of Halle (1812–18), the new Prussian University of Bonn (1818–21), and the Protestant University of Erlangen (1821–57). While at Halle he must have got to know Liebig's father as a supplier of chemicals, for in 1818, Johann Liebig contributed a paper to Kastner's periodical, *Die deutsche Gewerbsfreund* (The German Tradesman) on the best time of year to apply liquid manure. This little-studied but significant

32. C. Renger, *Die Gründung und Einrichtung der Universität Bonn* (Bonn, 1982). J. Bargon, "August Kekulé and the Kekulé Building at the University of Bonn," in J. Wotiz, ed., *The Kekulé Riddle* (Clearwater, Fla., 1993), pp. 33–47.

journal was used by Kastner to argue that it was the business of chemistry and of chemists to further trade and the prosperity of nations.[33] This was precisely the message that Liebig was to elaborate upon in the 1840s.

It is possible that Kastner passed through Darmstadt on his way to Bonn in 1818 and was sufficiently impressed by the academic keenness of Justus that he offered to take him on as his personal assistant and to train him in chemistry. Matriculation at the local University of Giessen, where chemistry was desultorily taught to medical students, was out of the question because Liebig had no *Abitur.* The situation at the new University of Bonn was either more relaxed, or Kastner, as its first professor of chemistry, had sufficient clout to engineer Liebig's matriculation. The major problem was financial, but Johann overcame this through his friendship with the grand-duke's chancellor, Ernst Schleiermacher (1755–1844), who arranged for Liebig to receive a small stipend. He, too, must have been strongly persuaded of the young man's potential. In practice, Liebig seems to have worked his way through his first semester at Bonn in 1820 as a kind of unofficial and extremely youthful Privatdocent, offering lessons to other undergraduates. He started up a student group to read and discuss contemporary chemical publications and worked extremely hard. One of his fellow students was the son of the then well-known German chemist, C. F. Buchholz, who became a close friend. Buchholz (1770–1818) was professor of pharmacy at Erfurt.

Liebig's wonderfully informative letters to his parents can be liberally quoted from to illustrate his life and activities as an undergraduate who did not drink and who devoted himself entirely to his studies.

> This semester I have taken experimental chemistry, experimental physics, and pharmaceutical chemistry. I take these courses and I also audit meteorology, and encyclopaedia of the natural sciences. I am rather busy this winter. In summer I may take mineralogy, and so forth. Kastner promised me that I may be present as an assistant or Famulus in his private experiments. This would be a wonderful opportunity if he keeps his promise. I see him from time to time. I told him also about my treatment of cobalt ores and the very peculiar phenomena which happen thereby. He desires now to analyze this ore in my presence and he asks me to bring him more of it. Therefore, send me at once at least four ounces and add at least two lots of the mineral green from Gotha and two lots of our self-produced myrin and my solution of the cobalt ore and the yellow residue from the

33. For Kastner's journal, see O. P. Krätz, "Die Chemiker in den Gründjahren," in E. Schmauderer, ed., *Der Chemiker im Wandel der Zeiten* (Weinheim, 1973), pp. 259–84, esp. pp. 262–63.

> evaporated nitric acid solution and a small sample of Vienna green which I have made and the calcinated cobalt ore.
>
> Dear God, though, one needs money here! I have already spent out 20 Gulden this month and am in dire straits. You must send me next month's allowance at once. . . .
>
> I take great pleasure in my studies. The more one gets into them, the more it pleases one. Now I perceive how little I know and how much more I must learn so that I can say, "I know something."[34]

Kastner was by inclination a pharmacist and technical consultant rather than a chemist, and it is doubtful whether the laboratory facility in the Poppelsdorfer castle at Bonn was any more than a kitchen workshop. Liebig later implied that Kastner failed to teach him the principles of inorganic analysis because he did not know them himself. More likely Kastner simply did not have the facilities to teach it adequately, and the foregoing letter certainly does not suggest lack of willingness. Other letters to his parents mention that he had become friendly with a student named Wöllner, whose father owned a Prussian blue factory at Dunewalde. In January 1821, Wöllner and Losen (the son of a wealthy Cologne pharmacist) took Liebig on holiday to see their parents' factories, and Liebig, "without spying," sent his father pages of detailed notes on the technical operations he had seen. Liebig must have made a considerable impression on Wöllner's father, for he offered to obtain Liebig a position in Cologne. These and other technological visits that he made from Bonn and on his journey to Erlangen led to another publication, on the manufacture of sulphuric acid, in 1823.[35] That Liebig was interested in the procedures of pharmaceutical chemistry is also clear:

> Yesterday we made lead chromate, glowing iron chromate with saltpetre, extracted it, and with dissolved potassium chromate, we precipitated a solution of lead acetate, but so that the acid was in excess. The precipitate settled magnificently. Also we produced morphine, this special alkaloid from opium which may be an essential part of the Aqua Tofana. We ground one ounce of opium with concentrated acetic acid until the mixture was uniform, filtered the acetate of morphine, added ammonium, whereby it did precipitate. The filtered liquid contained meconic acid, which was precipitated with acetate of barium, as the barium salt of meconic acid and was precipitated afterwards with [Ms damaged by mice]. . . . K. described in his *Zeitschrift* how to produce oxydized calcium chloride which he recommends as a remedy to promote the growth. At Schönebeck a plant

34. Liebig to parents, 1 November 1820, in Berl, *Briefe von Justus von Liebig*, p. 11.
35. J. Liebig, "Über die Bereitung der Schwefelsäure," *Buchner's Repertorium der Pharmacie*, 15 (1823), 199–222.

> for producing fertilizer has been established. I will furnish you the description of its preparation and several other matters when I return in six to eight weeks time. . . . A friend of mine, a Catholic theological student, is teaching me Greek, which will be useful when I take my doctorate.[36]

The need for Greek is puzzling, as only Latin would have been required for promotion at a German university. In another letter Liebig reveals that he is studying French because Kastner had told him, quite rightly, that it was the current language of science. He also aspired to acquire algebra, which he must have ignored at school. A third letter shows his continuing interest in practical dyestuffs of commercial potential and reveals his unhappiness at his brother Louis's refusal to return to his father's business:

> Together with Kastner I have prepared the new metal, cadmium,[37] from the Silesian zinc, also iodine from the sodium salt of hydro iodine acid, cinnebar and the blue molybid acid, tin oxide, which is known as Richters Blue Carmin. To produce blue carmin from indigo, one needs an addition of alum earth. . . . One now gets many earthern vessels from London which are covered with a beautiful metallic glaze of platinum and which are not attacked by any acid, even vitriol oil. It is used for coffee things, candlesticks, and so forth . . . Platinum has the same value as silver. If experiments would give good results, one could make enormous sums of money because such utensils are in great demand. . . . I'm sorry that Ludwig [Louis] persists in his attitude. . . . Together we could improve the [family] business and manufacture chemicals. It would be a pleasure to work together since when one of us was absent the other could be there.[38]

By then Kastner had already received the call to Erlangen, and Liebig was preparing to join him for the summer semester of 1821. He promised his parents that costs would be cheaper in Bavaria and that he would be able to attend all of Kastner's lectures without fee.

If Liebig had been the hard-working student at Bonn, determined to learn to make a living, Erlangen was to open his eyes to a social world. If his sole aim in life in 1820–21 was to serve his father's goals by turning his father's business into a chemical factory with his brother Louis he would now aspire to be a natural philosopher and academic who would teach others to be pharmacists and manufacturers.[39]

36. *Ibid.*, 1 February 1821, p. 17.
37. Cadmium had been discovered in 1817 by Stromeyer in the impure zinc carbonate used by pharmacists.
38. Liebig to parents, 20 February 1821, in Berl, *Briefe von Justus von Liebig*, p. 19.
39. This is implied by the letter of 20 February 1821.

The University of Erlangen

The old town of Erlangen, which lies on the River Regnitz a few miles to the north of Nuremberg, had founded its Protestant Friedrich-Alexander University (the only Protestant university in Catholic Bavaria) in 1743. It was particularly successful in attracting students from both the north and south, and like Oxford and Cambridge in Britain, it numbered a considerable body of rich and aristocratic students. Quite why Kastner, as a member of the Leopoldina, which had migrated to Bonn, was attracted to Erlangen is unclear. In principle, his laboratory facilities were better, for Bonn had no sizable laboratory until the rebuilding of the 1860s. During the tenure of Georg Hildebrandt (1764–1816) as professor of chemistry at Erlangen from 1796, an institute for the practical training of students in chemistry and physics had been opened in the Karlstrasse. This laboratory was poorly lit and had seating for only ten students to watch lecture demonstrations.[40] Although his first laboratory at Giessen was to be little better, well might Liebig have recalled later that the majority of the so-called laboratories in German universities before Giessen's were really workshops filled with ovens and equipment for carrying out metallurgical and pharmaceutical operations rather than analysis. Erlangen was not to have a purpose-built modern laboratory until Kastner was succeeded by Liebig's Munich disciple, Eugen von Gorup-Besanez (1817–78). The latter was succeeded in 1879 by another of Liebig's students, his biographer, Jacob Volhard. There was also to be a Liebig pharmacy tradition: In 1863 the first professor of pharmacy, Martius, was succeeded by Liebig's pupil, Hugo Zöller. Thus, although Liebig remained at Erlangen for barely two semesters, he left his mark on the university.

Liebig's letters to his parents are equally revealing about his academic activities at Erlangen, where he arrived on 9 May 1821. As Kastner planned no lectures until the winter semester, Liebig threw himself into other subjects:

> I take botany this summer with Professor Schubert [who commended him for "outstanding industry, attentiveness and progress"], and technology, stoichiometry, crystallography, and physics. I shall do a lot of botanising this summer because I need this very badly [for pharmacy]. . . . I have bought Kurt Sprengel's "Science of Herbs."[41] (Erlangen, 20 May 1821)

40. T. Kolde, *Die Universität Erlangen unter dem Hause Wittelsbach 1810–1910* (Erlangen, 1910).

41. K. Sprengel, *Anleitung zur Kenntniss der Gewächse*, 3 vols. (Halle, 1802–4). Liebig to parents, 20 May 1821, Berl, *Justus von Liebig*, p. 22.

> During this summer I have studied physics, botany and technology diligently. Whether I shall accomplish anything in these fields only time will tell. . . . The future is no longer dark before my eyes. I have made up my mind to devote myself completely to teaching. Kastner himself has encouraged me in this. I awake, so to speak, to a new life, because there lies before me the goal that I want to reach. This winter, owing to a suggestion of Kastner's, I shall make an analysis of a fossil. This is intended to be printed in a journal so that I may be known a little in the learned world. Kastner will send a copy of this publication directly to the Grand Duke whom he knows personally. He will recommend me and, at the same time, ask for a travel scholarship for me. Then if the Government is on my side, and I have no doubt that this will be the case because chemists are rare in our country [i.e., Hessen-Darmstadt], I will have made my fortune.

And he went on to speculate that because chemistry and physics were not taught independently of medicine at Giessen, he would plan to start a course of experimental chemistry in Darmstadt:

> If I get approval, I shall easily get a call to a university in association with other men, according to Kastner's wishes. I could establish an Institute like that of Trommsdorff which certainly would be supported by the state. Therefore, during the winter, I study chemistry and analysis with great diligence and zeal. I am also studying languages and mathematics, and I hope that I shall meet your expectations some day.[42]

In the event, there was never a published analysis of a fossilised mineral; perhaps it did not prove necessary. As at Bonn, Liebig also organized a student society to discuss the scientific literature of the day.

What was Trommsdorff's institute that Liebig and Kastner were setting such store by? To answer this, a short digression is necessary.

In recent years historians have come to doubt the tradition that Germany's hegemony in chemistry during the late nineteenth and early twentieth centuries owed its origins solely to its investment in academic research, competition between universities, or state policy, or to Liebig's "invention" of practical training. Bernard Gustin has argued:

> While this historiographical line contains some important elements of truth it nevertheless seriously distorts both Liebig's innovativeness and his role in the institutionalization of laboratory training in chemistry in Germany, and the role played by the German states in the patronage of chemistry. Far from being the simple result of Liebig's unique genius or of state support, or of the two combined, the efflorescence of chemistry in the nineteenth century was in its essence

42. Liebig to parents, 18 November 1821, in Berl, *Briefe von Justus von Liebig*, pp. 29–30.

> the consequence of developments already in motion before Liebig's birth.[43]

The consensus is that innovation was brought about in the Germanies (just as in Britain) by individual enterprise in the face of much government indifference. Although, as Hufbauer has shown,[44] it is possible to identify a German chemical community by the end of the eighteenth century, the discipline of chemistry remained marginal and subordinate, or at least coupled to, other academic or practical concerns such as medicine, mining and economics. Gustin concluded decisively:

> even though the hallmarks of the institutionalization of chemistry as an autonomous discipline – its establishment in the [inferior] philosophical faculty – was foreshadowed at the end of the eighteenth century, this step could not occur, paradoxically, until it coincided with the emergence of an explicit need and demand for training in chemistry.[45]

The crucial constituency that came to demand this practical training, just like the English chemists and druggists later in the 1840s, were the German apothecaries who were interested in the improvement of the scientific content of pharmacy and in raising the social status of its practitioners.

The importance and significance of pharmacy for the production of chemists and the creation of laboratory facilities has now been fully recognized by historians, and it is the context in which Liebig's early career and entrepreneurship must be set. For much of the eighteenth century pharmacists were socially the most inferior of the *Gelehrtenstand* (learned professions), for their learning was acquired, not by schooling or university certification, but through apprenticeship. The tightening up of the policing of apothecaries by the medical profession during the second half of the eighteenth century, campaigns to modernise the old-fashioned pharmacopoeia, and the need to come to terms with Lavoisier's chemical revolution raised the question of the more efficient training of pharmacists and led to the private initiatives of Andreas Marggraf in Berlin, Johann Wiegleb at Langensalza, Johann Trommsdorff at Erfurt, and their many imitators. In these initiatives, training institutions were set up within the premises of apothecary shops. Such private institutes for the training of apothecaries were increasingly established in university towns

43. B. H. Gustin, *The Emergence of the German Chemical Profession 1790–1867* (PhD thesis, University of Chicago, 1975), p. 24.
44. Karl G. Hufbauer, "Social support for chemistry. Germany during the eighteenth century: How and why did it change?" *Historical Studies in the Physical Sciences* 3 (1971), 205–31.
45. Gustin, *Emergence*, p. 50.

(where lecturers could be easily obtained), and though not immediately financed by the state, increasingly they gained government approval.[46] Trommsdorff's pharmaceutical school is known to have attracted more than 300 students between 1795 and 1828. (It is interesting to note that the son of the Birmingham manufacturer, Mathew Boulton, studied at Wiegleb's institute in 1789; he may well have been the first English student of practical chemistry in Germany.)

As we have seen, Liebig and Kastner hoped to go into partnership with a private pharmaceutical and manufacturing school – whether at Erlangen, or Darmstadt, or elsewhere is not stated. It was on this private understanding that Kastner engineered a travel stipend for Liebig from the duke of Hessen-Darmstadt. As we shall see, the terms of Liebig's appointment at the University of Giessen in 1824 were that he should teach pharmaceutical chemistry. It might be said, therefore, that Kastner's plans for a private institute were institutionalized at Giessen, partly because of explicit government approval and partly because of Liebig's desperate need for income. Kastner, however, remained at Erlangen. By 1833, in response to Liebig's growing reputation and to his pleas and threats to resign, the government formally recognized Liebig's private institute as an official public function of the University of Giessen.

Erlangen also broadened Liebig's mind socially. Although he allowed some hints of this to drop in his letters to his parents, Liebig was probably naturally reluctant to expand on his extracurricular activities. Two of these interests, however, were responsible for his abrupt departure from Erlangen without a degree. One of the features of German university life after the end of the Napoleonic Wars was the creation of the *Burschenschaften* (students' association), beginning at Jena in 1815. Although these student fraternities indulged in the fairly harmless drinking, dueling, and horseplay of the traditional *Landsmannschaften* (students' organizations with local affinities), they also had a serious political purpose: to further nationalism and democracy against the continued absolutism of German rulers. Inevitably they attracted the more militant and radical student. At the University of Giessen in 1819 a militant group known as the *Schwarzen* (Blacks) emerged under the leadership of Karl Follen, who was later exiled to America. When one of his fanatic followers stabbed the poet Kotzebue to death in 1819, the authorities took action. At a meeting of states in Carlsbad that year, it was agreed to tighten up the administration of universities, including the banning of the *Burschenschaften.* In practice, they were merely driven underground, often as a secret element

46. The key account is Dieter Pohl, *Zur Geschichte der pharmazeutischen Privatsinstitute in Deutschland von 1779 bis 1873* (unpub. thesis, University of Marburg, 1972).

in the still officially tolerated *Landsmannschaften.* Such groups flourished at Erlangen when Liebig was a student.

Seemingly, it was Kastner who suggested that Liebig should start drinking and eating with the *Burschenschaften* in the Zum Lämmle tavern. Finding that several members were, like himself, Rhinelanders, Liebig helped form their own corps, the *Korps Rhenania,* with the motto *virtus firmat fortes.* The earliest surviving likeness of Liebig, by Ernst Fries (the son of Kastner's brother-in-law), shows a dashing and wickedly handsome eighteen-year-old moustached and bearded Liebig wearing the Rhenish ribbon over his chest.[47] Unfortunately, after a boisterous New Year's Eve drinking bout at the Gasthaus, Liebig offended "persons of quality" by knocking the hat off a local counsellor and being involved in a skirmish between students and town louts. These skirmishes went on for several weeks and became sufficiently serious for the authorities to call in the militia. In disgust a majority of students boycotted classes and left Erlangen for Altdorf, a neighbouring town where the first home of the eighteenth-century university had been, and threatened to reestablish a seat of learning there. Now thoroughly alarmed by the loss of income this move would entail, the Erlangen citizenry successfully pleaded for the students' return. Their march back into Erlangen headed by a wind band was triumphant. (There was to be a curious replay of this incident at Giessen in 1846 when, during a student strike and protest and a threat to reestablish the university at Darmstadt, Liebig advised Eben Horsford to carry on working in the laboratory.[48])

There was still the little matter of deciding who should pay for the militia's services, the university or the city. A neat compromise was realized by levying a small tax on every tankard of beer sold in the following weeks. Students and townspeople vied with each other to pay off the debt as quickly as possible![49] Told in retrospect, these events appear like the plot of a Viennese comic opera; but they had serious consequences for Liebig. For a start he was punished with three days' imprisonment for being "particularly active in this disturbance and for making scurrilous remarks about those in authority." Inevitably that authority began to keep a close eye on Liebig and the activities of the *Rhenania.* His rooms were searched by the police, who reported finding 50 yards of the *Rhenanian* coloured ribbon; more seriously, they found incriminating evidence that Liebig was in correspondence with "foreign" (i.e., non-Bavarian) students

47. The original drawing was lost, but a copy made by Liebig's daughter Nanny is in the Bayerische Staatsbibliothek.

48. Rezneck, "Horsford," p. 376.

49. Volhard, *Liebig,* vol. i, p. 25; R. E. Oesper, " Justus von Liebig – student and teacher," *J. Chem. Educ.* 4 (1927), 1461–76.

and that he seemed to be planning a coordination of the separate university corps of Rhinelanders. If the Erlangen authorities seem not to have taken the Carlsbad Decrees seriously before, they now took fright and summoned Liebig before them on 20 March 1822.

This alarming development – the possible punishment was rustication without a degree, or, worse, exile – could not have come at a worse moment for Liebig, for Kastner was about to open negotiations with Grand Duke Ludwig for Liebig's Paris stipend. What happened at Liebig's interrogation is not known; but significantly, the semester being at an end, Liebig returned immediately to Darmstadt, where, either literally or metaphorically, he was placed, or placed himself, under house arrest.[50]

In the event, Liebig got off scot-free, though at least one of his associates who had fled from Erlangen at the same time was expelled from the university. The lack of reprisal against Liebig seems to have caused ill feeling among the *Rhenia,* who openly suspected that Liebig had given the names of corps members to the authorities in exchange for his travel grant.

If these political events were not bad enough, Liebig had also become sexually involved with the poet August Graf von Platen (1796–1835), an older fellow student at Erlangen.[51] Platen, a member of the aristocratic family of von Platen Hallermund in Ansbach, had been a page at the court of King Maximilian in Munich before reluctantly entering military service during the Napoleonic Wars. In 1818 he had decided upon a diplomatic career and to this end had begun studies of philosophy, literature, and political history at the University of Würtzburg and from 1821 to 1826 at Erlangen. Not needing to earn his living, he devoted himself to poetry and to a series of homosexual affairs. He died in Italy at the early age of thirty-nine.

From boyhood Platen kept a diary as a record of "the weakness of the human heart, and the history of my own impressions." His love affairs inspired hundreds of sonnets, many in imitation of the technically difficult Persian ghazels. Despite public knowledge of his homosexuality (notoriously exposed by his rival Heinrich Heine in a literary dispute[52]), on his death his family refused to allow publication of the seventeen volumes of notebooks, and they remained quietly deposited in the Royal Library at Munich until published on the centenary of Platen's death in 1896. By then most of his lovers, including Liebig, were dead, and no embarrass-

50. He uses the term "house arrest" in a letter to Platen, though he may have meant only that he was lying low. See Peter Bumm, *August Graf von Platen. Eine Biographie* (Paderborn, 1990), p. 289.
51. *Ibid.;* Xavier Mayne [pseud.], *The Intersexes* (priv. pr. 1908; repr. New York, 1975), Chap. 13; Jeffrey L. Sammons, *Heinrich Heine* (Princeton, N.J., 1979), pp. 141–47.
52. Sammons, *Heinrich Heine,* Chap. 8.

ment could be caused. In fact, since Platen went to considerable lengths to be discreet in his notebooks by using nicknames, avoiding male pronouns, and effusing the whole with Romantic and Oriental aesthetics and sentiments, it would have been hard to take offence, even at the end of the nineteenth century when attitudes towards homosexuality were hardening. The physical side of Platen's feelings was never made explicit, and his emphasis was always upon romantic longing and intellectual companionship. Not until his sojourn in Italy after leaving Erlangen did Platen seemingly indulge in anything more than embraces and kissing.

Even before the publication of Platen's notebooks and the subsequent publication of his extensive correspondence, knowledge of the strange friendship between chemist and poet had became available when Liebig's son-in-law Moritz Carrière published a discreet account in the *Allgemeine Zeitung* in June 1873. A more explicit account was published in 1908 by the Uranian advocate Edward Stevenson (1868–1942), writing in English under the pseudonym of Xavier Mayne. Because only 250 copies of his *The Intersexes* were published, the erotic nature of Liebig's friendship with Platen remained hidden. When Liebig's first biographers, Kohut and Volhard, came to deal with the affair, they treated it as an example of deep male friendship and comradeship. Consequently, although the matter has been noted by a few Liebig scholars, the full story of the extraordinary relationship and of Platen's obsession with Liebig has been known only since 1990 when Peter Bumm published a detailed biography of Platen.

Platen had already had a series of affairs with fellow students at Würtzburg and Erlangen when in September 1821 he was abandoned by his current lover. For a while he returned to his studies, which included visits to Kastner's lectures. Kastner (with whom Liebig lodged) lived next door to Platen, who one Sunday made a courtesy call on the professor, where, he wrote:

> I came upon him with a student who is eager to study science with him and who is a handsome young man. He withdrew immediately I came, which annoyed me.[53]

It says something of the assiduity of Liebig's chemical studies that Platen had no further opportunity of meeting Liebig until 12 March 1822, under three weeks before Liebig's abrupt return home. Their first meeting consisted of lunch and a long country walk when Liebig gave Platen "proof of a sudden and decided affection" that surprised them both. Liebig, with his still limited knowledge of literature and the wider world beyond Germany, was undoubtedly bowled over by the older man's charm, sophis-

53. Volhard, vol. I, p. 26.

tication, wealth of knowledge, and poetic imagination, though Bumm suspects, too, that there was also an element of satisfaction that he, a shopkeeper's son, should be conversing intimately with a count. Whatever the intensity of their meetings and conversations, they were to be extremely short-lived. The most dramatic moment probably occurred on Liebig's departure from Erlangen when Platen accompanied him as far as Nuremberg, where they spent the night pledging love and exchanging kisses in a hotel.

Inspired by his new but departed lover, Platen poured out his feelings in six remarkable "Liebig Sonnets." That Platen had shrewdly probed the depths of Liebig's character is suggested by one of them:

> Of thy enchantment who is unaware?
> Looking on thee who is there hath expressed
> Regret? How freely nature doth invest
> With scattered charms these cheeks, this lovely hair!
> Thou art so young, so tender, and so fair,
> With all the favours of good fortune blessed;
> Yet he who fears the cunning in thy breast
> Of thee by day and night may well beware!
> The trees still boast of many a green bough,
> The leaves still build many a broad arcade;
> Yet 'neath the bark destruction works e'en now.
> Shall cold reflection rob me of the shade
> Wherein, in sweet repose, I cool my brow?
> Ah no! Beauty bids faith be undismayed![54]

In another he refers more intimately to their secret:

> How can this frigid world participate,
> With its vain bustle, in our happiness?
> Can it imprison it, can it suppress?
> Can it unite us, can it separate?
> We see the things about us dissipate,
> Things which but held our love in close duress,
> Yet true affection never must confess
> Nor shall a third our secret penetrate.
> Those who behold us mid the gay crowd straying
> Little suspect what amorous misery
> Upon our hearts in secrecy is preying.
> I will forget every strange vagary,

54. Reginald B. Cooke, trans., *The Sonnets of Karl August Georg Max, Graf von Platen-Hallermünde* (Boston, 1923), Liebig sonnet 3, p. 54.

> When, thy embraces all my grief allaying,
> My wayward will submits only to thee.[55]

On reaching home, Liebig wrote to Platen, clearly wishing to continue the affair:

> I'm in Darmstadt, alone, as lonely as a man on an uninhabited island, for all that I love apart from my parents I have left behind me in Erlangen. Write to me soon because I can't bear the silence . . . I look forward to Easter when I shall see you again.[56]

Apart from these sentiments, Liebig also enclosed some mundane news, only to be upbraided by Platen in reply for the curt, business-like tone of his letter. Platen clearly wanted only romantic effusions from the besotted youth and acted rather like a wife berating her husband for not writing romantically in his letters. Nevertheless, Liebig's letter caused him to drop his own holiday plans and propose instead a joint trip along the Rhine.

Liebig's tone now suddenly altered as if he realised the predicament he was leading himself into. He confessed that another reason that he had left Erlangen abruptly was that he had been having an affair with a married woman whose husband had discovered the adultery and was furious. Liebig painted himself as the innocent victim of a *femme fatale;* but Platen was not deceived, realizing that Liebig was using this true or false story to indicate a cooling of his ardour. Nevertheless, Platen, still very much in love, decided to begin the journey to Aschaffenburg, to the east of Darmstadt, where they would meet to begin their holiday.[57] It was at this point that Liebig made the already quoted reference to his house arrest as an excuse for his not being able to leave Darmstadt for a holiday. Liebig's position is perhaps understandable, though hardly defensible if he had led the older man along: Each day he was waiting to hear from the Ducal Palace whether he would be able to go to Paris. Had the Erlangen affair put paid to his chances? To disappear suddenly on holiday with an aristocrat might have looked odd and was, in any case, forbidden if he had been truly placed on probation by the grand duke or his officials. Instead, then, Platen came to Darmstadt for three days and met Liebig's parents. It was on the second day, 24 May 1822, that Liebig learned from Schleiermacher, the cabinet minister, that the stipend had been granted him. According to a Liebig family tradition that Volhard heard, it was Liebig's quick-witted

55. *Ibid.,* Liebig sonnet 4, p. 55.
56. Bumm, *August Graf von Platen,* p. 289.
57. *Ibid.,* Liebig's previous letter implies that he had used the hotel at Aschaffenburg once before for a private assignation.

mother who ended the affair. On seeing her son embrace Platen to console him on his departure from their home, she said loudly to her husband, "Liebig, Liebig, Platen is a young girl" [*Mädche*].[58]

Although they were never to meet again, it was not completely the end of the relationship. Platen lived on in hope, writing from Cologne:

> My soul is with you . . . I try to express the inexpressible. Feel it yourself in your heart, love me friend, love me still with the thousandth part of my love and then no man could be as happy as I, as long as the sun shines.[59]

Fortunately for both men, no sooner had he arrived back in Erlangen in July than Platen fell in love with a law student named Hofmann. What Liebig did between news of receiving the stipend in May and moving to Paris in November 1822 is unknown. Bumm for one has speculated that he may have been forced to stay at home for fear of physical reprisals from irate members of the *Rhenia* who were looking for him.

In Paris, however, he must have had time to reflect on the way he had treated Platen's friendship. But instead of openly admitting that he no longer wished to pursue a possibly physical relationship, he continued to play with Platen's feelings. The several letters he wrote to Platen from Paris were informative about his exciting life there, but at the same time he also tried to excuse his conduct and to rekindle the embers of Platen's infatuation. For example, in one letter he drew an embarrassing metaphor from his studies with Biot and Gay-Lussac.

> I wish I could play the harmonica. I would play, and you would perhaps hear the notes which could tell you how I wholeheartedly love you . . . Gay-Lussac, the discoverer of the gas laws, has in his lectures contributed even less towards that end. I wish I were a gas that could expand without end. In an instant I would be satisfied at last, for I would expand myself only towards Erlangen, where I would surround you like an atmosphere.[60]

Unimpressed by this "inferior lover," Platen penned four lines of a ghazel in his notebook:

> Wie, du fragst, warum dein Wohlgefallen
> Mich erwählt, umschlossen hält vor Allen?

58. Volhard, *Liebig,* vol. i, p. 32.
59. Bumm, *August Graf von Platen,* p. 292.
60. Munday, *Sturm und Dung,* p. 38. See August Graf von Platen, *Der Briefwechsel,* 4 vols., ed. P. Bornstein (München, 1921), vol. 2, p. 310.

> Nur Verwundrung kann der Niegeliebte,
> Selten treue dir entgegenlallen.[61]

Although their desultory correspondence continued for a further two years, moving Platen from hope to polite resignation, the final blow came from Liebig in June 1825:

> I hate myself and must box my ears for not having replied to your letter of 10 March. What do you think, you exquisite singer of love, your dear Liebig is a bridegroom and head over heels in love. When I speak of love, as to you my dearest friend, my good Jettchen loves me with a sincerity and fervour of which I had no presentment; my dried-up heart believes this feeling cannot be described, and I conclude that I love you with the same sincerity still . . . You lucky man that you possess the delight of another man's heart . . . Write to me soon. We will write regularly every fortnight . . . I embrace and kiss you and love you heartedly.[62]

Unless Liebig was bisexual, it would be hard not to sympathise equally with both Platen and Liebig's fiancée at the tone of this letter. Sensibly, Platen never replied; but when in 1864 a committee was established in Munich to repair Platen's gravestone, Liebig was a subscriber.

It has been plausibly suggested that the vehemence of Liebig's attack on *Naturphilosophie* in 1840 may have been a mask for his feelings of repugnance for homosexuality after his marriage, and even for the bitter and unfair comments on Kastner that he expressed in his autobiographical notes. In the latter he claimed that Kastner's lectures were "without order, illogical, and arranged just like the jumble of knowledge which I carried about in my head."[63] It is true that Kastner was touched by the spirit of *Naturphilosophie,* a system that looked for answers to largely metaphysical questions and sought connections between all physical and spiritual phenomena; but he was undoubtedly also a good practical chemist. As Löw has emphasized, Kastner rarely made metaphysical statements in his practical writings; although he may have found comfort in such explanations they did not form part of his working practice.[64] And after all, without Kastner's professional guidance, Liebig might well have remained a provincial shopkeeper. The same defence cannot be made for philosophers like Lorenz Oken and Friedrich Schelling, whom Liebig also had in

61. Bumm, *August Graf von Platen,* p. 313. The verse may be translated roughly: "Why do you choose to tell me of pleasures in other's company/ Merely to wound the person you have never really loved."
62. *Ibid.,* p. 372.
63. Liebig, "Autobiography," p. 260.
64. R. Löw, *Pflanzenchemie zwischen Lavoisier und Liebig* (München, 1977).

his sights at Erlangen. But there is no evidence that Liebig, who attended some of Schelling's lectures at Erlangen (as did Platen), was ever in any way influenced or misled by them. As Liebig himself said, "Schelling possessed no thorough knowledge in the province of the natural sciences, and the dressing up of natural phenomena with analogies and images which was called exposition, did not suit me."[65]

Meanwhile Kastner, using the pretext of delivery of a personal copy of his latest book to the grand duke, took the opportunity on 12 April 1822 to praise Liebig's achievements and potential:

> Allow me, Your Highness, to present you with a copy of my *Vergleichende Übersicht des Systems der Chemie* through the hands of my zealous former pupil, Liebig of Darmstadt. The diligence with which the young Liebig has studied Physics and particularly Chemistry under my instruction, first in Bonn, and then here in Erlangen, as well as the evidence provided by the experimental work he has published, leads me to anticipate great things from him for his country [Hessen-Darmstadt] and for science. Moreover, it makes me desire that he should become one of the growing number of my students who now teach Chemistry in universities and similar institutions of higher learning with approbation and profit. A six months' stay in Paris, which would suffice for economical preparation and revision before he returned to Darmstadt would, in my opinion, be enough to help develop the young Liebig into a future chemistry teacher who, for example, Darmstadt would be glad to have if he were appointed as professor of Chemistry at the Military School, and capable also of founding a private training school for manufacturers and apothecaries.[66]

To this, Liebig added his own petition, which stressed that his parents, who still had seven unprovided-for children, could not afford his further studies.

It should not be thought that such petitions were unusual. We know of at least one other Hessian student who had been sent to study with Fourcroy in Paris at the beginning of the century – W. L. Zimmerman. Ironically, this was the chemist Liebig was destined to displace in 1824. Nor were scholarships peculiar to Hessen. In 1824, the young philosopher Ludwig Feuerbach was sent to study in Berlin by King Maximilian I of Bavaria.

Kastner's and Liebig's petitions worked and, on 24 May 1822, Liebig was awarded 330 gulden.

65. Liebig, "Autobiography," p. 262.
66. Berl, *Briefe von Justus Liebig*, pp. 34–35.

Paris

Liebig's grant of 330 gulden was supposed to be sufficient for six months' study in Paris, but when he arrived in early November 1822, within a week he was pleading for more money to Ernst Schleiermacher (1755–1844), cabinet secretary to the duke of Hessen-Darmstadt. His travel costs, his lodgings in the rue Harley, and food all proved much more expensive than he or the Court had anticipated. Most of the letters to his parents and to Schleiermacher, which were published by Berl in 1928, are filled with references to his financial worries. On the other hand, he was overjoyed by the atmosphere of scientific dedication that he found in Paris. In a letter to an Erlangen *Rhenian* friend, August Walloth, he wrote:

> Vain in my very little knowledge, I came here, expecting my life would only be a promenade; but all at once I found myself among people of whom I am the most insignificant. It struck me like a bolt from the blue sky, but what a beneficial thunderbolt! The old tree of prejudice with its deep roots is smashed, and in place of the old there is growing up a slowly-developing new tree with green foliage; and I hope, because the destruction is at the same time the blossom, a fruit-bearing tree will result. You see what a metamorphosis has taken place in me in the meanwhile. The lectures of Gay-Lussac, Thenard, and others are the cause of this. Science is no longer an old nag (*Gaul*) which one only has to saddle so that one can ride on it. It is a winged horse – the more I try to reach it, the more it flees from me . . . I thought I had worked hard in Darmstadt, but in Paris the daily song is from seven in the morning until midnight and later, and I enjoy it.[67]

The cause of this metamorphosis from an unsophisticated bucket chemist into a natural philosopher (the word *scientist* was not coined until the 1830s) were the lecture courses Liebig was able to attend at the Faculty of Science.

Joseph-Louis Gay-Lussac (1778–1850), a protegé of Berthollet's group of research-oriented chemists and physicists at Arcueil, had been appointed a professor of physics at the faculty of science in 1809.[68] The faculty shared premises and students with the Ecole Normale, and his task, along with other teachers, was to prepare students for public examinations for the secondary school licence, for under Napoleon's reforms and ahead of the rest of Europe, science teaching formed part of the national curriculum. Although the numbers of students varied from

67. *Ibid.*, p 49.
68. M. P. Crosland, *The Society of Arcueil. A View of French Science at the Time of Napoleon I* (London, 1967).

course to course and semester to semester, by Liebig's time, Gay-Lussac and his fellow professors were commonly lecturing to audiences of 300 to 400 students. Liebig would have experienced nothing comparable at Bonn or Erlangen, where such figures would have represented the entire university population.

As a young man in 1808, following a vigorous mathematical training at the Ecole Polytechnique, Gay-Lussac had discovered the simple integral law of combining gas volumes, which played an important role in the development of the atomic theory and the determination of atomic weights. The Scots toxicologist Robert Christison, who had attended his lectures in 1820, recalled:

> Gay-Lussac was perhaps the most persuasive lecturer I have ever heard. His figure was slender and handsome, his countenance comely, his expression winning, his voice gentle but firm and clear, his articulation perfect, his diction terse and choice, his manner most attractive, and his lecture was a superlative specimen of continuous unassailable experimental reasoning.[69]

Gay-Lussac had collaborated with Louis Jacques Thenard (1777–1857), a pharmacist who had been trained by Louis Vauquelin and Lavoisier's disciple Antoine Fourcroy. Together Gay-Lussac and Thenard developed an important technique for analyzing organic compounds, using potassium chlorate as an oxidising agent to produce carbon dioxide and water, volumes and weights of which could be determined. Later they had modified this rather dangerous procedure by using copper oxide as the oxidiser. Already adept at mineral analysis, this was the technique that Liebig needed to learn. His own later modification of the procedure was to be a key event in his career as a chemist and teacher.

In his first semester, Liebig attended the lectures of Gay-Lussac on physics and Thenard on chemistry; in the second semester, they gave way respectively to Jean-Baptiste Biot (1774–1862), another protegé of the Arcueil group who had done outstanding work on optics, and Thenard's assistant Pierre Dulong (1785–1838), another student of Berthollet's who had achieved fame for his work on the law of atomic heats with A. T. Petit. Liebig soon realised that if he was to gain the maximum benefit from Gay-Lussac's and Biot's lectures, he would need to improve his knowledge of mathematics. The quantitative and mathematical approach to science was another big difference he found between German and French teaching. He had, however, not entirely lost sight of the practical nature of his intended career as a pharmaceutical and manufacturing chemist and teacher. He

69. M. P. Crosland, *Gay-Lussac* (Cambridge, 1978), p. 148.

Crayon drawing of Liebig as a student in Paris made by Christian Julius Portmann (1799–1861). (Courtesy Liebig-Museum Giessen)

also attended the interesting lectures on industrial chemistry that were offered by Nicolas Clément (1779–1841), who, together with his father-in-law, Charles Désormes, owned a chemical factory at Verberie, where they had made important contributions to the understanding of the mechanism of the lead-chamber process for the manufacture of sulphuric acid. As Liebig told his government patron Schleiermacher on 17 February 1823:

> Clément's application of chemistry to arts has a purely practical tendency. One finds used in his lectures everything in relation to life and, furthermore, useful things newly discovered or invented for factories. Because Clément is known as an excellent chemist one always finds new and interesting results in his lectures. It would be a very great pleasure to me if you would allow me to tell you those things in the future.[70]

70. Berl, *Briefe von Justus von Liebig*, pp. 47–49.

And he promised to send his father, who was apparently thinking of preparing alcohol for sale, details of Clément's discussion of distillation techniques after they were delivered. If these lectures and the extra mathematics coaching he took were not enough, Liebig also rekindled his Latin and began to learn Italian and English. When we bear in mind that he was also having to listen, learn, and think in French, this was a formidable schedule of activity for any student. Useful friendships were also forged with his roommate Karl Oehler (1797–1874), who in 1842 was to join Sell in distilling coal tar at Offenbach; and a young professor of philosophy from the University of Giessen, Friedrich Schulz (1797–1829), who may possibly have replaced Platen in Liebig's affections.[71]

Liebig's diligence in attending lectures began to be rewarded when Thenard arranged research space for him in the private laboratory of Gaultier de Chaubry, an independent investigator and former pupil of Vauquelin's, who had frequently collaborated with Gay-Lussac and Thenard. The facility cost Liebig's parents another 600 francs.

The Translation to Giessen

In Gaultier's laboratory Liebig had the bright idea of extending his earlier work on fulminates, using his new knowledge of organic analysis. By July 1823 he had prepared a paper that he asked Thenard to present to the French Academy of Sciences on his behalf. Thenard, as the current president, was unable to do this and asked Gay-Lussac to do it instead on 28 July 1823, while Liebig performed the demonstrations. In the audience was the sixty-year-old naturalist and polymath Alexander von Humboldt, who was always on the lookout for talented young men. Whether or not Gay-Lussac needed persuasion, Humboldt urged him to take Liebig into his private laboratory at the Arsenal. Here Gay-Lussac, who had a longstanding interest in cyanogen, to which the fulminates were related, and Liebig jointly extended the work on fulminates, publishing the results in the *Annales de chimie* the following year.[72] Forty-four years later, in an afterdinner speech at a banquet of jurors of the Paris Exposition of 1867, Liebig told the extraordinary story of his collaboration with Gay-Lussac:

> Never shall I forget the hours passed in the laboratory of Gay-Lussac. When we had finished a successful analysis (you know without my telling you that the method and the apparatus described in our joint

71. Schulz, who had been sent to Paris to learn Persian, was murdered by robbers in Kurdistan. See Michaud's *Biographie Universelle* (Paris), vol. 38, p. 467. For Oehler, see *Hessische Biographien*, vol. I, p. 196.

72. J. Liebig, "Analyse du fulminate d'argent," *Annales de chimie* 25 (1824), 285–311.

> memoir were entirely his), he would say to me, "Now you must dance with me just as Thenard and I always danced together when we had discovered something new." And then we would dance.[73]

Another well-known anecdote, first reported by Liebig's English student and close friend E. K. Muspratt, is that Liebig forgot to ask for Humboldt's name and address when the old man first spoke to him at Liebig's Academy début and asked him to dinner.[74] It was only after he had rudely missed this engagement that a friend identified Liebig's intense questioner as the great Prussian traveller and savant. Horrified by his *faux pas,* Liebig had rushed round to apologise and found the old man amused by Liebig's lack of social grace. This is a wonderful story; unfortunately, correspondence reveals that Liebig had already been introduced to a sitting of the Academy by Humboldt's co-worker Knuth in January 1823, six months before the *Knallsilber* paper was laid before the Academy. During the interim, he had continually met academicians socially, so that it would be reasonable to suppose that he must have met Humboldt before July.[75]

Whatever the truth of Liebig's gaffe, while Liebig was dancing with Gay-Lussac, Humboldt took it upon himself to write in French to the grand-duke of Hessen-Darmstadt through Schleiermacher. He urged that this bright young Hessen chemist be given a post as soon as possible.

> We have the privilege, my Lord, that we have among us one of your subjects who by the superiority of his talent, his extensive knowledge of chemistry, and the sagacity of his spirit has quickly brought him to the attention of the Royal Institute of France. Dr Liebig combines with these his qualities of mind a gentleness of character and such goodness of manners which is uncommon among savants of his age. If my feeble voice could be of some weight, I would ask your Royal Highness to continue your efforts to support him. As a Professor he would give honour to your country and the gratitude that I shall feel towards a Sovereign who deigns to honour such a distinguished talent will be shared by my Academy colleagues, Mons. Gay-Lussac, Thenard, Dulong and Vauquelin.[76]

73. C. A. Browne, "The 'Banquet des chimistes,' Paris, April 22, 1867," *J .Chem. Educ.* 15(1938), 253–59.
74. First told by Muspratt to J. Campbell Brown, who reported it in his translation of Liebig's autobiography, *Chemical News* 63 (1891), 265. See W. A. Shenstone, *Justus von Liebig* (London, 1901), pp. 181–89; printed in E. K. Muspratt, *My Life* (London, 1917), pp. 158–60.
75. See, however, Liebig's dedication of *Die Chemie in ihrer Anwendung auf Agricultur* (Braunschweig, 1840) to Humboldt.
76. Berl, *Briefe von Justus Liebig,* p. 68.

Liebig had long ingratiated himself with Schleiermacher, who had a private long term interest in creating a great natural history museum and collection in Darmstadt. Aware that Cuvier was active in Paris, he had encouraged Liebig to make the acquaintance of the great palaeontologist, who supplied fossils to Schleiermacher with Liebig as intermediary. Since Schleiermacher could have done this just as easily through the Hessen envoy in Paris, Baron von Pappenheim, it shows the complete confidence that the state already had in Liebig's abilities. It was also to Liebig's advantage. While he was in Paris, Liebig worked on a comparative analysis of fossil and modern mammalian teeth, although it was never published.[77]

As Liebig's correspondence shows, for most of 1823 he still imagined that through Kastner's and Schleiermacher's patronage he would obtain a teaching appointment in Darmstadt, perhaps at the Military Academy, from which he would be able to branch out privately into a school for teaching pharmacy and manufacturing. Indeed, when his father mentioned the University of Giessen in passing in a letter in March 1823, Liebig sharply rejected the establishment and in another letter referred to Giessen as offering only "restricted opportunities" for chemistry.[78] A difficulty about obtaining such an appointment there was that having left Erlangen in such a hurry, he had no doctoral degree, and such a form of promotion was unavailable to foreigners in Paris.

It was Kastner who came to Liebig's rescue with a plan to use his personal influence at Erlangen to obtain the necessary degree. As Munday has brilliantly shown, the outcome was to all intents and purposes an honorary degree, or more bluntly, a paid degree, *in absentia*.[79] All Liebig's biographers have been puzzled by the fact that there is no surviving copy of Liebig's Erlangen thesis. Given its fascinating title, "On the Relationship between Mineral and Plant Chemistry," with its hint of Liebig's future book *Agricultural Chemistry* (1840), Volhard was reduced to suggesting that a Liebig admirer had pilfered the thesis from the university archives. The arrangement was that Liebig's parents were to send Kastner 66 thaler (a sum equivalent to Liebig's first stipend of 330 gulden), the sum normally required to print a doctoral diploma and administration fees being only 66 ordinary gulden. (The thaler, or gold florin, was worth five gulden, or ordinary florins.) In May 1823, Kastner assured Johann

77. Cited by Munday, *Sturm und* Dung, p. 39. There are also several references in Berl, *Briefe von Justus Liebig*.
78. Liebig to parents, 30 March and 28 April 1823, Berl, *Briefe von Justus Liebig*, pp. 52, 53.
79. Munday, *Sturm und Dung*, pp. 40–47. German scholars will not necessarily agree with this interpretation, but unless a copy of Liebig's dissertation is discovered, Munday's interpretation seems to fit the evidence.

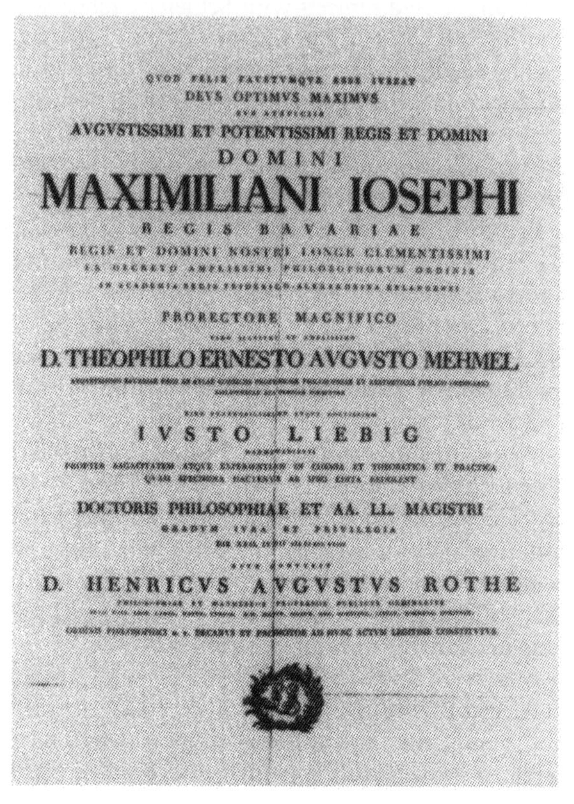

QVOD FELIX FAVSTVMQVE ESSE IVBEAT
DEVS OPTIMVS MAXIMVS

AVGVSTISSIMI ET POTENTISSIMI REGIS ET DOMINI
DOMINI
MAXIMILIANI IOSEPHI
REGIS BAVARIAE
REGIS ET DOMINI NOSTRI LONGE CLEMENTISSIMI
EX DECRETO AMPLISSIMI PHILOSOPHORVM ORDINIS

PRORECTORE MAGNIFICO

D. THEOPHILO ERNESTO AVGVSTO MEHMEL

IVSTO LIEBIG

DOCTORIS PHILOSOPHIAE ET AA. LL. MAGISTRI
GRADVM IVRA ET PRIVILEGIA

D. HENRICVS AVGVSTVS ROTHE

Liebig's doctoral diploma, University of Erlangen 1823. (Courtesy W. Lewicki)

Liebig that everything was proceeding happily and that the doctorate would be awarded to his son in a few weeks. As Liebig had written no dissertation Kastner cajoled the Erlangen faculty into accepting in lieu of it one of Liebig's published papers, a Latin petition, a brief answer to a question that Kastner had supposedly asked Liebig – "How does mineral chemistry relate to plant chemistry?" – and a curriculum vitae. When asked by the faculty about the lack of a formal dissertation, Kastner pleaded Liebig's absence in a foreign county and his urgent need for an extension to his grant, which would come only if he had the doctorate. As the minutes reveal, on the understanding that Liebig promised "great things to come" it was agreed to interpret the single-line answer to Kastner's question as a thesis. On that basis the doctorate was conferred on 23 June 1823 for 66 gulden; the remaining 264 gulden were presumably pocketed by Kastner as a managerial fee.

Liebig was exultant about his doctoral success and immediately began signing himself, "Dr. J. Liebig," even in letters to his parents.[80] His title, the collaboration with Gay-Lussac on Knallsilber, and the combined weight of recommendations by Kastner and Humboldt ensured preferment. Exciting plans to travel to England with Humboldt before the summer semester came to nothing, and Liebig left Paris quietly for Darmstadt in April. On 26 May 1824, Liebig was duly made extraordinary professor of chemistry at the University of Giessen.

At Bonn Liebig was an earnest student with no higher goal than to learn practical chemistry to improve his father's business. Continually interested in technical processes, on which he frequently reported to his father, Liebig's aim seems to have been to collaborate with his brother, once Louis had qualified in pharmacy, and to extend his father's business into a pharmacy and chemical manufactory. At Erlangen, however, he entered a new social world and blossomed both socially and sexually. Turned down by his brother, Liebig now aspired, like Trommsdorff and other pharmacists before, to make a career as an independent teacher of pharmacy in an academy that would instruct pharmacists in the chemical principles of their work. At the same time, by teaching the principles of industrial chemistry, he would fulfil Kastner's long-standing claim that chemistry was the backbone of a nation's prosperity in the modern age. Such an enterprise seemed even more important as he read more chemical literature at Erlangen and learned about Germany's backwardness as a scientific nation, Britain's industrial revolution, and Paris's status as the Mecca of scientific intelligence. He trained at Erlangen to fulfil the pharmaceutical dream and industrial ambition of his own country.

Paris had been a revelation in the quality and quantity of its scientific teaching. Liebig was always to remember the collaboration with Gay-Lussac as the turning point in his life and an experience that had put him in touch with the very roots of modern chemistry through the pupils of its great founder, Antoine Lavoisier. Although he was later to quarrel extensively with other French chemists, he always remained on the best of terms with Gay-Lussac. His first foreign student at Giessen in 1833 was none other than his master's oldest son Jules. And when Gay-Lussac published his account of a titrimetric or volumetric method for the analysis of silver as a result of his work in the Paris Mint, Liebig translated it into German.[81] Paris also provided Liebig with the opportunity to meet and mix

80. Berl, *Briefe von Justus Liebig*, p. 61.

81. J.-L. Gay-Lussac, *Instruction sur l'essai des matières d'argent* (Paris, 1832); translated by Liebig as *Vollständiger Unterricht über das Verfahren Silber auf nassem Wege zu probiren* (Braunschweig, 1833).

with some of the highest and mightiest of European science, and the Erlangen doctorate (dubious though its origins had been) and the presentation he made before the Academy of Sciences provided him with an excuse to make himself known to other famous men, including Döbereiner, Berzelius, Oersted, and Thomas Thomson. Through Humboldt's patronage and the continued faithful support of Kastner, the twenty-one-year-old son of a shopkeeper was transformed into a professor of chemistry. The cost to his parents and the state, more than 5,000 gulden, was to the benefit of European chemistry, and on a personal level it raised Liebig from the respectable working class to the respected middle-class intelligentsia.[82]

82. Volhard, *Liebig*, vol. I, p. 33.

2

Organic Analysis and the Giessen Research School

> In this apparatus there is nothing new but its simplicity and thorough trustworthiness. [Liebig, J. (1831), 1]

Giessen, the principal town of Upper Hessen, had a population of only 5,500 people when Liebig arrived at the university in May 1824. There was no rail connection with Marburg and Kassel to the north or Frankfurt to the south until the 1850s, and Giessen could be reached only by the *Familienwagen,* or horse omnibus.[1] The town, which was almost destroyed in World War II, still retained its medieval appearance in Liebig's day. In 1527, the then Landgrave Philipp, a strong supporter of the Reformation, created the first Protestant University of Hessen at Marburg. When he died in 1567, he willed that Hessen should be partitioned among his four sons into four *Länder:* Hessen-Kassel, Hessen-Marburg, Hessen-Rheinfels, and Hessen-Darmstadt. Although the Augsburg Edict of 1555 allowed official toleration of both the Protestant and Catholic religions, all four *Länder* remained Protestant, with Hessen-Kassel largely subscribing to Calvinism and the other three states to Lutheranism.[2] Because of his Lutheranism and anti-Calvinism, Ludwig VIII (1678–1739), the sixth Landgrave of Hessen-Darmstadt, decided to create a Lutheran University, the Ludoviciana, at Giessen in 1607. Until the 1830s this university remained a small provincial backwater of some 300 to 400 mainly local students and was divided into the traditional faculties of philosophy, medicine, law, and theology. It was scarcely renowned for its scholarship. An attempt between 1775 and 1788 to create a faculty of economics, or cameralism, to train state officials for the Treasury and its commercial ventures proved to be a disastrous failure. As late as 1827, the university library opened only four days a week for one hour.[3]

Carl Vogt, who was born in Giessen and whose father was a professor

1. E. K. Muspratt, *My Life* (London, 1917), pp. 21–23.
2. Donald B. Tower, *Hensing, 1719* (New York, 1983). Chapter 4 provides a coherent English account.
3. Peter Moran, *Kleine Geschichte der Universität Giessen 1607–1982* (Giessen, 1982).

of medicine at the university, is one of our principal sources of information on life at Giessen in the 1820s and 1830s.[4] He described Giessen as a town of twisting narrow streets whose earlier fortification ditches had long ago become gardens and tree-lined walks. The one humpbacked bridge over the River Lahn sufficed only as a pedestrian crossing; horse- and ox-drawn traffic was forced to ford the river. Like Marburg and other classic German "university towns," Giessen was economically dependent upon the biannual influx of students and professors for its livelihood.[5] Indeed, apart from a tobacco factory, the university was the sole source of revenue. Knowing this, the government was always able to suppress any political agitation by the inhabitants by threatening to move the university to Darmstadt. Situated on a bend in the picturesque River Lahn as it turns westwards towards Wetzlar (*giessen* referred to the streams that *pour* into the Lahn), the university had no building other than the *Kollegiengebäude,* which functioned as a library, registry, and lecture hall. With one or two exceptions, most teaching was done in individual professors' homes, and professors, like the townspeople, commonly took in several student lodgers during the two teaching semesters.

Liebig's appointment has to be seen as part of a deliberate ploy by Ernst Schleiermacher and the grand duke to modernise the university and make it more attractive both to Hessens and to German students from other *Länder.* We have already seen that Liebig had preferred to go to Bonn and Erlangen rather than to his local university and that he initially held no very high opinion of the institution. Apart from Liebig, Schleiermacher had also recently appointed the philosopher Joseph Hillebrand (1788–1871), the forestry scientist Johann Hundeshagen (1783–1834), the philologist Johann Adrian (1793–1864), the mathematician Hermann Umpfenbach (1798–1862), a lively writer of geometrical textbooks, and an anatomy demonstrator Friedrich Wernekink (1798–1835), who worked on the nervous system before switching completely to his real interest in 1826 when he was appointed professor of mineralogy.[6] All these men were at least five years older than Liebig. Carl Vogt's father Philipp (1787–1861), a Giessen physician, had been appointed to a chair of medicine in 1817 and concentrated on materia medica. There was also the oriental scholar E. F. Schulz, who had become a close friend of Liebig's in Paris, where he had been sent for further training following his appointment as professor of philosophy in 1822.

4. C. Vogt, *Aus meinem Leben. Errinnerungen und Rückblicke* (Stuttgart, 1896). A section from this concerned with Liebig is translated into English in H. G. Good, "The Early History of Liebig's Laboratory," *J. Chem. Education* 13(1936), 557–62.
5. E. W. Gilbert, *The University Town in England and West Germany* (Chicago, 1961).
6. For Umpfenbach and Wernekink, see entries in *Allgemeine Deutsche Biographie.*

Alongside these more forward and research-oriented professors, there was much deadwood. In the medical faculty Bernhard Wilbrand (1779–1846), a professor of anatomy, physiology, and natural history, as well as director of the Botanic Gardens, became a particular bugbear of Liebig's, because of both his overt commitment to *Naturphilosophie* and the embarrassing fact that he did not believe in the circulation of the blood. Wilbrand was to be mercilessly satirised by Georg Büchner in his posthumous drama *Woyyzeck*. Büchner had studied medicine at Giessen in 1835 alongside Carl Vogt. There was also the incumbent professor of chemistry when Liebig arrived at Giessen, Wilhelm Zimmermann (1780–1825), who had proved a bitter disappointment to the state, which had sent him to train in Paris at the beginning of the century. Following that, he had taught in a Giessen high school for many years. Although he knew Kastner well, he lacked the latter's drive and ambition, publishing little and teaching chemistry to medical students without enthusiasm and with little reference to anything that had happened since Lavoisier.

Although Giessen was one of the earliest German universities to have a chair of chemistry – Johann Hensing, the first professor, had discovered phosphorus in the human brain at Giessen in 1719 – the attempt to teach applied chemistry within the economics faculty in the 1770s through a second chair of chemistry was as short-lived as the faculty. Between 1779 and 1815 chemistry was taught solely within the medical faculty by K. W. C. Mueller, whose duties also included the administration of the Botanical Gardens. On his death in 1815, Philipp Vogt senior, the professor of medicine, was given the additional responsibility of teaching chemistry and pharmacy. Three years later, Zimmermann was appointed to the philosophy faculty to teach general chemistry and mineralogy, leaving medical chemistry and pharmacy to Philipp Vogt in the medical faculty. Because he was a friend of Kastner's, the state perhaps hoped that he would resuscitate the cameralist experiment of the 1770s.

As professor *extraordinarius*, a poorly paid position compared with a full *ordentlicher* professorship, Liebig was immediately faced with the Zimmermann problem. There was no question of their mutual collaboration and respect; the state had virtually thrown sand in Zimmermann's face by appointing a very young man to replace him.[7] No one can blame Zimmermann for defending his position, his machinations to monopolise the available students and the limited laboratory facilities that he had in

7. G. Weihrich, *Beitraege zur Geschichte des chemischen Unterrichtes an den Universität Giessen* (Giessen, 1891); B. H. Gustin, *The Emergence of the German Chemical Profession* (PhD, University of Chicago, 1975), pp. 90–102; P. Munday, *Sturm und Dung* (PhD, Cornell University, 1990), pp. 68–80.

Wilbrand's building in the Botanical Gardens. Both because Liebig had indicated to Schleiermacher his interest in establishing a private institute for the training of pharmacists and manufacturers, and to prevent conflict, Liebig's initial commission was only to teach pharmacy. Zimmermann was to retain a monopoly of teaching general chemistry. As a further gesture towards harmony, pharmacy students were transferred with Vogt's approval from the faculty of medicine to the faculty of arts; this turned out to be an important move for the future that was widely copied at other universities. It meant that chemistry, emerging from pharmaceutical chemistry, could become an autonomous discipline and not merely one of medicine's many collateral sciences.

The state's thoughtfulness did not solve the problem of where Liebig was to teach practical pharmacy, and Zimmermann immediately made it clear that he would not share his small laboratory space with Liebig. The matter was of great urgency. Because Liebig's salary was only 300 gulden, less than the stipend he had had for six months in Paris, he urgently needed laboratory and lecture fees from students. When Liebig found teaching space elsewhere within the university, Zimmermann immediately used his weight in the Senate to force Liebig to abandon it. It took three months of pleading with Schleiermacher before the government found space outside the university's jurisdiction where a new laboratory could be created – the guardhouse of a relatively new (1817–19) but disused barracks on the south side of the town.[8] As Liebig told an Erlangen friend, August Walloth:

> They have decided to build a new laboratory. Since the barracks offered enough space, one of the annexes with colonnades was chosen. Before they made up their minds three months had passed, in spite of my constant urgings. Imagine, three months! At this moment I am busy with the internal installation. Whether this can be finished within five weeks, only Heaven knows. But in five weeks I have to start my lectures. Imagine, after great difficulties I could only get a donation [subsidy] of 100 gulden for the laboratory and for buying instruments, reagents and materials. May Heaven have mercy on me! How can I manage with so little. But I could stand all of this if I could only be spared the endless tricks of Zimmermann and his clique. They embitter my life here and spoil every pleasure.[9]

Indeed, we know from a cash book kept by Liebig's father that Johann had to continue to subsidise his son's expenses for some further years even though Liebig was now a professor.

8. The garrison had moved in 1821 and turned over for use as a hospital. Only the guardhouse was made available to Liebig.
9. E. Berl, *Briefe von Justus Liebig nach neuen Funden* (Giessen, 1928), p. 75.

View of Giessen garrison block that housed Liebig's original 1824 laboratory facilities and living quarters on the first floor. From a photograph taken in the early 1950s. (Courtesy Liebig-Museum Giessen)

Following the opening of the winter semester in November 1824, it became a tussle between Zimmermann and Liebig as to who could obtain the most students from whatever faculty. Liebig began with twelve medical students – more than Zimmermann, who immediately cut his own fee to 9 gulden, half of Liebig's fee of 18 gulden. When this 9 gulden loss leader failed to entice students back to his classes, Zimmermann offered free tuition. When this ploy also failed, he asked the government for a period of study leave (probably to look for a position elsewhere). Obviously severely depressed and drunk, he drowned himself in the Lahn on 19 July 1825.

Liebig's reaction may seem callous, but he had had to put up with a great deal of enmity and deliberate obstruction. He wrote to Schleiermacher immediately following Zimmermann's death:

> As painful as the unexpected death of Professor Zimmermann has been for all of us, yet I cannot allow myself to be blinded to the good which can come out of it for me. Although since my coming here, Zimmermann has advertised free lectures on chemistry each semester, he never actually held them, as the students preferred to attend my

> classes [for a fee] on the same topic. Up to now, I have held the *ausserordentliche* teaching position in chemistry. I believe, therefore, that, now that the position has been made vacant by his death, I am entitled to be a candidate for the *ordentliche* teaching position in chemistry.[10]

Since, by unhappy coincidence, Blumhof, the professor of technology and mining had also recently died, Liebig suggested that the government should combine both positions, which would enable him better to emphasize industrial chemistry in his lectures. Accordingly, he resurrected plans for a private institute of pharmacy and manufacturing, for which he gained the promise of cooperation from Wernekinck (who would teach mineralogy and geology), Umpfenbach (who would offer mathematics), and Georg Schmidt (1768–1837), the professor of physics who agreed to lecture on his subject. Unfortunately, the prospectus for this institution that he sent Schleiermacher on 30 July 1825 has not survived. Liebig's letter stated:

> You will see from the enclosed prospectus that Professor Wernekinck, Umpfenbach and I have united to establish a chemical-pharmaceutical institute. In Germany there are only two such institutions: one at Erfurt under the direction of Professor Trommsdorff, and the other in Jena, erected by Professor Göbel [there were, in fact, others]. The number of students who have applied every year as freshmen at these institutes is so great that both can accept hardly one-sixth of the applications. Accordingly, they fix their numbers at twenty. But I believe that we could take many more at Giessen if one of us alone directed the institute, because a personal interest will be increased considerably by a joint action. We have no doubt at all that our institute will be a great success, especially since Professor Schmidt has decided to lecture on physics. It promises fame for Giessen and the university. There will always be about twenty or thirty students, which will increase the numbers of students at Giessen overall.[11]

Liebig's and his colleagues' strategy was obviously to allow the state to underwrite the venture with a laboratory subsidy in exchange for the promise of its bringing extra students to Giessen. The strategy worked, and within two years Liebig was able to tell Schleiermacher the intriguing news:

> My chemical institute has been supported more this year [1826–27] from abroad than from this country. Among my pupils are French and

10. *Ibid.*, p. 78.
11. *Ibid.*, p. 79.

> Dutch. I hope it will increase much more next year and that the students will be convinced that something can be learned here.[12]

Schleiermacher had duly made the recommendation of Liebig's promotion to the *ordentlicher* chair to the Giessen senate. With the support of the philosophical faculty (only Zimmermann's ally, Heinrich Pfannkuche, the professor of Hebrew, dissented) and strong vocal support from Vogt in medicine and Schmidt in physics, Liebig was duly elected to the full professorship of chemistry on 7 December 1825 at the increased salary of 800 gulden and a laboratory allowance of 120 gulden. On the other hand, the senate categorically refused to allow Liebig's proposed pharmaceutical institute to be incorporated within the university. It was "the state's [i.e., the university's] task," they minuted, "to train civil servants, *not* apothecaries, soapmakers, beer-brewers, dyers, and vinegar-distillers."[13] Faced by such academic unanimity (only Vogt had spoken in support of the incorporation), Schleiermacher ruled on 17 December 1825 that Liebig's institute would have to be a private venture. This decision explains how Liebig apparently got away with admitting nonmatriculated students (i.e., students without the *Abitur*) at Giessen. As it was a private venture, he could do what he liked; the muddying of the waters came insofar as he offered the same lecture and laboratory space in the Giessen barracks to properly matriculated students in the medical and philosophy faculties. This intellectual and social mixing of students clearly remained a thorn in the side of some of his professorial colleagues for the next decade, even though Liebig's students (like the distinctive forestry students) remained withdrawn from other students. Despite the politically radical atmosphere of the university during the 1830s and 1840s, Liebig and his students were to remain outsiders. Only in 1848 did Liebig become seriously involved with politics when he volunteered for the Giessen home militia.[14] Liebig's other disappointment in December 1825 must have been the discovery that Zimmermann's apparatus, which had been locked away in his house to prevent Liebig's using it, was so out of date as to be of no use.

With his increased salary, a payment of 60 gulden for a *Diener* or *Familus* (a laboratory steward), confidence that his institute would succeed and that any homosexual inclinations were banished, Liebig decided to marry. His work as a chemical consultant for state enterprises (see Chapter 4) had brought Liebig into contact with a Darmstadt official named Michael Moldenhauer (1772–1860), whose family he may well have known as a teenager. It was Moldenhauer's daughter Henriette [Jet-

12. *Ibid.*, p. 83.
13. Volhard, vol. i, p. 58, translated by Gustin, *Emergence*, p. 95.
14. Volhard, vol. i, 176. For Liebig's political views, see Chapters 11 and 12 of this book.

tchen] Moldenhauer (1807–81), whom he wooed and married in May 1826. Volhard, who knew her from his childhood in Giessen, described her as a pretty woman who had received the typical poor education of middle-class girls of her generation. Although not particularly clever or able to understand her husband's work, she had a healthy human intelligence that made her a tender, loving wife and excellent mother. Whatever Liebig's sexual activities and preferences may have been as a student, there is no evidence that he was ever anything but a devoted husband and father.[15] As their correspondence shows whenever Liebig was travelling, their mutual love remained undiminished over the years, and she became a magnificent hostess after 1837 when, influenced by what her husband had experienced in upper- and middle-class households in Britain, the Liebigs began to entertain regularly. Because his wife loved gardening, Liebig bought a plot close to the barracks laboratory when his finances were more secure. This garden became a social gathering place for professors and students on Sunday afternoons. In Munich they were to entertain on an even wider and more lavish scale.

The fact that Jettchen was a pious Catholic and Liebig was nominally Lutheran seems to have caused no friction. Their two sons Georg (1827–1903) and Hermann (1831–94) were brought up as Lutherans; their three daughters Agnes (1828–62), Johanna [Nanny] (1836–1925), and Marie (1845–1920) as Catholics. Georg was trained as a doctor, and after serving in India for a few years, he became a spa doctor at Reichenhall near Munich. Hermann took up his father's interest in agricultural chemistry and had a varied career as practising farmer and landowner. Agnes married the philosopher Moritz Carrière but died tragically young. Johanna, whose diaries have been recently discovered, married the eminent surgeon Karl Thiersch and settled in Leipzig. Marie, following the death of her fiancé Karl Knapp, remained a spinster. Other marriages on the Moldenhauer side also enriched the Liebig intellectual circle. Jettchen had two brothers, each of whose daughters married into the Liebig circle. Johanna Moldenhauer married Liebig's student and Giessen colleague Heinrich Buff (1805–78), and her sister Elise was the first wife of Liebig's student Wilhelm Hofmann. Helene Moldenhauer, the daughter of another brother, became Hofmann's third wife after he had failed to persuade Johanna Liebig to marry him.

As Gustin has shown, Liebig's new institute was widely advertised in the leading German pharmaceutical journals. It even received the imprimatur of Trommsdorff when he closed down his own institution at Erfurt in

15. See discussion in R. Blunck *Justus von Liebig* (Berlin, 1938), 137–44.

1828 and later, when as editor of the *Neues Journal der Pharmacie* he referred to Liebig's "excellent institution." We know that when the institute was opened in 1826 the curriculum included mathematics, botany, mineralogy, and instruction on reagents and analysis – all of which seem to have been taught by Umpfenbach and Wernekink. Liebig taught experimental chemistry, knowledge of pharmaceutical wares, and tests for the purity of drugs. From 1829 Liebig handed over some of the pharmacy teaching to a former pupil, Johann Mettenheimer (1802–64), who had become the official university pharmacist at the Pelikan Apotheke.[16] In the second semester of the course, there was more mathematics teaching from Umpfenbach, a practical physics course from Schmidt, and a rigorous course of chemical analysis from Liebig, with the laboratory open to students from early morning until the evening. It cannot be a coincidence that the curriculum seems remarkably similar to the one that Trommsdorff had been offering since 1795. The major difference was that Liebig did not have an apothecary's shop the sales of which could underwrite the laboratory teaching expenses.[17] All this private enterprise was in addition to the formal university courses that Liebig offered: "general experimental chemistry" in the summer, "agricultural and forestry chemistry" in the winter, together with the year-round course in analytical chemistry that students could attend as many days a week as they wished. One modern touch was that each week's lecture course was followed by a Saturday examination, which, rather as the mathematics tripos was used at Cambridge, Liebig used to weed out the weak and the ordinary (Good, 1936; Schierz, 1929; Vogt, 1896; Volhard, vol. i, p. 60). Those, like Karl Ettling, Carl Vogt, and Wilhelm Hofmann, who survived the tests were commonly invited to work with Liebig personally or to serve as his assistant. At a later date, when his course of qualitative analysis had been formalised, Liebig rewarded students who had successfully done "the alphabet" with a platinum crucible.[18] Liebig's English student Frank Buckland left an interesting account of this teaching method:

> When a young man begins here, he generally goes through the course of analysing a set of one hundred bottles, which takes him sometimes a year. These bottles are various compounds, which he must find out – viz., in the first ten he has only to find one metal, etc; in the second,

16. C. Billig, *Das Pharmaziestudium an der Universität Gießen* (unpub. PhD thesis, Universität Gießen, 1994).
17. Gustin, *Emergence*, 71–72.
18. The "alphabet" involved the analysis of "the hundred bottles" of unidentified solids or solutions.

two metals or substances, etc; till at last the highest bottles contain six or seven substances, all of which he must find out. Liebig thought I had better not begin these, as I had to learn German, and to stay only a short time here. However, I think I know more chemistry than when I came. About 3 o'clock Liebig comes into the laboratory, and he seems to be able to tell everybody, whatever they may be doing, what to expect, and how to proceed.[19]

Similarly, after only three weeks in the laboratory, the American, Oliver Wolcott Gibbs, noted:

> We have already finished our course of analytical exercises having analysed sugar, Ether and Oil of Turpentine and Uric Acid of each several Analyses or combustions. We have made besides 4 or 5 determinations of Nitrogen according to the 3 different methods in use, and it only remains for us to determine the Spec[ific] Grav[ity] of a vapour according to Dumas' method to be ready to undertake some investigation. You see from this that the mere learning how to make a combustion does not take much time.[20]

Liebig recalled this heuristic method somewhat idyllically in his autobiography:

> Actual instruction in the laboratory, with experienced assistants in charge, applied only to the beginners; my special students learned only in proportion to what they brought. I gave them assignments and supervised the execution; like the radii of a circle, everybody had a common centre. There was no actual guidance; every morning I received from each one a report on what he had done the day before as well as his views on what he intended to do; I either agreed or made my objections, everyone was obliged to seek his own way. In companionship and constant intercourse, and in which each one participated in the work of all, everyone learned from the others. . . . We worked from daybreak until nightfall, distractions and diversions were not available at Giessen. . . . The memory of their stay in Giessen awakens in most of my pupils, as I have often heard, the pleasant sense of satisfaction for a well-spent time.[21]

This quiet, workmanlike atmosphere is hard to reconcile with Liebig's known shortness of temper, and other student reminiscences suggest he

19. G. H. O. Burgess, *The Curious World of Frank Buckland* (London, 1967), p. 34.
20. O. W. Gibbs to W. F. Channing, 22 November 1846, Gibbs Family Papers, State Historical Society of Wisconsin. My thanks to J. B. Morrell and J. Sturchio for this reference.
21. J. Liebig, *Autobiography*. I have used Fruton's translation. See J. S. Fruton, *Contrasts in Scientific Style* (Philadelphia, 1990), p. 34.

could also adopt a bullying, hectoring, impatient tone when he was in a hurry for results.

The mystery about Liebig's institute is how long the collaboration with his colleagues Umpfenbach, Wernekink, and Schmidt lasted, and what their financial role was. In the absence of any contrary information, we must assume that they faded from the picture rather rapidly and that for many years all the teaching was done by Liebig and Mettenheimer, with any financial profits going directly to them. Such a success did he make of it, however, that in 1833 he was able to persuade the university's chancellor Justin von Linde (1797–1870) to absorb the institute into the university.

In the event, although large numbers of Liebig's students were destined for Germany's apothecary shops and emerging chemical industries, he was to become chiefly renowned as the director of a model institution for the teaching of practical chemistry. Historians are well aware that, whatever impression Liebig himself gave to the contrary later in his lifetime, he was not the first person to teach practical chemistry. Besides the efforts of pharmacists like Trommsdorff, with their private institutes established to make pharmacy more scientifically respectable through instruction in chemical analysis, the cameralist tradition had also demanded skills in mineral analysis of mining inspectors. Such needs had been met, notably by Johann Döbereiner at Jena and Friedrich Stromeyer at Göttingen, from about 1816. A few British pupils, for example, were learning inorganic analysis from Stromeyer several years before they had heard of Giessen. Nevertheless, with Liebig's headquarters in an unheated and disused barracks, the Giessen laboratory, or laboratories as they were to become by 1840, were to become the most famous in the world for practical instruction in chemical analysis and especially for a surefire method of organic analysis.

From 1825 until its first enlargement in 1835, Liebig's laboratory was the guardroom of a disused barracks on the town's boundary – now the Ludwigstrasse – close to the railway station that was to arrive in the 1850s. Apart from a spectacular open colonnade in Greek style, where dangerous reactions could be performed in the open air, Liebig had an internal laboratory space of little more than 38 square metres. This was divided into a small room that served for lectures, an unheated broom cupboard that served as a balance room and store, and the principal space, which was filled with ovens and work tables that served for both Liebig and the eight or nine students who might work there at any one time. On the floor above, in equally cramped conditions, lived Liebig, his wife, and the first three of their five children. The family's laundry was done in the laboratory in the rare moments it was empty.

Organic Analysis Perfected

Nineteenth-century organic chemists faced tremendous technical difficulties. But the intellectual problems of organic chemistry would never have been promoted or solved without the analysts' dogged search for adequate foundations. Each practical chemist searched for the organic analysts' philosopher's stone – the perfect method. William Prout, for example, spent twelve years looking for an apparatus and a technique that would produce accurate analyses of organic materials. The research was time-consuming. Although Andrew Ure could write optimistically that his own method allowed him to make six determinations in a single day, even Liebig (in recommending the time saved by his method) could claim only 400 analyses per year with an army of research assistants.

Because all methods for determining the carbon, hydrogen, and oxygen content of organic substances were to depend on measurements of the volumes of carbon dioxide and water produced, final accuracy was dependent not only on the accurate determination of the quantities of these substances but also on accurate knowledge of their composition. Organic analysis a specialized branch of gasometry.

Most of Lavoisier's organic analyses remained unpublished at his death, and his use of oxidizing agents remained largely unknown. Although some success using his cumbersome, dangerous, and rather inaccurate method of oxidation was employed by the Swiss chemist Saussure, most chemists continued to use methods of proximate analysis by destructive distillation. Organic substances had been analyzed by distillation for centuries, and by the beginning of the nineteenth century the products were usually collected, and often weighed, as fractions of gas, oil, phlegma, and residue. Such distillation techniques continued to be used in vegetable analysis long after Gay-Lussac and Thenard, frustrated by Berthollet types of distillations, revolutionized organic analysis in 1810 by adopting an oxidizing agent.

Not that their use of potassium chlorate was a satisfactory solution to the problem. It proved to be time-consuming, dangerous, and very much dependent on the operator's skill for its accuracy. Certainly, the two Frenchmen obtained important results, but other chemists did not adopt their apparatus.

Jöns Jakob Berzelius, whose untranslated Swedish textbook of animal chemistry appeared between 1806 and 1808, continued to rely mainly on proximate analysis until 1812 and seems to have been unaware of the French method until his return to Sweden from England that year. In the hands of Berzelius, the Frenchmen's cumbersome apparatus was trans-

A view of the interior of Liebig's 1839 teaching laboratory as restored in the early 1950s. (Courtesy Liebig-Museum Giessen)

formed into a safer and simpler horizontal arrangement, and uncertain volumetric estimations were replaced by the direct weighing of carbon dioxide and water by absorption and condensation. The combustion process took about two hours, though Berzelius's proper insistence on purification and the drying of the substances over sulphuric acid made the complete process of analysis much longer.

Berzelius's technique underwent modifications at the hands of various chemists, most of whom quickly adopted copper oxide as an oxidizing agent when it was introduced by Gay-Lussac in 1815. The final version of Berzelius's apparatus and method was that of Liebig in 1830, which used a coal fire instead of dusty charcoal or a cool spirit lamp. Water (the oxidation product of the substance's hydrogen content) was absorbed in a bulb of hygroscopic calcium chloride, which could be weighed directly. Carbon dioxide (the oxidized product of the substance's carbon content) was similarly weighed directly by absorption in a solution of potassium hydroxide in an ingeniously arranged array of five glass bulbs that became known as the *Kaliapparat*. Oxygen was determined by difference. The full details of Liebig's method were published in the student manual, *Instructions for the Analysis of Organic Bodies*. This, like most of Liebig's books, was published in both German (Liebig, 1837) and English (Liebig, 1839). The method proved to be so nearly perfect that few changes had to be made when Liebig published a second edition in 1853, by which time his pupil and translator A. W. Hofmann had begun to experiment with gas heating to replace the spirit lamp. Apart from the adoption of the Bunsen burner in the 1860s, Liebig's method remains in current use.

Liebig's perfection of combustion analysis arose in the context of his discovery of hippuric acid in horses' urine in 1829. Hippuric acid (empirical formula $C_9H_9O_3N$) contains 8% nitrogen, 5% hydrogen, and a massive 60% carbon. The would-be analyst was therefore faced with the dilemma of, on the one hand, burning only small amounts of acid because of the very large volume of carbon dioxide produced on oxidation, and on the other hand requiring large amounts of acid to increase the volumes of nitrogen oxides and water for accuracy in determining nitrogen and hydrogen content. In practice, this meant making time-consuming separate estimations for carbon, nitrogen, and hydrogen, using different sample weights. (Oxygen was always estimated by difference.) Pharmaceutical work was also demanding improved methods for analysing the large amounts of carbon in the important alkaloid compounds that were of interest in the Giessen laboratory. By weighing carbon and hydrogen directly as carbon dioxide and water, Liebig was able to maximise the amount of material he combusted, to increase the accuracy of estimates of contained carbon and hydrogen because they were being done gravimetrically rather than volumetrically, and at the same time to optimise the chance of accurately estimating any nitrogen volumetrically.

It is worth emphasizing that large numbers of students flocked to Giessen, particularly overseas students, because in the *Kaliapparat* Liebig had a wonderful technique for analysis. (Wöhler was once to ask Liebig jokingly whether he had a Greenlander working with him, 8 May 1839.) His triangular series of five glass bulbs filled with potassium hydroxide was first described in Poggendorff's *Annalen* at the end of 1830. The three lower bulbs ensured the total absorption of carbon dioxide by potash; the two upper bulbs were effectively safety devices to ensure that potash could not flow back into the combustion tube or forward to the calcium chloride trap. Although Liebig had learned glassblowing in Paris, the making of the bulbs demanded very considerable skill and it was his assistant Carl Ettling (1806–56) who perfected the apparatus. It was Ettling, on a visit to the British Association in 1840 at Glasgow, who showed a Scottish glassblower the necessary technique. As glassblowers became more skilled, the inelegant prototypes were transfigured into objects of beauty, so much so that the bulbs, like Kekulé's benzene hexagon, have become part of the heraldry of chemistry. According to Pettenkofer, Liebig's students were marked out at Giessen by a potash-bulbs brooch or button that they wore on their jackets (Pettenkofer, 1877).

Individual chemists who had not trained at Giessen personally found the Liebig method easiest and best only after much hard work. What was the best form of heating? How should substances be purified? Should the products of combustion be estimated volumetrically or gravimetrically? What could be done to diminish the hygroscopic nature of copper oxide, or its perverse habit of absorbing air when warmed? Should substances be triturated? Were rubber tubing or cork stoppers best for connecting the parts of apparatus, and which would alter least in weight during a combustion? How could nitrogen be accurately estimated? How were the determined elements arranged in the molecule?

No wonder Prout exclaimed: "To conquer these, every means that could be thought of, as likely to succeed, were tried, but without effect, and I was obliged to relinquish the matter in despair."[22] The great advantage of Liebig's method, as he was the first to admit, was not that he introduced anything radically new over Berzelius, or that he offered new heights of accuracy, but that he refused to be trapped by such questions. As he said, defeatism merely means you do not do any research. Here, he suggested, was a simple, cheap, reliable method that answered well enough for everyday determinations of carbon, hydrogen, and oxygen.

The determination of nitrogen remained a separate issue and was always to be the subject of separate elaborate determinations along with other, more rare organic elements such as sulphur. In October 1840, Wöhler suggested to Liebig that nitrogen might be determined accurately by heating an organic compound with slaked lime to form ammonia, which could then be absorbed in an acid of known strength.[23] Liebig appears to have passed this information to his assistant, Heinrich Will (1812–90), who with the student Franz Varrentrapp (1818–77) perfected and published this technique.[24] Wöhler was not pleased when Liebig on 28 June 1841 sent their paper to him with the comment, "I believe you have also once employed this method."[25]

22. W. Prout, "On the ultimate composition of simple alimentary substances," *Phil. Trans.* 1827, p. 360.
23. A. W. Hofmann, ed., *Aus Justus Liebig's und Friedrich Wöhler's Briefwechsel* (Braunschweig, 1888), p. 165.
24. F. Varrentrapp and H. Will, "Neue Methode zur Bestimmung des Stickstoffs," *Annalen* 39(1841), 257–96.
25. Hofmann, *Liebig-Wöhler Briefwechsel,* 28 June 1841.

The *Annalen der Chemie*

Because of his early interest in fulminates and cyanates but mainly because of his commitment to the training of state pharmacists, Liebig's research during the 1820s was turned towards organic chemistry. Following their meeting at the *Versammlung deutscher Naturforscher* at Hamburg in 1830, he became a close friend of the Heidelberg pharmacist Phillipp Geiger (1785–1836), who interested him in the alkaloids that dozens of German pharmacists, including Emanuel Merck, were busily analyzing.[26] In 1831, aware that Liebig was desperately short of money, Geiger persuaded him to co-edit a journal Geiger had been editing since 1824, the *Magazin für Pharmacie.* As Liebig told Berzelius:

> I have, recently, been burdened with a heavy load, by joining Geiger as co-editor of his journal, all for the sake of the damned money involved. At this small university where I live, I am almost on the verge of starvation.[27]

In point of fact, it proved to be one of the most important steps in his career.

Geiger had been using the *Magazin* to bolster the scientific foundations of pharmacy, which he did by experimentally checking authors' reports. The presence of a rising chemist on the journal's editorial board was part of Geiger's attempt to stress its scientific orientation. To this end, Geiger and Liebig set up the Southern German Association of Pharmacists, who were to submit samples of their quality pharmaceuticals to an "exchange," or depot, managed by Emanuel Merck in Darmstadt. The idea was that Geiger and Liebig (and, presumably, Merck) would test the samples free of charge for their quality and purity, and if the samples were up to standard, would award them the association's seal of approval (Thomas, 1988; Wankmüller, 1973). It is doubtful whether the scheme ever got off the ground, for to have run it properly would have involved considerable time and effort.

The journal was retitled the *Annalen der Pharmacie* in 1832, following its amalgamation with the North German *Archiv des Apothekervereins.*[28] Following Geiger's death in 1836, Liebig was appropriately joined by Emanuel Merck, who with Trommsdorff and Rudolph Brandes, the latter a Westphalian pharmacist, made up the editorial team. Brandes withdrew

26. U. Thomas, "Philipp Lorenz Geiger and Justus Liebig," *Ambix,* 35(1988), 77–90.
27. J. Carrière, ed., *Berzelius und Liebig ihre Briefe* (München, 1898), p. 11.
28. See contract with the Heidelberg publisher Christian Winter transcribed by Wankmüller (1973).

in 1836, disapproving of Liebig's critical stance towards contributors. On Trommsdorff's death in 1837, Liebig replaced him with his close friend, the Coblenz pharmacist Friedrich Mohr (1816–79).[29] In 1837, following his first exciting tour of the British Isles, where he had met Thomas Graham, Liebig floated the idea of transforming the magazine into a truly international journal, with Graham and Jean-Baptiste Dumas in Paris as his co-editors. To this end he asked Mohr to stand down, much to the latter's irritation, although it did not destroy their friendship. Wöhler thoroughly disapproved of this plan, as he told Berzelius:

> I was greatly annoyed to see that Liebig since the beginning of this year published his *Annalen* with the co-operation of Dumas and Graham. Although I know for sure that this cooperation is merely a trick in favour of the publisher and the sale of the journal, this alliance seems to me not only ridiculous, but fatal, because apart from everything else, it ignores shamefully all nationality and it humiliates us in the eyes of these Frenchmen. I told Liebig what I thought of it and in order to re-establish to some extent a kind of equilibrium, and also to show that he not going to change his scientific partners every minute, I proposed to have my name also placed on the title page. Of course, I promised him not only my name like the other two, but also my real cooperation, and this, naturally, without any remuneration.[30]

This new arrangement suited Liebig perfectly, for he had already fallen out with Dumas, and as Wöhler had suspected, the arrangement with Graham had been only window dressing. Nonetheless, their names remained on the masthead until 1842, even though the effective editors were Liebig and Wöhler. It was Wöhler, too, who made the sensible suggestion that, in order better to reflect the journal's emphasis on organic chemistry, the title should be changed to *Annalen der Chemie und Pharmacie* (Phillips, 1966; van Klooster, 1957). Three numbers made up a volume, so there were four volumes a year, and sometimes, when copy became brisk, a supplementary volume was necessary. From 1839 Liebig also took a leaf out of Berzelius's book by introducing an annual summary (*Jahresbericht*) of progress in chemistry, physics, and geology. For this purpose he had the help of various Giessen colleagues such as Heinrich Will, Hermann Kopp, Heinrich Buff, and Ernst Dieffenbach. An attempt by Hofmann and various London colleagues to create an English edition of the summary in

29. See W. A. Kahlbaum, *Justus von Liebig und Friedrich Mohr in ihren Briefen* (Leipzig, 1904); and J. M. Scot, "Karl Friedrich Mohr, 1806–1879," *Chymia* 3(1950), 191–203.
30. Wöhler to Berzelius, 28 July 1838. See J. Braun and O. Wallach, eds., *Briefwechsel Zwischen J. Berzelius und J. Liebig*, 2 vols. (Leipzig, 1901).

(*Above and opposite*) Two unfamiliar versions of Wilhelm Trautschold and Hugo von Ringen's lithograph of the Giessen laboratory in 1842, reproduced by Friedrich Schödler in *Illustrierte Deutsche Monatshefte* 1875. (Courtesy The British Library)

the late 1840s was not successful, mainly because the translation was always two or three years out of date.[31] When Liebig left Giessen in 1852, his colleague and successor Hermann Kopp (1817–92) took over day-to-day control of the journal.

Geiger had originally needed Liebig to check that authors' chemical statements were accurate – a task that Liebig accepted with relish. His bitter editorial denunciations of German and overseas colleagues soon made him enemies, including Eilhard Mitscherlich in Berlin, while at the same time making the *Annalen* indispensable reading. Above all, it gave Liebig a stage for developing his talents as a publicist and writer, for which his contact and friendship with the *Annalen*'s Heidelberg publisher Christian Winter (1773–1858) proved to be indispensable. One of Winter's largest publishing projects was Geiger's plan to edit a huge *Handbuch der Pharmacie,* for which he had also sought Liebig's help. Liebig generously saw this project through the press after Geiger's sudden death in

31. A. W. Hofmann, ed., *Annual Report of the Progress of Chemistry and the Allied Sciences,* vol. 1 (London, 1849), vol. 2 (1850), vol. 3 (1852), vol. 4 (1853).

order to support his widow and her five children.[32] (Liebig later took special responsibility for the education of Geiger's son Friedrich.) Liebig was embarrassed in 1851 when his English pharmacy student William Bastick translated Geiger's introductory essay to the *Handbuch* on "the nature, import, history, and literature of pharmacy" and misattributed it to Liebig.[33]

Winter was to publish several of Liebig's books, including the important *Chemische Briefe* in its several editions. Another valuable contact Liebig had made in Paris as a student was with the young Braunschweig publisher Eduard Vieweg, who became a partner in his father's business in 1825.[34] On a visit to Berlin in 1832, Liebig met the physicist J. C. Poggendorff (1796–1877), who had edited the *Annalen der Physik und Chemie* since 1824. Together they drew up a plan for a great dictionary, or encyclopaedia, of chemistry to be called a *Handwörterbuch der reinen und angewandten Chemie,* which Liebig persuaded Vieweg to publish. It was a colossal and time-consuming venture for both editors. As Liebig quickly realised, the letter "A" alone involved discussion of some of the thorniest theoretical problems of the day – the nature of aldehyde, alcohol, ether

32. P. L. Geiger, *Handbuch der Pharmacie (1824),* 5th edition edited by Liebig, 2 vols. (Heidelberg, 1843), amounting to 1411 pages.
33. See *Pharmaceutical Journal* 11 (1852), 384.
34. M. and W. Schneider, eds., *Justus von Liebig Briefe an Vieweg* (Braunschweig, 1986).

(Äther) and "ethyl" (Äthyl) – quite apart from the need to say something systematic and methodical on analysis. Not surprisingly, considering that Liebig was simultaneously editing the *Handbuch* as well as *Annalen*, progress was extremely slow; Poggendorff was equally dilatory. Consequently, the first fascicle of the *Handwörterbuch* did not appear until the autumn of 1836, and the first complete volume not until 1842 (by which time some of the earlier fascicles had to be re-edited because they were already out of date).[35] When by 1847 the project had got only to the letter "E," both Liebig and Poggendorff had had enough, and they happily handed the entire project over to Hermann Kolbe, who energetically disposed of "F" to "R" in four further volumes by 1859. By then, even he had begun to flag, and the final two volumes of the dictionary were overseen by Hermann von Fehling at the University of Stuttgart. Apart from the contribution of the essay on fermentation (*Garung*) in the third volume, Liebig did no more writing for the *Handwörterbuch*. The significance of his contributions to the first two volumes must not, however, be underestimated. On the theoretical level, they forced him to express his views on controversial matters; on the research front, articles on, for example, blood and endosmosis, pushed him and his students towards the field of physiological chemistry that was to preoccupy him from 1840. Last, but by no means least, articles could be tried out first in the *Annalen* or, like the important entry on organic analysis, lead to a separately printed monograph.[36]

The Giessen Research School

We are now in a position to explain how Liebig built up a great teaching and research school that became the model for others in Germany and overseas. In a seminal article, J. B. Morrell suggested that the necessary conditions for Liebig's success were intellectual, institutional, technical, psychological, and financial.[37] In the first place, Liebig had a definite programme of research, the analysis of organic compounds, and of instruction, the practical teaching of qualitative and quantitative analysis.

35. M. and W. Schneider, "Das *Handwörterbuch* in Liebigs Biographie" in W. Dressendörfer and W. D. Müller-Jahncke, eds., *Orbis Pictus. Kultur – und pharmaziehistorische Studien* (Frankfurt, 1985), pp. 247–54; A. J. Rocke, *The Quiet Revolution* (Berkeley, 1993), pp. 68–71.
36. J. Liebig, *Anleitung zur Analyse organischer Körper* (Braunschweig, 1837).
37. J. B. Morrell, "The chemist breeders," *Ambix* 19(1972), 1–46; G. L. Geison, "Scientific change, emerging specialities, and research schools," *History of Science* 19(1981), 20–40; Fruton, *Contrasts in Scientific Style;* J. W. Servos, "Research schools and their histories," *Osiris* 8(1973), 3–15.

By 1831 he had more than adequately established a national and international reputation for the prosecution of these related programmes, and the *Annalen* was to become a mouthpiece for him and for his students.

Unlike many later directors of research, Liebig did not add his name to a student's published work. "Nothing enthuses young people more than to see their names in print," he told Vogt (1896). This reputation, together with the growing awareness that he had a skill and a technique to impart that could be extremely useful in the prosecution of pharmacy and medicine (and by the 1840s, in agriculture and chemistry teaching also) ensured that another necessary condition of success – an adequate supply of students – was fully met. As we have seen, this position was mainly advanced through Liebig's management of a private pharmaceutical institute separate from his state-supported teaching at the university, though there must have been some overlap of activities. Unfortunately, no details have survived concerning student numbers and how financially successful the institute was. All that is known is that it did succeed in attracting students from other German states and, significantly, from overseas, particularly Britain. Some, like the young Francis Galton, found German or chemistry, or both, too difficult and stayed only briefly.[38] Others spent one or two semesters in Giessen, or like William Gregory, returned after obtaining their doctorates to do more research with Liebig.

In 1833, following some bargaining with the government, Liebig amalgamated his private school with his official university course, by which time he was teaching ten to fifteen pharmacy students and three to five chemistry students per year. Over the next two decades, the number of students reading chemistry was to exceed those in pharmacy.[39]

	1830–35	*1836–40*	*1841–45*	*1846–50*	*Totals*
Chemistry	15	75	174	143	407
Pharmacy	53	63	74	62	252

By 1852, when he left Giessen for the University of Munich, more than 700 students of chemistry and pharmacy had passed through Liebig's hands.

Such large numbers could be dealt with only in a much bigger laboratory than the premises Liebig had been granted in the 1820s. Leaving aside the small extension built in 1835, it was not until 1839 that he and the state architect Paul Hofmann (whose son was soon to become Liebig's favourite German pupil) obtained funds from the government to enlarge his facilities to include a lecture theatre and two separate laboratories for

38. Galton travelled to Giessen with William Allen Miller in 1840. Miller stayed, Galton did not. See F. Galton, *Memories of My Life* (London, 1908), p. 48.

39. Fruton, *Contrasts in Scientific Style*, p. 25.

pharmacy and chemistry students. The chemistry laboratory – familiar from the famous engraving by Trautschold – was fitted with glass-fronted cupboards from which the fumes from dangerous reactions could be vented directly into the outside air through a special chimney. Such "fume cupboards," which hitherto had been found only in the private laboratories of Gay-Lussac in Paris and the pharmacist W. H. Pepys in London, soon became standard laboratory furniture in laboratories all over the world.

The extension was commemorated in 1842 in a portfolio of eight plates published jointly by Winter in Heidelberg and Taylor & Walton in London. This atlas contained what were to be two of the most frequently reproduced illustrations in the whole history of chemistry: an 1841 engraving by J. M. Bayrer showing a horse and carriage passing the exterior of the building and the other by Wilhelm Trautschold (engraved by Heinrich von Ritgen) of the interior of the new analytical laboratory showing students at their work places.[40] Because owners and dealers have broken up copies, the complete folio is now extremely rare. The exterior view by Bayrer, a professor of drawing at Giessen, is frequently misattributed to Trautschold, as is a later 1848 view by the Finnish artist Friedrich Soldan – probably because of the splendid portrait of Liebig that Trautschold painted in 1840 and his execution of the laboratory scene in the atlas. Born in Berlin in 1815, Trautschold's early paintings were of historical subjects, animals, and landscapes. After executing Liebig's portrait, he settled in Giessen in a house close to Liebig's laboratory, where he made a living from portraiture before becoming the university's official drawing master in 1843. As this did not pay, he left Giessen in 1846 to try his fortune in London, where he married an Englishwoman and died in 1876.[41]

The enlargement of Liebig's laboratory also suggests another vital feature of his success: state patronage and its corollary, financial support. As Morrell has shown, all previous teaching laboratories had failed because of a catch-22 situation: If a teacher charged a low fee in order to encourage large numbers of students, he was unable to meet the expenses of running a laboratory of sufficient size; if, on the other hand, he charged realistic fees to cover laboratory expenses, few students (other than those motivated by a lucrative career as a German apothecary) were attracted. The result of this situation, especially in the Scottish universities, had been

40. I. P. Hofmann, *Acht Tafeln zur Beschreibung des Chemischen Laboratoriums zu Giessen* (Heidelberg and London, 1842).

41. C. Rauusch, "Trautschold," in *Ludoviciana Festzeitung* (Giessen, 1907). Ironically, he is remembered for nothing but his association with Liebig. See M. T. Hayford, *C. F. Trautschold 1815–1877*, priv. print. (1980).

that university teachers were discouraged from running practical classes; they opted for low fees, very large classes, and chemical entertainment by lecture demonstrations.

Liebig had the advantage that from the very beginning the state provided the University of Giessen with a modest annual subsidy for laboratory expenses over and above the professor's salary and such fees as he was able to take from students. By being prepared to use some of his salary and student income to subsidize the laboratory costs for several years, Liebig was able to bargain his success in attracting students to the University of Giessen, as well as his own growing fame as a chemist, to obtain not only increases of salary but also increases in laboratory expenses.

All this was done at very considerable personal cost to Liebig's energy and health. By the summer of 1833, he was on the verge of a nervous breakdown. A painting made at this time by Engel portrays a prematurely aging thirty-year-old man. Recuperating at Baden-Baden in August, he wrote a long, angry, and passionate letter to the university chancellor:

> The resources of the laboratory have been too small from the beginning. I was given four bare walls instead of a laboratory; despite my solicitations, nobody thought of a definite sum for its outfitting [and] for the purchase of supplies. I needed instruments and [chemical] preparations and was obliged to use 3–400 gulden per year of my meagre salary for the purchase of preparations; I have needed in addition to the attendant paid by the state [since 1828}, an assistant, who costs me 320 gulden; if you subtract these two expenditures from my stipend [800 florins] not much remains to clothe my children.[42]

Better that Linde should sack him, Liebig declared, than that in resigning he should appear ungrateful to a state whose support of him as a student had enabled him to become famous. But if nothing was done, he would, for his own peace of mind, be forced to work only half the year at Giessen, spending the other half at Darmstadt, where he would found a new private institute.

By implication, this action would have drawn students away from Giessen to the detriment of the recruitment that Liebig's presence had brought about. This extraordinary, passionate letter, together with the real danger that Liebig would be poached by another university to the great loss of Hessen-Darmstadt, persuaded the government to satisfy Liebig's demands. Throughout the nineteenth century (indeed, even to this day), German professors commonly traded their reputations to gain rises

42. Liebig to Linde, 12 August 1833. See E. M. Felschow and E. Heuser, eds., *Universität und Ministerium im Vormärz* (Giessen, 1992).

in salary. Liebig had already made semiserious approaches to Dorpat in 1826 and in 1835 had received the first of several external calls, to Antwerp. The following year he had the galling experience of actually being declined by Göttingen on the grounds of his fractious personality.[43] Although he was pleased that his friend Wöhler was appointed, it was a mortifying experience, as he confessed to Berzelius:

> I have had a painful experience. I have battled against everything that is bad in the chemical literature for the past eleven years, while burning for the truth and full of enthusiasm for science. Free of selfishness, I have spoken my mind, and this could not be done without hurting many people. Far from my intentions being recognized, I have instead acquired a reputation as an intolerant person who is envious of others' accomplishments. Had I not behaved so, I would have received a call to Göttingen and a large rise in salary from my government [to make him stay in Giessen]. Thinking about this makes me very unhappy and depressed. Everything grates on me and I have no zest for my dear work. Word of this dammed opinion of my character will soon get around, and every future opportunity in the country I live in will be closed to me, for in Germany betterment only comes from being "called." This is the only measure of merit, since bureaucrats can neither read nor understand chemical work.[44]

Although this episode does not seem to have tempered Liebig's editorial style in the *Annalen,* his new awareness of his reputation may have stunned him sufficiently to cause him to patch up his quarrel with Dumas in 1837 and, through his British and French contacts, persuaded him temporarily to assume the mantle of an international cooperationist.

In fact, his career did not suffer, and he went on to receive calls from St. Petersburg in 1837, Vienna in 1840, London in 1845, Heidelberg in 1851, Munich in 1852 (which he accepted), and Berlin in 1865. The financial effects of these calls is best shown in tabulated form (see Table 2.1).

Not surprisingly, the government provided generous funds for the enlargement of the laboratory between 1835 and 1839, so enabling Liebig to expand his student population still further. In 1835 funds were provided to convert the tiny lecture room of the barracks into a balance room, and a new auditorium and small private laboratory sufficient for Liebig and two or three assistants were added. This work was preliminary to the major extensions and reconstruction made by the architect Paul Hofmann in 1839. As Gustin has shown, by 1843, when the number of students auditing chemistry and pharmacy peaked at sixty-eight, Liebig was "wholly responsible for the attraction to Giessen of no less than 15% of the

43. Volhard, vol. i, 130.
44. Braun and Wallach, *Briefwechsel Berzelius und Liebig,* p. 112.

Table 2.1

	Liebig's Salary (gulden)	Event	Laboratory Expenses, Paid by the State (Including Laboratory Servant) (gulden)
1824	300	Associate professor	100
1825	500	April: raise	400
1826	500	January	438
	500	March	446
	800	July: full professor	446
1827	800		600
1833	880	Complains and threatens to resign	619
1835	1,250	Invited to Antwerp; promised larger lab.	714 plus Ettling's salary
1837	1,650	Invited to St. Petersburg	
1840	3,200	Invited to Vienna	1,500
1843			1,900

students in the university."[45] Students were being attracted from overseas – particularly from Britain. Liebig's arrangement and the recognition by a government that it was unreasonable to expect a large science school to be financed entirely from a professor's own pocket soon became widely recognized by other German states. Although not the first to teach practical chemistry in a German state, Liebig played a direct role in this expansion process through polemical essays published in 1839 and 1840. Even so, in 1843 such was the need for student places that Liebig was forced to invest about 1,500 gulden to buy a house in the neighbouring Seltersweg that he converted into an elementary analytical laboratory for Henrich Will (1812–90). This was abandoned only in 1852 when Liebig left Giessen.[46]

Since Morrell's essay was published in 1972, there has been much discussion of what constitutes a research school or group. On one level, Liebig's school certainly fits Geison's definition: "Small groups of mature scientists pursuing a reasonably coherent programme of research side-by-side with advanced students in the same institutional context and engag-

45. Gustin, *Emergence German Chemmical Profession*, pp. 99–100.

46. Volhard, vol. i, 83–4; *Liebig-Wöhler Briefwechsel*, October 1843; Felschow and Heuser, *Universitäten und Ministerium*, 14 May, 18 July, and 18 August 1843.

ing in direct, continuous social and intellectual activity."[47] This camaraderie is beautifully captured in a letter Gibbs wrote to Channing in 1846:

> Of Liebig' talent there can be no doubt . . . and Giessen is not a bad school for one who has learnt accuracy and careful manipulation elsewhere. It is a great advantage to be where so many are at work. One's mind gets well stirred up. One converses continually on scientific subjects and during those little intervals which so offen [sic] recur in one's work in the laboratory, while something is evaporating or filtering, one may be finding out what others are doing and how they do it and so pick up many small scraps of experience.[48]

On the other hand, it must be stressed that up until the late 1830s, at which time Liebig moved into physiological chemistry, there was little coherence in his programmes; research was an eclectic mixture of inorganic chemistry, qualitative and quantitative analyses, the theory and practice of organic chemistry, and pharmaceutical chemistry. Moreover, it is difficult to set forth an exact definition of a "Liebig student" if we bear in mind that only a few took doctorates with him (the modern criterion of postgraduate training in a research school), some stayed less than two semesters with him, many published nothing in *Annalen,* many never graduated beyond the analytical "alphabet," having gone to Giessen purely to acquire a skill, and others who were attracted from other university disciplines came only to listen to a famous professor. Yet others, such as the Englishmen Penny and de la Rue, were awarded purely honorary degrees and never seem to have set foot in the Giessen laboratory. More controversially, the very considerable roles of Liebig's assistants in both elementary teaching and the advanced training at Giessen must be emphasized. Liebig possessed considerable talent in choosing these men. Ettling (1806–56), who assisted from 1837 until becoming *ausserordentliche* professor of mineralogy at Giessen in 1849, was an able teacher, demonstrator, and glassblower.[49] Heinrich Will, who succeeded Liebig at Giessen in 1852, virtually ran his own laboratory class from 1843 in an annex to the laboratory, so large had Liebig's enterprise become. Two of Liebig's most famous pupils, Kekulé and Erlenmeyer, actually made their doctoral studies under Will's supervision.

Nevertheless, what is striking and impressive is the extent to which many of the leading chemists of the later nineteenth century were products of the Giessen school. This was recognised by later nineteenth-century

47. G. Geison, "Scientific change, emerging specialities and research schools," *History of Science* 19(1981), 20–40.

48. Gibbs to Channing, 22 November 1846. A similar point was made by Bergemann; see Appendix 1 in this book.

49. See Vogt's description of Ettling, *Aus meinen Leben,* in Good (1936), p. 558.

chemists who saw Liebig as the *Stammbaum,* the root of the family tree of chemistry.[50] At least two dozen of Liebig's students obtained full professorships in German and Austrian universities and technical high schools. A similar number were to be found in Britain. Fruton has similarly identified nearly 100 students who went into pharmacy and more than fifty who became leading industrialists and manufacturers. The numbers of pupils entering pharmacy actually increased during Liebig's tenure at Giessen, despite the growth in numbers of students taking chemistry rather than pharmacy. Fruton's examination leads to the inescapable conclusion that "the main educational function of the Giessen laboratory throughout the period 1830–1850 was the training of future pharmacists and industrial chemists."[51] Ernst Homburg (1993) has, however, sounded a note of caution about too readily assuming from this kind of evidence that Liebig and Giessen were totally responsible for the rise to prominence in Europe of German chemistry and industry; he emphasises the important role played by the Technische Hochschule, to which Liebig, with his emphasis on "pure science," was indifferent or even opposed. On the other hand, through the judicious appointment of Heinrich Buff to a chair of physics in 1834, and of Friedrich Knapp to a chair of technological chemistry in 1841, Giessen was effectively functioning as a polytechnic, or technical, university by the 1840s, a position it did not relinquish until the foundation of the Hochschule in Darmstadt in 1877. Whereas other universities had experienced a decline in student numbers, a newspaper noted in 1846, Giessen's recruitment was continually increasing. The explanation, the newspaper noted, lay in the facilities "for the thorough study of the natural sciences." If only the university would appoint a first-rank botanist and zoologist, it concluded, there would be few other universities "where doctors, manufacturers, men with technical professions, pharmacists, and farmers could prepare themselves more thoroughly for their specialty than here."[52] This was exactly Liebig's own view, and much of his correspondence with Linde concerned his attempts to persuade the government to recruit to Giessen such stars as Theodor Bischoff, the anatomist, Alexander Braun (1805–1877), the botanist, and Carl Vogt, the zoologist.

As the youngest professor in the philosophical faculty, it was not until 1846, when Liebig was forty-three, that his colleagues elected him dean.

50. A geneaological chart of chemists "descended" from Liebig, drawn up by Fritz Kröhnke, is displayed in the Liebig Museum, and reproduced in *Giessener Universitätsblätter* 6(1973), endpaper.

51. Fruton, *Contrasts in Scientific Style,* p. 17. See Billig, *Das Pharmaziestudium.*

52. Borscheid (1976) as quoted by C. Jungnickel and R. McCormmach, *Intellectual Mastery of Nature,* vol. 1 (Chicago and London, 1986), pp. 218–19.

Table 2.2. *Students of Chemistry and Pharmacy at Giessen 1830–50*

	1830–35	1836–40	1841–45	1846–50	Total
Initial Matriculation					
Chemistry	15	75	174	143	407
Pharmacy	53	63	74	62	252
Natural Science	1	1			2
Medicine	2		1	2	5
Philosophy	1	1	9	5	16
Public Affairs			1	1	2
Law		1		1	2
Architecture				1	1
Not in Matriculation Lists	4	7	11	9	31
TOTAL	76	148	270	224	718
Dr. phil. (Giessen) awarded to students of chem. or pharm.	10	12	49	70[a]	141
Postdoctoral persons in Liebig's laboratory		5	18	6	29
Persons from other than German or Austrian states					
United Kingdom	2	20	36	25	83
Switzerland	1	11	11	15	38
France	6	9	8	4	27
United States			2	14	16
Russia		2	8	3	13
Poland	1		1		2
Italy			2	3	5
Norway		1	1		2
Denmark				1	1
Netherlands			2		2
Belgium			1		1
Luxembourg			1	1	2
Spain			1		1
Mexico		1			1
TOTAL	10	44	74	66	194

[a]Includes degrees awarded after 1850. From Fruton (1990, p. 26), with acknowledgements to the American Philosophical Society.

(He had been sub-dean in 1841.) The job was hardly onerous, mainly consisting in signing matriculation certificates and compiling an annual register of successful promotions to doctorates, migrating colleagues, and new appointments. The position, which he held again in 1851, did, however, give Liebig powers of patronage and the ability to promote distinguished scholars to the doctorate *in absentia* and *honoris causa*.[53]

The State of Chemistry in Vienna and Prussia

In a Munich lecture of 1864, Liebig recalled:

> In 1838 and 1840 I discussed the state of chemistry in Austria and Prussia, and my unfavourable descriptions were accorded a remarkably different reception.[54]

In the same lecture he reiterated his belief that instruction in the principles of pure chemistry should come before learning and thinking about applications and, therefore, that it would be unwise for Bavaria to invest in the Technische Hochschulen that were becoming fashionable in other parts of Germany, Austria, and Switzerland.

Like Kastner, Liebig consistently maintained that chemistry was the basic discipline underlying the economic success of modern societies and the health of human beings. In one of his first independent *Annalen* editorials in 1835, he had strongly argued that chemical understanding was essential for apothecaries and doctors, and this position was extended into a discussion of chemistry's value for industry in an essay "On the position of chemistry in Austria" in 1838, soon after his first visit to Great Britain. The reason he focused on Austria is not entirely clear, for he had not yet visited Vienna. On the other hand, he had been inspired by what he had seen of British industrial activity, particularly the way chemistry was being applied in Glasgow, and he seems to have been dismayed by a Viennese book on technical chemistry that he had read. Accordingly, he accused Austrians of failing to participate in any of the advances made in chemistry during the last twenty years. This, he claimed, was in dramatic contrast to Austria's important contributions to mathematics and physics. This lack of participation could not be blamed on the political situation, which had once held up progress in France and Germany; rather, the poor quality of Austrian chemistry and pharmacy was owing to the wretched quality of the teaching personnel. Tactlessly, he put the blame squarely on

53. H. G. Gundel, *Giessener Universitätsblätter* 6(1973), 58–80.

54. J. Liebig, "Die Bayerische Landwirtschaft und das technische Schulwesen in Bayern" (1864), *Reden* (1874), 37–47.

the type of specialised instruction that was being developed in the Viennese Polytechnic by the professor of technical chemistry Paul Meissner (1778–1864). He saw the Viennese kind of instruction as an asylum for the inept and slow rather than as an institution that would encourage deep understanding and individual talent. Inevitably, a few years later, Meissner seized the opportunity to publish a fierce denunciation of Liebig's *Agricultural Chemistry*.[55]

The unexpected result of Liebig's attack was that the Austrian government took his criticism seriously. When Liebig sent a copy of his *Agricultural Chemistry* to Austrian government officials, including the chancellor, Prince Metternich, they took it as a hint that he might be willing to come to Vienna to reform the teaching of chemistry along Giessen lines. Semiofficial approaches were backed up by a letter from the German chemist, Carl Reichenbach (1788–1869), who, following his extraction of paraffin wax and creosote from coal tar in the early 1830s, had become an industrialist in Austrian Moravia. Liebig's pupil Joseph Redtenbach (1810–70), who had just been appointed professor of chemistry at Prague – then part of the Austrian empire – also wrote. It is clear from Liebig's correspondence with these Austrian-based chemists and with Linde in Giessen that Liebig thought quite seriously about the invitation. Negotiations continued into 1841. The sticking points were not entirely financial, for a salary and laboratory expenses far higher than his existing ones were offered, but the enormity of the task and its accomplishment. He also foresaw the difficulty of educating his children in a Catholic country. In the event, following a private meeting with Prince Emil von Hessen (1790–1886), the son of the grand duke of Hessen-Darmstadt, who warned Liebig that he might well suffer more political supervision in Vienna than he did from Darmstadt, Liebig decided to stay put. For his consideration, however, his salary was raised to 1,500 gulden.

That was not quite the end of the matter; the Austrian government did take action to improve the teaching of chemistry by sending a number of promising students to Giessen in the following decade, including Heinrich Jesowitz, Friedrich Rochleder, Theodor Schlosser, Gustav Kayser, Franz Ragsky, and Erwin Waidele. None of these chemists, apart from Rochleder, achieved much distinction, and the irony is that it was the very institution Liebig had attacked, the Vienna Hochschule, that became the principal "breeder" of Austrian chemists later in the century and a model for German polytechnic education.[56]

55. P. T. Meissner, *Justus Liebig, Dr der Medicin und Philosophie, Professor der Chemie zu Giessen, analysiert* (Frankfurt, 1844).

56. See E. Homburg, *Van beroep 'Chemiker'* (Delft, 1993).

Liebig's friendship with Reichenbach did not survive the latter's growing commitment to the paranormal. In a supplement to the *Annalen* in 1845, Liebig had enthusiastically followed up Reichenbach's work on creosote in 1833 and had published his own preliminary account of the observation that some sensitive people (though not Reichenbach himself) were able to detect the presence of hidden magnets and crystals in darkened rooms. Wöhler, his co-editor, immediately complained about this inclusion of such nonsense in a professional chemical journal, but Liebig defended his decision on the grounds that the observations were of medical interest and, more mysteriously, that "insupportable pressure" had been put upon him.[57] It is difficult to imagine Liebig being leaned upon, but possibly his Scottish pupil William Gregory, who was to translate Reichenbach's work and who was fascinated by animal magnetism, was the source of the pressure. Gregory's support for Liebig's agricultural chemistry was crucial in the British context. Once he had moved to Munich, however, Liebig used the opportunity of a lecture on the philosophy of science to dismiss Reichenbach's work on "the odylic force" as completely false.[58]

Liebig's polemic on Austrian chemistry had been followed up in 1840 by an even more startling *ad hominem* attack on "the position of chemistry in Prussia," which the Prussian government, unlike the Viennese, did not take to at all kindly. Liebig explained to both Wöhler and Berzelius that this polemic was a publicity stunt and prolegomenon to his book *Agricultural Chemistry,* due out later that year and designed to show how chemistry lay significantly at the heart of practical agriculture and vegetable physiology. The article would awaken the Prussian government to the importance of chemistry, which his own small state of Hessen had fully understood, as shown by its investing in Giessen.[59] The book, in other words, would show the consequences of investing in chemical education. There may, however, have been private reasons for the attack, namely, that Liebig suspected that the Prussian government was deliberately preventing its students from studying with him at Giessen. In his usual tactless manner, Liebig exaggerated the situation in Prussia by claiming that there was not a single chemical laboratory in that country, and that too much emphasis was placed instead upon literary and philosophical studies, including that "false Goddess" and "Black Death," *Naturphilosophie:*

57. Hofmann, *Liebig-Wöhler Briefwechsel,* 30 December 1844 and 14 March 1845.

58. J. Liebig, "Über das Studium der Naturwissenschaft" (1852), *Reden* (1874), pp. 156–71; see Maria Habacher (1964).

59. Schneider, *Liebig Briefe an Vieweg,* 17 May 1840; *Liebig-Wöhler Briefwechsel,* 1 June 1840.

> From this product of obdurate presumption and ignorance no progress for the science [of chemistry] is to be expected; with what haste, with what concupiscence they grasp onto the false Goddess of German *Naturphilosophie* with its straw-stuffed and rouge-painted dead skeleton. It promises them light, without troubling them to open their eyes; it gives them results without observation or experiment, and without acquainting them with nature and form, purpose and activity, which one wants to explain; with life-force, dynamic, specific, with loud, and in their mouths, senseless words, which they do not understand, they explain experience, which they likewise do not understand. The life-force of *Naturphilosophie* is the *horror vacui,* the *Spiritus rector* of ignorance.[60]

Such speculative philosophy, he claimed in adumbration of *Agricultural Chemistry,* had led agriculturists and plant physiologists to "ascribe the most important function [in plant nutrition] to an imaginary substance, humus, which exists in their imagination in a form that occurs in no type of soil,"[61] and to an educational bureaucracy that had failed to understand the significance of science. Liebig's remedy, as in the case of Austria, was for the Prussian government to invest in chemistry for its own sake and to appoint properly trained chemists to teach doctors, pharmacists, physiologists. and manufacturers.[62] The future intellectual development of Prussia depended upon chemistry, from which, so long as it continued to be taught only as an auxiliary science, no new scientific blood would emerge.

Unfortunately, because Liebig explictly derided the lack of facilities for teaching practical chemistry in the six Prussian universities, as well as making rude comments about named teachers such as Mitscherlich at Berlin,[63] the Prussian government was able to ignore his essential message concerning the centrality of chemical education and concentrate instead on the appropriateness of his criticism that Prussia's university teaching facilities were inadequate. The result of Liebig's hyperbole was a twenty-year-long debate in Prussia that was closed only in the 1860s with the large-scale expansion of chemistry teaching at Bonn and Berlin.[64]

60. J. Liebig, "Der Zustand der Chemie in Preussen," *Annalen* 34(1840), 97–136; translation adapted from Munday, *Sturm and Dung* (1990), p. 175.
61. *Ibid.*, p. 132.
62. Munday, *Sturm und Dung* (1990), pp. 173–8; R. Zott "The development of science and scientific communication," *Ambix,* 40(1993), 1–10.
63. See H. W. Schutt, *Mitscherlich* (München, 1991), Chap. 22.
64. S. Turner "Justus Liebig versus Prussian chemistry," *Hist. Stud. Phys. Sciences,* 13(1982), 129–62; Zott, "The development of science," Felschow and Heuser, *Universität und Ministerium.*

The initial reaction of Johannes Schulze, the long-serving secretary of the Prussian ministry of education, was to order an immediate inquiry at each of the medical and philosophical faculties of the six universities he commanded. Because Schulze himself was the civil servant most closely connected with Karl von Altenstein, the humanist whose philological and philosophical ideology of university education had been the target of one of Liebig's principal criticisms, Schulze's correspondents were placed in a catch-22 situation. Should they please the ministry of education by defending humanistic education against Liebig's claim that chemistry was no longer an art or craft but the science that underlay an understanding and exploitation of nature? Or should they agree with Liebig that their facilities were poor and use his polemic as a lever to argue for the improved funding of science? The texts of the replies, which have been published by Felschow and Heuser (1992) provide fascinating details of the low-key and service status of chemistry teaching in Prussia compared with its high profile at Giessen. The replies also point to the dilemma – echoed in Britain in the 1860s and 1870s – over whether universities were institutions for the liberal education of gentlemen civil servants or utilitarian institutions for training scientists and technocrats. One result was that Carl Bergemann (1804–84), a pharmacist from the University of Bonn, personally visited Giessen in order to view Liebig's facilities. His twenty-two-page account, which he sent to Schulze, is a most interesting contemporary description of Liebig's newly expanded laboratories. (See Appendix 1 for a translation.) Most of the other replies from medical faculties also queried whether the kind of Giessen training in which students worked all day in a chemical laboratory was not far too specialized for medical students. In any case, they insisted, pharmacy and physiology were more important, thereby ignoring Liebig's point that chemistry was the scientific basis of these sciences. Prussian philosophical faculties, on the other hand, although deploring Liebig's hysterical tone and rebuking his attack on the role of the humanities in university education, tended to agree that more practical training in chemistry was desirable. They fought shy, however, of Liebig's image of huge chemical institutes, seeing this aspect of training as the task of trade schools and polytechnics, not universities. Clearly, Liebig did not succeed in eradicating from the Prussian mind chemistry's image of the workshop and of "bread studies."

The affair also had British repercussions. Liebig's Scottish pupil William Gregory (1803–58) regarded Liebig's arrangement at Giessen as a model for similar schools in the British Isles.[65] But when the Royal College of

65. W. Gregory, *Letter to the Right Honourable Earl of Aberdeen on the State of the Schools of Chemistry in the United Kingdom* (London, 1842).

Chemistry was finally opened in London in 1845, despite bearing most of the hallmarks of the Giessen laboratory and being privately financed, it lacked the vital financial stability of its German model. Although amazingly successful because its director A. W. Hofmann fulfilled the Liebigian conditions intellectually, technically, and psychologically, it was only after 1853, when the British government took over the management of the college, that the institutional and financial conditions of Liebig's success were fully met.

Teaching Style

There are many contemporary descriptions of Liebig's teaching strengths and weaknesses. According to Volhard, who was obviously referring to the Munich era:

> Liebig's delivery was not fluent, nor had he a finished style; at times it was almost halting. There were frequent pauses for which the audience could see no reason, and to those not used to his lectures, they were quite painful [embarrassing?]. At such times, he would stand thinking about the subject he had just been discussing, mentally pursuing his train of thought into regions far beyond the bounds of the lecture, peering before him apparently absent-minded until he suddenly realised that he had an audience before him. He would sigh and resume the broken thread of his discourse. However, his remarks were thoroughly to the point and so free from rhetorical flourishes, and he brought out the essential features of the subject matter so clearly, that the hearers had the feeling that they were actually witnessing the discovery of new facts. This directness aroused the keenest interest of the audience; one felt that the speaker believed his science to be of sacred importance, and this devotion soon infected the listeners.[66]

Such absorption in the matter at hand extended to his lecture demonstrations. The consequences could be hilarious, as Vogt recalled from 1835, when Liebig picked up two test tubes, one containing colourless lead acetate and the other a yellow solution of potassium chromate, and poured them together to explain the principle of double decomposition. The expected beautiful precipitate of the yellow dye, chrome yellow, was not, however, produced:

> Liebig shakes the glass and goes, constantly shaking it, up and down, along the front row of students, all the time repeating: "Chrome yellow! A beautiful yellow precipitate! You see, gentlemen, you see!"

66. Volhard, vol. 1, pp. 89–90; translation in R. Oesper, "Justus von Liebig, student and teacher," *J. Chem. Education* 4(1927), 1472.

> At last, he raised the glass and held it up in front of his own eyes. "That is, you see nothing, the experiment has failed." (In a rage, he throws the glass into a corner.) Ettling, the assistant, shrugs his shoulders without speaking and points to a glass still standing on the table as a way of telling the students: "The Professor in his zeal has again used the wrong solution."[67]

Episodes like this must have endeared Liebig to his students and formed part of his personal magnetism, charisma, and charm. Gregory, although admitting that Liebig lacked eloquence or fluency as a lecturer, believed him unsurpassed as a teacher because he never used superfluous words and because of his ability to adduce apposite and beautiful illustrations.[68] In letters to his mother, Horsford admired Liebig's "simplicity of character, his noble generosity, his enthusiasm, his self-sacrifice for his friends, and I venerate him for his measureless acquisitions."[69] And Horsford took to heart precepts that he heard from Liebig's lips: "never to give up anything taken in hand, and immediately to put to the proof any new idea that occurs." Such bullish impatience was very much a key to Liebig's success as an investigator, but it also accounted for his rush to judgement of others.

The outburst of literary activity in the 1830s, together with the increases of salary he inveigled from the government, and the improved living quarters his family was able to enjoy over the enlarged laboratory from 1839, rescued Liebig from the financial worries that had plagued him since he was a student at Bonn. Although he had returned to France in 1828 at his government's expense to look at French industries, he had otherwise stayed in Giessen. He could now afford to travel farther afield – in 1837, at the invitation of one of his first English pupils, William Charles Henry, he visited Britain. By then he held a European reputation as an organic chemist, a subject to which we now turn.

67. Vogt, *Aus Meinen Leben*, pp. 122–33; translated in Good (1936), p. 57.
68. W. Gregory, "The cerebral development of Dr. Justus Liebig," *Phrenological Journal* 18(1845), 54.
69. Quoted by S. Reznek, "The European Education of an American chemist," *Technology and Culture* 11(1970), 366–86.

3

Liebig the Organic Chemist, 1820–40

> The loveliest theories are being overthrown by these damned experiments; it is no fun being a chemist any more.[1]

We have seen that when he lived in Paris Liebig attended the lectures of Gay-Lussac and Thenard, from whom he learned how to analyse animal and vegetable materials. Using their technique, Liebig was able to analyse the explosive silver cyanate that he had first prepared at Erlangen and to show that it was a derivative of an unknown organic acid, fulminic acid, which was an oxide of cyanogen. This was actually a correction of his earlier assumption that the fulminates were metallic acids. He published his analyses with Gay-Lussac in 1824, just when Friedrich Wöhler (1800–82) was analysing silver cyanate in Berzelius's home laboratory in Stockholm, where he had gone to perfect his analytical technique. Wöhler showed that silver cyanate was a salt of another new acid, cyanic acid, identical in composition with Liebig's fulminic acid. Because fulminates and cyanates had quite different properties, it was assumed by their contemporaries that one of the two young men was an incompetent analyst. Liebig, whose silver cyanate preparation was probably impure, was quick to make this charge against Wöhler; but after they had met at Frankfurt in 1826 and had gone through their respective analyses together, each agreed that their original findings had been justified. Liebig graciously admitted that he had blundered in describing Wöhler's analyses as incorrect.

Not only did their little conflict lead to the greatest friendship and partnership in the history of chemistry (more than 1,000 letters exchanged over their lifetimes have survived), but it was also one of the principal factors that led Berzelius to announce the doctrine of isomerism in 1831: that two (or more) substances might have the same composition, yet different properties, because their atoms were differently arranged. Both Dalton and Gay-Lussac had already speculated about this geometri-

1. Liebig to Berzelius, 22 July 1834; see J. Carrière, ed., *Berzelius und Liebig ihre Briefe* (München, 1892), p. 94.

cal extension of the atomic theory. By the time of Berzelius's announcement Wöhler had made an even more sensational discovery. In 1828, as he told Liebig, he had discovered that the urea extracted from a dog's urine had exactly the same composition as ammonium cyanate. Between them and in the space of a decade, Liebig and Wöhler had revealed the source of the richness, fascination, and difficulty of organic chemistry, namely, that the simple elements of carbon, oxygen, hydrogen, and nitrogen could combine in myriad different ways to produce millions of different compounds.

Friedrich Wöhler, the son of a veterinary surgeon, had attended the Gymnasium at Frankfurt. His chemical interests had been kindled by a family doctor who encouraged him to experiment in a small home laboratory. As a medical student at Marburg, Wöhler carried on experimenting and published his first work at the age of twenty-one. Although he took a medical degree at Heidelberg in 1821, the professor of chemistry Leopold Gmelin easily persuaded him that his future lay in chemistry and that he should spend a year studying with Berzelius in Stockholm. Wöhler struck up a warm friendship with the great Swedish chemist, whose textbook and annual reports on the progress of science Wöhler translated into German.[2]

It was Wöhler, then teaching at the trade school (*Gewerbeschule*) in Berlin, who first suggested on 8 June 1829 that he and Liebig might occasionally collaborate on research. The impetus was again to resolve an analytical difference of opinion between them over the composition of an organic compound, picric acid:

> The contents of your last letter to Poggendorff [on picric acid] have been communicated to me by him, and I am glad that they afford me the opportunity of resuming the correspondence we began last winter. It must surely be some wicked demon that again and again imperceptibly brings us into collision by means of our work, and tries to make the chemical public believe that we purposely seek these apples of discord as opponents. But I think he is not going to succeed. If you are so minded, we might, for the fun of it, undertake some chemical work together, in order that the result might be made under our joint names. Of course, you would work in Giessen and I in Berlin. When we are agreed upon the plan, we could communicate with each other from time to time as to its progress. I leave the choice of subject entirely to you.[3]

2. J. J. Berzelius, *Jahresbericht über die Fortschritte der physischen Wissenschaften* (Tübingen, 1822–48).
3. A. W. Hofmann, ed., *Aus Justus Liebig's und Friedrich Wöhler's Briefwechsel,* 2 vols. (Braunschweig, 1888), vol. 1, p. 4.

The immediate result of this suggestion was a joint paper on the mellitic acid they had extracted from aluminium mellitate (then known as "honeystone," or mellite, from the honeycombed shape of its crystals).[4]

A further fourteen joint papers were published, their final collaboration being on thialdine and selenaldine in 1847. Many of these productions do not seem to have been genuine collaborations but more in the nature of chemical "gifts" (Liebig's term in his autobiography) to one another, which one or the other would investigate and the other merely check before joint publication. One of Liebig's last letters (31 December 1871) to Wöhler is usually quoted in illustration of their mutual affection and altruism:

> When we are dead and long-since decomposed, the ties which united us in life will always hold us together in the memory of mankind, as a not too common example of two men who faithfully, without envy and ill-feeling, strove and disputed in the same field and yet remained throughout closely united in friendship.[5]

This reflection seems to acknowledge that such collaboration between two potential rivals did not entirely prevent disagreement. It will be recalled from Chapter 2 that Wöhler was much upset by the way Liebig allowed Will and Varrentrapp the honour of working up a method of nitrogen analysis that Wöhler had first suggested to Liebig; and more than once Wöhler had to reprimand Liebig for publishing their joint work under Liebig's sole name.

Nevertheless, the partnership undoubtedly had a restraining and calming effect upon Liebig. In Hofmann's excellent characterisation:

> Liebig, fiery and rash, seizing a new idea with enthusiasm, readily giving free rein to his imagination, tenacious of his opinions, yet open to the recognition of error, sincerely grateful, indeed, to those who made him conscious of it. Wöhler, calm and deliberate, approaching a new problem with temperate consideration securely guarded against over-hasty conclusions; but both equally inspired by the same invariable love of truth. Liebig irritable, easily offended, hot-tempered, hardly master of his emotions, which often found vent in bitter words that involved him in long and painful quarrels. Wöhler, unimpassioned, even under the most malignant provocation, disarming the bitterest opponent by the sobrety of his speech, the sworn foe of quarrels and dissension, yet both animated by the same unerring sense of right.[6]

4. *Poggendorff's Annalen,* 18(1830), 161–64.
5. Hofmann, *Liebig's und Wöhler's Briefwechsel,* vol. 2, p. 324.
6. *Ibid.,* vol. 1, pp. xlviii–xlix.

Another of Wöhler's famous letters to Liebig (9 March 1843) reinforces this judgement:

> To make war against [Richard] Marchand [a Berlin chemist], or, indeed, against anybody else, brings no contentment with it and is of little use to science . . . Imagine that it is the year 1900, when we are both dissolved into carbonic acid, water and ammonia, and our ashes, it may be, are part of the bones of some dog that has despoiled our graves. Who cares then whether we have lived in peace or anger; who thinks then of your polemics, of the sacrifice of your health and peace of mind for science? Nobody. Your good ideas, the new facts which you discovered – these, sifted from all that is immaterial, will be known and remembered until the end of time. But how comes it that I should advise the lion to eat sugar?[7]

On examining Liebig's list of research publications between 1824 and 1830, which are conveniently listed in Paoloni's (1968) bibliography, it is difficult at first to see any pattern. After an explosion of interest in fulminates, Liebig seems to have jumped erratically between applied chemistry (on pigments for paper manufacturers and artists, 1824); salts in mineral waters (1825); instruments (Howard's new thermometer, 1825); inorganic chemistry (iodine and bromine salts, 1826); six papers on indigo (1827, which included the isolation of carbazotique acid, later renamed picric acid); atomic weights (of bromine, 1828); the composition of organic acids (1828); the analysis of horses' urine and the identification of hippuric acid (1829); the properties of spongy platinum (1829); and an industrially significant method for purifying nickel and cobalt ores from arsenic (1830). It was not until about 1828 that there appears to have been a growing commitment to organic chemistry and not until 1830 that all or the majority of his annual output of publications fall definitely into this area. Before then, Liebig seems to have been almost indiscriminate in his choice of subjects for investigation. Wöhler perceived Liebig's fault only too well in a private letter to his mentor, Berzelius (26 March 1830):

> Liebig is exceedingly diligent, but it is truly a shame that he works in the Franco-German manner of L. Gmelin and, as a rule, produces a mass of little new pieces of information without giving certain, reliable results and completed researches. He dwells on far too many trivialities, and places far too much weight on the certainty of his analyses than on the probability that nature may have established the composition of a compound.[8]

7. *Ibid.*, vol. 1, p. 224.
8. J. Braun and O. Wallach, eds., *Briefwechsel zwischen J. Berzelius und F. Wöhler*, 2 vols. (Leipzig, 1901), vol. 1, p. 291.

The last remark was presumably a reference to their amicably-resolved difference over fulminic and cyanic acids.

Prior to about 1830, therefore, it would be incorrect to speak of Liebig's having a definite research agenda. Before then he was an opportunist, seizing upon any chemical issue that came his way. Such an approach is not unusual in the annals of chemistry. Lavoisier, for example, had spent many years of apparently aimless research before homing in on the problem of combustion. Like Lavoisier, Liebig took time to gain a reputation. Before 1830, his publications did not mark him as in any way out of the ordinary; the dispute with Wöhler had shown Wöhler to be the more reliable analyst, and it was left to Gay-Lussac and Berzelius to explain the real significance of fulminic and cyanic acids possessing the same composition. Liebig's haste and impetuosity also caused him to miss discoveries. In his analysis of mineral waters, Liebig had, to his chagrin, missed the opportunity of discovering a new halogen, bromine.

Liebig's maturation as an organic chemist seems to have come only gradually after 1826 when he met Wöhler and began to interest himself in organic compounds, starting with the composition of certain vegetable acids in 1828. He also began to reveal his experimental dexterity when he demonstrated that the remarkably absorptive spongy platinum first prepared by Edmund Davy in Ireland could be prepared in purer form by boiling platinum chloride with potash and precipitating the platinum black (*Platinschwarz*) with alcohol. He found, after carefully washing and drying the platinum, that it would absorb up to 250 times its volume of oxygen and spontaneously ignite combustible gases.[9] Although Liebig's exceptionally pure preparation was known for many years among chemists as "Liebig's Platinum Black," its significance for the theory and practice of catalysis remained for Berzelius and others to explore.

As we saw in Chapter 2, difficulties with the analyses of hippuric acid and alkaloids directed Liebig's attention to the need to speed up and to simplify the basic copper oxide technique of organic analysis that he had learned in Paris from Gay-Lussac and Thenard. Although the potash bulb apparatus remained to be perfected by Liebig's assistant Ettling, the new method was first announced by Liebig in 1830. The perfection of the technique led to a decade of intensive investigation of organic compounds by Liebig and his students, Liebig himself publishing an average of thirty papers each year between 1830 and 1840. Several of these, to be reviewed, were of the very greatest significance for the theory and practice of

9. J. Liebig, "Ueber Edmund Davy's schwarzen Platinederschlag," *Poggendorff's Annalen der Physik* 17 (1829), 101–14; W. H. Brock, "Liebig buys platinum from Janety the younger," *Platinum Metals Review* 17 (1973), 102–4.

organic chemistry, namely, an extensive series of papers on the nitrogen content of bases; the joint work with Wöhler on the benzoyl radical (1832) and on the degradation products of urea (1837); the discovery of chloral (1832); the identification of the ethyl radical (1834); the preparation of aldehyde (1835); and the polybasic theory of organic acids (1838).

Liebig's new method of combustion analysis was put to immediate pharmaceutical use with the alkaloids, morphine, strychnine, quinine, and cinchonine, which had been isolated by Sertürner, Pelletier and Caventou earlier in the century. In 1831 he investigated the chloride salts of these alkaloid bases and deduced that their power of neutralising acids was directly proportional to their nitrogen content.[10] More detailed analyses of narcotine, peperine, and atropine, as well as caffeine, confirmed the proposition, though when challenged by Regnault in 1838, Liebig admitted that the relationship between basicity and nitrogen content was not completely one to one.[11] By then he had perfected a new and accurate method for analysing organic bases, using platinum salts.[12] Liebig's method of determining the equivalents, or molecular weights, of organic bases by estimating the platinum content of double salts (platinic chloride plus the base hydrochloride) soon became (and remains) a standard analytical practice. Among many consequences, it gave a cast-iron formula for quinine, leading Liebig into an uncertain commercial venture in the 1840s (Chapter 5).

In June 1832 Liebig learned that Wöhler's first wife had died soon after giving birth to their child. He wrote a noble letter (15 June 1832) of consolation to his friend:

> My poor dear Wöhler. Who could have predicted such a dreadful misfortune after so happy a confinement; my poor friend, how empty are words of consolation after such a loss. I cannot tell you, I cannot express the feelings I had on receiving the news; it was as if I had actually experienced the loss myself. When I think to myself how satisfied and lucky you were in your domestic arrangements and of the affection and love you had for one another, and now this shocking dismemberment of all your hopes, this foundering of all your wishes! The good wife, so young, so full of life and goodness, and for her parents and yourself so irreparable. Come to us, dear Wöhler if we can

10. J. Liebig, "Neue Versuche über die elementare Zusammensetzung organischer Salzbasen," *Magazin für Pharmacie* 33 (1831), 143–44.
11. J. Liebig, "Über den Stickstoffgehalt der organischen Basen," *Annalen der Pharmacie* 6 (1833), 73–74.
12. J. Liebig, "Über die Constitution der organischen Säuren," *Ann. Pharm.* 26 (1838), 41–60, in reply to his own translation of Regnault's paper, "Neue Untersuchungen über die Zusammensetzung der organischer Basen," *ibid.* 26 (1838), 10–40.

> provide any consolation and help to heal your grief. To stay in Cassel at such a time will be injurious to your health. We can busy ourselves with something. I've bought some amygdalin from Paris [from Pelouze] and I shall have some 25 pounds of bitter almonds to work on. You must not travel, you must get occupied, but not in Cassel. I feel your distress will disappear in work and, dear friend, it will also be better if your sorrow is shared with a friend . . . Come to us; I expect you by the end of the week.[13]

From this tragedy emerged one of the most important collaborative papers in the history of chemistry. Liebig and Wöhler were by no means the first chemists or pharmacists to have investigated the composition of bitter almonds. In 1823 the German pharmacist Carl Strange had identified the crystals formed when oil of bitter almonds oxidises in air as benzoic acid, which he also extracted from the cherry laurel. In France, Pierre Robiquet also extracted a crystalline nitrogeneous material, amygdalin, from the nut, which was also known for containing the poison, prussic acid. Using almond oil supplied by Gay-Lussac's favourite student Théophile Pelouze (1807–67) and by means of a rigorous distillation process, Liebig and Wöhler succeeded in obtaining a pure oil for analysis, though since its boiling point (179° C) lay beyond the scale of their thermometer, they had to rely on consistent specific gravity readings as a criterion of purity. The pure oil was then analysed, as was its oxidation product, benzoic acid. The value for the latter challenged an analysis performed by Berzelius in 1813, but the Swede was to agree with their new values.

The two chemists then transformed the pure oil into several halogenated products, which were further transformed in other reactions. The only trick they missed was a close investigation of benzoic acid itself, and it was left to Mitscherlich in Berlin to distil it with slaked lime and identify the product as benzene, or benzol, as Liebig proposed it should be called in 1834. In their joint work, to their evident surprise but delight at the prospect of classification that it afforded, they found in each case that a common multiple of $C^{12}H^{10}O^{2}$ consistently appeared in the derivatives after their composition had been accurately determined. If hydrobenzoyl (as they named the pure oil) was represented as (14C + 10H + 2O) + 2H [i.e., modern C_6H_5COH], then their paper implied the following reaction schema:[14]

13. Hofmann, *Liebig's und Wöhler's Briefwechsel,* vol. 1, pp. 53–54.
14. J. Liebig and F. Wöhler, "Untersuchungen über das Radikal der Benzoesäure," *Ann. Pharm.* 3 (1832), 249–82. The whole was translated by Wöhler's American student J. C.

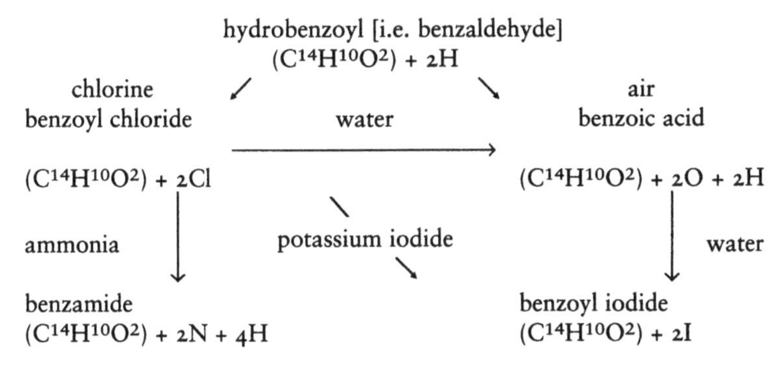

In reviewing these relationships, they concluded:

> We find that they group themselves about a single compound, which preserves its nature and composition unchanged in nearly all its associations with other bodies. This stability, this sequence in the phenomena, induced us to assume that this group is a compound element and hence to propose for it a special name, that of benzoyl. The composition of this radical we have expressed by the formula 14C + 10H + 2O.

In view of the later controversy over the substitution of an electronegative chlorine atom for an electropositive hydrogen, it is important to notice Liebig and Wöhler's explicit remark that[15]

> benzoyl hydride (the oil of bitter almonds) is composed of (14C + 10H + 2O) + 2H. By the action of chlorine, 2 atoms of hydrogen combine with 2 atoms of chlorine to form hydrochloric acid, which escapes. In the place of the hydrogen, however, 2 atoms of chlorine enter, according to the following formula: (14C + 10H + 2O) + 2Cl.

This brilliantly executed paper was forwarded to Berzelius in advance of publication, and his reply of 2 September 1832 praising the investigation was printed as an appendix. Berzelius was principally struck by the way a combination of the three elements carbon, hydrogen, and oxygen could act as if it were a single element. Using his knowledge of Greek, he proposed to call this first example of a ternary radical, *proin* (dawn); but then, realizing this would involve altering the familiar name of benzoic acid, he accepted Liebig and Wöhler's term, *benzoyl* and gave it the special symbol of Bz (i.e., Bz = $C^{14}H^{10}O^{2}$). He then gave lots of examples of how

Booth, in *American Journal of Science* 26 (1834), 261–85, and has been reprinted in O. T. Benfey, *From Vital Force to Structural Formulas* (1964; repr. Philadelphia, 1993), p. 35.

15. Benfey, *From Vital Force*, p. 25.

compositions could be represented by this formula, adding the caveat that such formulas should be used only "when the ideas they express are advanced in some measure to confirm truths, else they would lead to a Babylonian confusion."[16]

Six years later in 1838, Berzelius had second thoughts, arguing that the concept of an oxygen-containing ternary radical was against the spirit of Lavoisier's original definition that a radical was the part of a molecule that combined with oxygen. It was, as well, against his own electrochemical theory of composition. He now proposed that the benzoyl radical Bz should be defined as the hydrocarbon ($C^{14}H^{10}$). On the atomic weight scale of C = 12, this reduces to the familiar modern benzyl radical, C_7H_5. Unfortunately, although Liebig and Wöhler readily agreed with Berzelius, other chemists deployed benzoyl as an oxygen-containing radical and so muddled the clarification and order that the paper had brought. Later, in the 1840s, Berzelius clouded the issue further by reinterpreting bitter almond oil not as a hydride of benzoyl but as a dioxide of a radical, pikramyl, $C^{14}H^{12}$ (*pikros,* Greek for bitter). By then, however, Berzelius had lost his previous authority, and the suggestion was quietly ignored.

As Liebig and Wöhler claimed in the famous opening sentences of their paper:

> When in the dark province of organic nature, we succeed in finding a light point, appearing to be one of those inlets whereby we may attain to the examination and investigation of this province, then we have reason to congratulate ourselves, although conscious that the object before us is unexhausted.[17]

The notion that a radical group might persist unchanged through a series of reactions, as if it were an element, was not entirely new. It had been used by Gay-Lussac in his work on cyanides in 1815 and by Dumas and Boullay in their work on esters in 1827. Liebig and Wöhler's paper reinforced this taxonomic approach and ushered in a method of classifying organic compounds that contemporaries and historians came to call the radical theory.

In 1833 in Ireland, Robert Kane, who worked with Liebig in 1836, used an analogy with ammonium compounds to suggest that the common component of alcohol, ether, and their derivatives was the group ethereum C^4H^{10}. He formulated this as a hydride of a radical (C^4H^8), which Berzelius had named etherin in 1832 in recognition of a classification of the relationships between alcohol and ether made by Dumas and Boullay. Liebig renamed Kane's radical as ethyl in 1834. In the same year in France,

16. *Ibid.*, p. 39.
17. *Ibid.*, p. 15.

Dumas and Péligot published work on the derivatives of wood alcohol, interpreting them in terms of a methyl radical. So successful was the idea of radicals in imposing order on what Wöhler had memorably once dubbed the "dark jungle of organic chemistry" that by 1837, in their one and only joint publication, Dumas and Liebig were able to make a fundamental analogy between inorganic and organic chemistry:

> in inorganic chemistry the radicals are simple; in organic chemistry they are compounds – that is the sole difference.[18]

Later, Liebig's pupils C. Gerhardt (1816–56), A. W. Williamson (1824–1904), and Hofmann developed a complementary classificatory system based upon a compound's resemblance to the molecular types of hydrogen, water, hydrogen chloride, and ammonia. Both methods gave way in the 1850s and 1860s to the more profound notions of valency and chemical structure that were developed by Liebig's pupil Kekulé and an English visitor to Giessen, Edward Frankland (1825–99).[19]

A good example of Liebig's use of the radical theory was his investigation of the composition of alcohol in 1834, which involved him in bitter and futile controversy with French chemists. French research dating back to the time when Liebig was a student in Paris had suggested that alcohol (ethyl alcohol) and its acidification product, ether (ethyl ether), were compounds of water and olefiant gas (ethylene). The reaction (etherification) was by no means straightforward, and a number of side-products, including a sulphovinic acid, were also formed when sulphuric acid was used. In joint work with Wöhler in 1831 on the analysis of barium sulphovinate, Liebig suggested that sulphovinic acid was either anhydrous sulphuric acid combined with alcohol (i.e., an addition compound that then decomposed into ether) or hydrated sulphuric acid combined with ether. This suggestion was contested by Dumas and Boullay in the following year, using accurate vapour density measurements of both alcohol and ether. In their conclusions, they suggested analogies with ammonium compounds, thus supplying a way of classifying esters as salts.

	ammonia		olefiant gas
	NH_3		C_2H_4 (C = 6)
		alcohol	$4C_2H_4, 2H_2O$
		ether	$4C_2H_4, H_2O$
hydrochlorate	NH_3,HCl		$4C_2H_4, 2HCl$

where C = 6 and 4-volume formulas are used. According to this scheme

18. J. Liebig and J. B. Dumas, "Note sur l'état actuel de la chimie organique," *Comptes Rendus* 5 (1837), 567.
19. See W. H. Brock, *Fontana History of Chemistry* (London, 1992), Chaps. 6 and 7.

of rational formulas, then, alcohol was a double hydrate of etherin (Berzelius's term) and ether a single hydrate, and ether was formed from alcohol by the extraction of a molecule of water.

Liebig saw the matter very differently. Although he agreed with Dumas that alcohol could be considered a hydrate of ether, he was adamant that water did not enter into the composition of ether itself. Instead, building on Kane's suggestion, he formulated ether as an oxide of an electropositive ethyl radical $(C_4H_{10})O$ and alcohol as its hydroxide, $(C_4H_{10})OH$, and sulphovinic acid as ether plus sulphuric acid.[20] To complicate matters, Berzelius then disagreed with both Dumas and Liebig that the formulas of alcohol and ether were related by hydration or dehydration. Although retaining an ethyl radical C_2H_5 (C = 6) for ether, he explained alcohol as the oxide of a different radical C_2H_6. When one bears in mind that all three chemists were using different atomic weight values and different volume standards, as well as positing different radicals, it is understandable that the situation was ripe for confusion and argument. Today's chemists will recognize that, although Liebig's formulation of ether was correct, his expression for alcohol $C_4H_{12}O_2$ (i.e., $2C_2H_5O + H_2O$) was double the modern value. It was not until Williamson's work in the 1840s that the relationships among alcohol, ethers, and their esters were placed on firm foundations.

As Larry Holmes has observed, Liebig's ardent defence of an ethyl theory against the French etherin theory is difficult to reconcile with Liebig's professed claim that theories were merely stepping-stones to truth. His hostility towards Dumas owed much more to his jealousy of the Frenchman's dominance in European chemistry. As Holmes said:

> As an influential independent voice, Liebig felt he had a duty to oppose these tendencies and to rescue French chemistry from the "false route" on which Dumas was directing it. Liebig was also resentful because he thought that the younger French chemists still acted as though no important investigations were going on outside Paris. In seeking to overthrow Dumas's ether theory, Liebig was in part trying to make the French acknowledge him and his followers as a leading force in organic chemistry.[21]

The result of these taxonomic polemics was a shifting set of alliances between Liebig and Wöhler, Berzelius, Dumas, Pelouze, Laurent, and Gerhardt as these and other chemists strove to bring order to the labyrinth

20. For fuller details, see F. L. Holmes, "Liebig" in *Dictionary Scientific Biography*, vol. 8 (New York, 1973), pp. 329–50.

21. *Ibid.*, p. 336. For an elaboration, see Holmes's "Justus Liebig and the construction of organic chemistry," in S. H. Mauskopf, ed., *Chemical Sciences in the Modern World* (Philadelphia, 1993), pp. 119–34.

of organic chemistry. Slowly, agreement grew that since Liebig's benzoyl and ethyl radicals behaved as electropositive molecules analogous to metals or ammonium in inorganic salts, they illuminated organic chemists' ways through the darkness.

Radicals also enabled Liebig to understand chlorination, a subject that, according to Volhard, Liebig had first investigated as a student. In 1831, using his improved method of analysis, Liebig isolated a white crystalline product from the action of chlorine on alcohol that he named chloral. In devising this word from *chlor*inated *al*cohol, Liebig was following a precedent set by the French chemist Chevreul, who, unable to suggest a systematic name for what is now called cetyl alcohol, had named it *éthal* from the first two syllables of ether and alcohol. On adding an alkali to chloral, Liebig obtained a volatile liquid the composition of which he deduced to be a carbon chloride C_2Cl_5, so missing the presence of hydrogen. This sloppy piece of analysis may well have been due to his anxiety to beat Dumas to publication, though it also suggests that Liebig's new method of analysis still faced problems with volatile compounds. Berzelius, echoing Wöhler's criticism already quoted, told Wöhler that because Liebig tended to publish things too hastily, "much remains undeveloped and required later corrections."[22] Liebig was not alone in missing the discovery of chloroform that year. In Paris, the pharmacist Eugène Soubeiran distilled chloral with bleaching powder and gave the product he called ether bichloride the formula CH_2Cl_2. In 1834, however, as a result of his controversial theory of substitution, Dumas was able to show that the volatile product, which he named chloroform because of a supposed relationship with formic acid, corresponded to the formula $CHCl_3$. Not until 1847 were its anaesthetic properties explored by the Edinburgh surgeon James Young Simpson on the advice of Liebig's Scots pupils Gregory and Anderson – as Liebig proudly noted in discussing its narcotic properties.[23] At that stage, as with the missed discovery of bromine, Liebig was irritated that Dumas had the glory of identifying a chemical of humanitarian significance, and in one of his last publications he laid claim to being its real discoverer.[24]

Alcohol also produced another interesting reaction for Liebig. Earlier pharmaceutical investigations by Scheele in the eighteenth century and by Fourcroy at the beginning of the nineteenth century, and by Döbereiner,

22. Berzelius to Wöhler, 24 June 1832; Braun and Wallach, *Briefwechsel zwischen Berzelius und Wöhler*, vol. 1, p. 292.
23. J. Simpson, "Anwendung des Chloroform's bei chirurgischen Operationen," *Annalen* 65 (1848), 121–26.
24. J. Liebig, "Zur Geschichte der Entdeckung der Chloroforms," *Ann. Chem.* 162 (1872), 161–64; Max Speter, "Liebig oder Soubeiran?" *Chemiker Zeitung* 55 (1931), 781–82.

had drawn attention to an odiferous product often produced when ethyl alcohol was oxidised with manganese dioxide; it was usually described as an "ether." In 1832 Liebig, who had been in correspondence with Döbereiner, confirmed the latter's account of the preparation of this pungent-smelling "light ether." Döbereiner, in return, sent Liebig a sample of a crystalline compound the ether formed with ammonia. This gift spurred Liebig into a full-scale investigation of the oxidation of alcohol, from which he concluded in 1835 that the light ether was a mixture of an *al*cohol *dehyd*rogenatus, or "aldehyde," and acetal (another neologism). This aldehyde, later more formally known as acetaldehyde, formed distinctive crystals of an aldehyde ammonia when ammonia was added, which was to become a standard tests for aldehydes.[25] On tabulating the conjectured formulas, Liebig was certain that there must be an intermediate compound between aldehyde and acetic acid ($C = 6$):

C_6H_6	unknown radical or hydrocarbon, "acetyl" (Berzelius's term)
$C_4H_6 + H_2O$	aldehyde
$C_4H_6O_2 + H_2O$	aldehydic acid
$C_4H_6O_3 + H_2O$	acetic acid

Much later in the century it was shown that Liebig's hypothetical aldehydic acid was a mixture of acetic acid and aldehyde.

Liebig took umbrage a few months after publishing this paper when Döbereiner, with whom he had always got on well, published a paper on platinum in which he remarked in passing that the two products of alcohol oxidation, acetal and aldehyde, "were likewise prepared by J. W. Döbereiner and analysed by Liebig." Using his editorial privilege in *Annalen* Liebig published a polemical and sarcastic essay, "Who discovered aldehyde?" which opened with the aphorism, "If our friends say this, what worse can be done by our enemies?" By tabulating Döbereiner's boiling points and account of the properties of the "light ether" against his own determinations of pure aldehyde, the superiority of Liebig's findings were obvious. He concluded bitingly, "Döbereiner has had in the discovery of acetaldehyde about the same position as Newton's apple had in the discovery of gravity and the laws of free fall."[26]

Although Döbereiner had used a distillation retort to prepare "light ether" in 1822, one of the undoubted reasons Liebig obtained better results was that he used a superior form of distillation that employed a counter-current water-cooling system that erroneously came to be called

25. J. Liebig, "Ueber die Produkte der Oxydation des Alkohols," *Ann. Pharm.* 14 (1835), 133–67; F. Walker, "Early history of acetaldehyde and formaldehyde," *J. Chem. Educ.* 10 (1933), 546–51.

26. J. Liebig, "Wer ist der Entdecker des Aldehyds?" *Ann. Pharm.* 22(1837), 273–77.

the Liebig Condenser. The principle of a water-cooled jacket had been familiar to pharmacists in the eighteenth century, and was a familiar tool to Liebig's friend Mohr.[27] Its association with Liebig's name came about solely because it became such a common tool in organic preparations and purifications in the Giessen laboratory.

In 1839, in his last contribution to the theory of organic chemistry, Liebig incorporated both Dumas's etherin and his own ethyl radicals within the acetyl radical. If acetyl, $C_4H_6 = Ac$, then,

olefiant gas [ethane]	$= AcH_2$
ethyl	$= AcH_4$
ether	$= AcH_4O$
alcohol	$= AcH_4O + H_2O$

At one stroke he recognised that the question of the composition of organic compounds could not be settled on the basis of one investigation, that of etherification. He was no longer confident that there was any unique way of classifying organic compounds and deciding upon their composition from the point of view of Berzelius's electrochemical theory, which he now thought had become too inflexible. Berzelius, whom he had hitherto revered as a father figure, was failing to adjust to discoveries such as substitution reactions and, above all, to Liebig's ideas on acidity.

Lavoisier's new chemistry had largely revolved around the generalisation that oxygen was an acid-former, and that all acids contained oxygen. Although by 1820 chemists had been forced to accept that there was also a class of hydracids, such as hydrochloric acid and hydrocyanic acid, because oxyacids (unlike hydracids) could be thought of as containing the elements of water, simple molecules such as SO_2, SO_3, and CO_2 were usually called acid anhydrides in recognition of the fact that they formed acidic solutions with water. By analogy, organic acids were supposed to consist of an inseparable acid anhydride combined with water, which was simply displaced when the acid reacted with a base. Thus, acetic acid was formulated by Berzelius as an anhydrous acid $C_4H_6O_3$, combined with water. This preserved Lavoisier's original dualism.

In Scotland in 1833, Thomas Graham (1805–69) had explored the relationships among three of the four known phosphorus acids whose existence Berzelius had attributed to isomerism.[28] By very precise and careful analyses, Graham showed that two of these acids and their respective salts contained hydrogen, which he believed to be present in the form of water. On interpreting these findings according to Berzelius's prevalent

27. R. J. Forbes, *Short History of the Art of Distillation* (Leyden, 1948; reprint 1970); and F. Szabadvary, *History of Analytical Chemistry* (Oxford, 1966), p. 243.
28. Michael Stanley, "Thomas Graham," *Chemistry in Britain* 27 (1991), 239–42.

electrochemical dualistic system, the water was assumed to be acting as a base to the acidic phosphoric acid:

$3HO + PO_5$ (O = 8) or H_3PO_4 (O = 16)
$2HO + PO_5$ or $H_4P_2O_7$
$HO + PO_5$ or HPO_3

Graham's use of Berzelian symbols in his paper was unique; it was the first time such symbols had appeared in the *Philosophical Transactions of the Royal Society*.[29]

The perspective that, among the mineral acids some (like phosphoric acid) could be saturated by one, two, or more equivalent bases, overrode the long-standing view that acids were acid anhydrides or simple nonmetallic oxides, as Lavoisier had posited in the 1780s. For Graham, however, the presence of water remained an essential feature of their acidity. Nevertheless, as Partington commented, Graham "established the *fact* of the existence of polybasic acids in a masterly way."[30] Although Graham published his views on acids in 1833, it was not until Liebig met Graham during the latter's summer tour of Europe in 1836 that Liebig understood the significance of polybasicity for the composition of organic acids. An additional complication for traditional views of acidity was the production in the 1830s of large numbers of new acids by heating natural vegetable acids. These so-called pyrogenic acids were investigated in particular by Liebig's Parisian friend Pelouze, who spent the summer of 1836 examining some of them at Giessen. Neither Liebig nor Wöhler made any secret of their speculation that these acids might be organic hydracids, though Berzelius was not to be convinced.

On his return from Britain in October 1837, Liebig stopped off in Paris, where he signalled a truce with Dumas in the form of the joint statement that organic chemistry was based on radicals. This same paper also contained the suggestion that certain specified organic acids (e.g., citric acid) apparently contained fractional parts by weight of water, an anomaly that would be removed if their molecular weights were doubled or trebled. This would, however, have the further consequence that their salts would contain two or three molecules of base, and Graham's findings for mineral acids would also apply to organic acids. In a letter he sent directly to the French Academy of Sciences, Berzelius was adamant in his refusal to accept this suggestion. The letter was read publicly by Pelouze, who dramatically claimed conceptual priority of the suggestion espoused by Liebig and Dumas.

29. T. Graham, *Phil. Trans.*, 1833, 253; Liebig published a translation in *Ann. Pharm.* 12(1834), 1–12.
30. J. R. Partington, *A History of Chemistry*, vol. iv (London, 1964), p. 274.

The rift between Liebig and Berzelius had a potentially dangerous consequence for Pelouze's career in Paris when in the spirit of reconciliation in 1837 Liebig allowed Dumas to report on Liebig's views on acids to the French Academy of Sciences. Pelouze was convinced that Dumas (who reported the ideas from two letters he had received from Liebig) was appropriating Liebig's research on citric acid as his own. Following private protests to both Berzelius and Liebig, the latter became convinced that Dumas was not to be trusted, as did Berzelius, who called Dumas a Jesuit. However, since Pelouze was up for election to the Academy of Sciences, Liebig was forced to hold his fire against Dumas until Pelouze was safely elected.[31] The resulting skirmish pleased Gay-Lussac, because, as Pelouze's mentor, he had always disliked Dumas:

> Now, my dear Liebig, I congratulate you for having left the gallery that you entered. I did not understand your marriage; I did not think at all that it could last and your divorce does not surprise me. You two characters were too opposite, a great loyalty, a great openness, cannot at all agree with a fine finesse. You knew him very well, this colleague Ignatius, this Jesuit as you said and heard. It is natural that you were fascinated and susceptible as I didn't think you were. One should recognize him as a snake, a great power of fascination. Do not think there is another one in France![32]

More amusingly, amidst these squabbles, Liebig showed that sauerkraut contained lactic acid, commenting that "we are so accustomed in our investigations to occupy our attention with what is novel and complex, that we overlook what is near at hand, though of far greater interest."[33] Liebig was always a pragmatist.

In the following year, 1838, Liebig published his masterly seventy-six-page memoir on the constitution of organic acids.[34] This contained analyses of dozens of vegetable acids determined via their silver salts. Citric, meconic, and cyanuric acids were now seen as tribasic and certain derived pyrogenic acids as bibasic. For example, the opium-derived meconic acid (a pyrone) could be formulated as:

meconic acid	$C_{14}H_2O_{11} + 3H_2O\ [C_{12}H_4O_{10} + H_4]$
silver salt	$C_{14}H_2O_{11} + 3AgO$
acid salt	$C_{14}H_2O_{11} + 2H_2O$ KO

31. Mi Gyung Kim, "Constructing symbolic space," *Ambix*, 43 (1996), 1–31.
32. Gay-Lussac to Liebig, cited by Liebig in a letter to Berzelius, 28 June 1838, in Carrière, *Berzelius und Liebig ihre Briefe*, p. 170.
33. J. Liebig, "Milchsäure, die Säure des Sauerkrauts," *Ann. Pharm.* 23 (1837), 113–15.
34. J. Liebig, "Über die Constitution der organischen Säuren," *Ann. Pharm.* 26 (1838), 113–89.

potassium salt	$C_{14}H_2O_{11} + H_2O$ $2KO$	
and, meconic acid →	comenic acid	→ pyromeconic acid
$C_{14}H_2O_{11}$	$C_{12}H_4O_8 + 2H_2O$	$C_{10}H_6O_5 + H_2O$
	(forms two salts)	(forms one salt)

Although Liebig's findings and interpretations were by no means all correct, the paper contained the important suggestion that it would be better for chemists in future if they defined acids (whether inorganic or organic) as "particular compounds of hydrogen, in which the hydrogen can be replaced by the equivalent of a metal."[35] Neutral salts lost their acidity because all their hydrogen had been replaced; acidic salts correspondingly retained some acidic tendencies because not all of their hydrogen had been lost. Although Liebig still continued to write the formulas of acids and salts dualistically, Berzelius did not approve of Liebig's generalization, which he dismissed as irresponsible. The growing rift between the older and younger chemist was to be marked by a complete break in 1842 in response to Liebig's views on physiological chemistry.

The full impact of Liebig's theory of acidity became clear only in the 1840s when electrolytic data and the shift of organic chemists towards unitary theories of composition and classification were seen to be beautifully harmonised with the concept of polybasicity and hydrogen displacements and substitutions. By then, however, Liebig had had his fill of polemics over theories of organic classification and composition, and he had moved instead towards the applications of his chemical skills and insights to the problems of physiological chemistry.

His swan song on conventional organic chemistry, the second great joint paper with Wöhler in 1837, anticipated this move and brought him especial fame in Great Britain. The joint work on benzoyl had given Liebig and Wöhler the confidence to analyse urine and to identify and classify its constituents, such as urea, uric acid, allontoin, and uramil, which they believed to be produced by "innumerable metamorphoses" of uric acid – itself a degradation product of flesh and blood.[36] There had been many studies of the composition of urine before, notably by the English doctor and chemist William Prout. To Liebig and Wöhler, however, the isolation of individual substances was of little value if their interrelationships and mutual degradations and metamorphoses were not also studied. Only in this way could any significant relationship between their chemical com-

35. *Ibid.*, p. 180.

36. J. Liebig and F. Wöhler, "Untersuchungen über die Natur der Harnsäure," *Ann. Pharm.* 26 (1838), 241–340; abstracted earlier in *Poggendorff's Annalen der Physik*, 41 (1837), 561–69.

position and physiological function be deduced. "From this investigation," they declared,

> the philosophy of chemistry will draw the conclusion that the production of all organic substances no longer belongs just to living organisms. It must be seen as not only probable, but as certain, that we shall be able to produce them in our laboratories. Sugar, salicin, and morphine will be artificially produced. Of course, we do not yet know how to do this, because we do not yet know the precursors from which these compounds arise. But we shall come to know them.[37]

As if in recompense for some occasions when he had not given his friend due acknowledgement, Liebig went out of his way in his autobiography to recall that this work (as well as that on bitter almond oil) had really been Wöhler's achievement:

> I cannot sufficiently highly estimate the advantage which the association with Wöhler brought to me in the attainment of my own as well as of our mutual aims, for by that association were united the peculiarities of two schools – the good that was in each became effective by cooperation. Without envy and without jealousy, hand-in-hand, we plodded our way; when the one needed help, the other was ready.[38]

This investigation of urine, which first brought Liebig's name prominently before a wider British public, gave contemporary doctors a new insight into the pathology of kidney and bladder diseases. But before turning to this topic and to the second phase of his career as a physiological chemist, it will be convenient to mention his few other contributions to organic chemistry.

In 1842, while in the middle of a dispute concerning the nature of protein, Liebig found time to devise a method of preparing potassium cyanide from potassium ferricyanide and potassium carbonate. This cheap method gave chemists a reagent that soon proved invaluable in organic chemistry and in the new art and science of photography, which was becoming such an endless source of fascination in the 1840s. The silver electroplating industry also found uses for potassium cyanide, and later in the century this substance became an important agent in gold mining.

Although this preparation was uncontroversial, Liebig's analysis of another cyanogen derivative, melamine, which he had first investigated in 1829, brought him into a bitter and indefensible squabble. Melamine, later known as cyanuramide, is a trimeric cyclic form of cyanamide

37. Liebig und Wöhler, "Untersuchungen," p. 241.
38. W. A. Shenstone, *Justus von Liebig* (London, 1901), pp. 38–39.

View of Liebig's 1839 lecture room as restored in the early 1950s. Note doorway and hatch leading into the general laboratory. (Courtesy Liebig-Museum Giessen)

$C_3N_3(NH_2)_3$, which is today used in the preparation of formaldehyde plastics. In 1834, while investigating the reactions of potassium thiocyanate with ammonia, Liebig stumbled upon an insoluble honey-coloured product which he named *melam* and formulated as $C_6H_{10}N$. After heating melam with potash, two bases separated, which he called melamine, $C_3H_6N_6$, and ammeline, $C_3H_5N_5O$. From the reaction of melam with sulphuric acid, he separated a neutral material he called ammelide, $C_6H_9N_9O_3$. How might these products be classified by a radical theory? The answer seemed to lie in the action of chlorine on ammonium thiocyanate, which produced a yellow product he christened mellon, C_3N_4. Because this mellon formed a potassium compound C_3N_4K, if mellon was assumed to be a radical analogous to cyanogen, then the new products would be:

melamine	$C_3N_3(NH_2)_3$
ammeline	$C_3N_3(HH_2)_2(OH)$
ammelide	$(C_3N_3)_2(OH)_2$
cyanuric acid	$C_3N_3(OH)_3$

Without doing any preparative chemistry or analysis, Liebig's former French pupil Charles Gerhardt decided from the symmetry of these formulas that something was wrong. He proposed that ammelide was really a previously unknown melanuric acid $C_3N_3(NH_2)(OH)_2$. However, Liebig found it easy to show that this theory was impossible by preparing the real melanuric acid from urea and showing that its composition was quite different. He bitterly castigated Gerhardt for using theory (or rather intuition) instead of experiment to judge such an issue. Although Gerhardt's judgement was at fault in this case, he and his colleague Laurent were to demonstrate that theoretical models had an important function in the classification of organic compounds.

The dispute did not end here, for Gerhardt, joined by Laurent, also argued that Liebig's mellon formula C_3N_4 could not possibly be right. The theoretical reason for their argument concerned standardising the formulas of organic compounds as "two-volume formulae," where the formula for water is H_2O. On this standard, the numbers of atoms in an organic compound always formed an even number, which Liebig's mellon did not. They were able to show that Liebig had missed an atom of hydrogen in his analysis, and they reformulated mellon as $C_6H_3N_9$.

Liebig was livid, making one of his longest and most vitriolic replies in "Herr Gerhardt and organic chemistry" and "Illustrations of an investigation of mellon compounds by Laurent and Gerhardt."[39] In these articles he accused the Frenchmen of making a "monstrous alliance " against the Giessen school and wrote that Gerhardt reminded him of a highwayman – a figure that Laurent's rhetoric transformed into a brigand in his own dispute with Dumas. The row is piquant, considering that Gerhardt had been a student of Liebig's and that he had translated Liebig's work into French. Although a concord was struck between the two men in 1850,[40] the argument erupted again violently in 1855 over the same mellon compounds when Gerhardt (correctly) deemed that they were best interpreted as derivatives of cyanuric acid $C_3N_3(OH)_3$.[41]

Holmes, the leading authority on Liebig's organic chemistry, has seen Liebig's shifting alliances between Wöhler, Berzelius, and Dumas as fairly characteristic of "scientists as contenders in an agonistic field filled with social conflicts, disputes, alliances and maneuvers," as they set about trying to construct reality from nature's clues.[42] In such a rapidly develop-

39. J. Liebig, "M. Gerhardt et la chimie organique," *Revue scientifique* 23 (1845), 422; *Ann. Chem.* 57 (1846), 93–118; and "Beleuchtung einer Untersuchung von Laurent und Gerhardt über Mellonverbindungen," *ibid.* 58 (1846), 227–64.
40. E. Grimaux and C. Gerhardt, Jr., *Charles Gerhardt, sa vie, son oeuvre, sa correspondance* (Paris, 1900), p. 201.
41. Volhard, vol. ii, pp. 277–91; Volhard ignored the savagery of Liebig's rhetoric.
42. Holmes, *Chemical Sciences*, p. 129, quoting Bruno Latour, Steve Woolgar, *Laboratory Life. The Construction of Scientific Facts*, 2nd ed. (Princeton, N.J., 1986), pp. 105–49.

ing research field, however, no sooner was one clue followed up than it led on to others, so that in Holmes's striking metaphor, Liebig and others were "being swept along in a moving investigative stream that none of them could control."[43]

As Liebig told Pelouze in 1834, "with the impetus that organic chemistry has acquired, one truly does not know where it will stop, and where we shall be led. One becomes dizzy from so many discoveries."[44] Faced by such a whirligig, there were three options – to hold on tightly and inflexibly as Berzelius tried to do, to hold tightly but bend with the wind as Dumas did, or to abandon the course as Liebig decided around 1839. Disillusioned by Berzelius's attitude to new ideas, mistrustful of Dumas's imperial ambitions, and outraged by the newer unitary theories of Laurent and Gerhardt, which promised to undermine the classificatory scheme of radicals he had invested so much time and energy in, Liebig decided the field was an impossible one for leadership. Far better, then, to strike out in a new direction in which he could lead by arguing that chemistry was a fundamental science of broad significance for the modern age.

Essentially, after 1840, Liebig switched himself off from the theoretical debates that were engaging his contemporaries. As he told Mohr, whereas "formerly when a problem presented itself, it seemed as though I had but to thrust my hand into a full basket and draw out the answer, now I have to consider it a long time."[45] And he came to share Schönbein's old chemists' view that "today's obsession with atom complexes, substitutions, rational formulae, etc., is to me, I openly confess, not to my taste, and I'm afraid, I'm not the only one."[46] In another letter Schönbein asked Liebig what he thought of "today's type formulas," confessing that he saw them as " a Cartesian game, little likely to lead to insights into chemical constitution."[47] To which Liebig replied unequivocally, "You're right about types. Organic compounds are like calico patterns forced into three packets from which the simplest pattern is pasted on. H[ydrogen] is a type, CO_2 is not."[48]

Nevertheless, despite this distancing from the theoretical interpretation of organic compounds, although he worked in the laboratory only during the mornings, Liebig remained relatively productive in Munich. In 1851

43. Holmes, *Chemical Sciences*, pp. 129–30.
44. *Ibid.*, p. 130.
45. G. W. A. Kahlbaum, ed., *Justus von Liebig und Friedrich Mohr in ihren Briefen* (Leipzig, 1904), p. 72.
46. Schönbein to Liebig, 24 February 1860, in G. W. A. Kahlbaum and E. Thon, eds., *Justus von Liebig und Christian Friedrich Schönbein Briefwechsel* (Leipzig, 1900), p. 96.
47. Schönbein to Liebig, 11 July 1860, *ibid.*, p. 102.
48. Liebig to Schönbein, 16 July 1860, *ibid.*, p. 104.

he provided clinicians with a simple chemical procedure for the quantitative determination of the amount of urea in urine and of the oxygen content of air in an alkaline solution of pyrogallol. Similarly, he identified and prepared further mellon compounds and fulminuric acid, and he synthesised oxamide. These were examples of the broader significance and applications of chemistry that he had first drawn to international attention in 1837 when he had paid his first visit to Great Britain.

4

Liebig and the British

> More students from this country than from any other land beyond the bounds of Germany, have worked in the laboratory of Giessen, and have derived incalculable benefit from the institution there imparted, and from the noble example there presented to them of an elevated philosophical and scientific life. (Thomas Graham to Liebig, July 1854)[1]

Liebig lived through a transport revolution as the railway train and steamship came to replace the stagecoach and sailing ship. Although distances shrank, it was the generation of chemists after Liebig who made the most of travel opportunities and who truly began to internationalise their chemical meetings – as the Versammlung der Deutsche Naturforscher and Aerzte and the British Association for the Advancement of Science (BAAS) had begun to do in limited fashion by welcoming foreign visitors. Thus, whereas some of Liebig's pupils, such as Playfair and Hofmann, visited the United States, Liebig's sphere of a lifetime's travel was limited to Aberdeen in the north, Naples in the south, Vienna in the east, and Paris to the west. He paid six visits to the British Isles and six to France, as well as holiday visits to Austria, Italy, and Switzerland, very often in the company of close friends such as Wöhler and Buff. Apart from other holidays to well-known German health resorts, his wife and family seem rarely to have accompanied him on such excursions. Correspondence, rather than travel, remained of great significance in Liebig's world. He came face to face with Berzelius but once, with Dumas three or four times, and with his closest friend Wöhler scarcely a dozen times. The exchange of information through correspondence was vital to Liebig's lifestyle.

Liebig's visits abroad are all well documented by Volhard from Liebig's correspondence with his wife and friends. His trips to Britain deserve special comment, both because of the large number of English, Irish, and

1. The letter accompanied the testimonial from Liebig's British friends that marked his retirement from Giessen. See W. H. Brock, ed., *Justus von Liebig und August Wilhelm Hofmann in ihren Briefen* (Weinheim, 1984), p. 179.

A Liebig family photograph taken outside a garden house at Giessen c. 1844. Liebig and his wife, Henriette, are surrounded by (*left to right*) Hermann (age 13), Georg (age 17), Agnes (age 16), and Johanna (Nanny, age 8). Their other daughter, Marie, was born in 1845. (Courtesy Liebig-Museum Giessen)

Scottish students who studied with him at Giessen, and because of his particular attachment to the country. The first Englishman to study with Liebig was William Charles Henry (1804–92), the son of the Mancunian chemist William Henry, who was a friend of John Dalton's. The Henry family's fortunes were based upon the manufacture and sale of the pharmaceutical product milk of magnesia, which was used in the treatment of dyspepsia. Charles Henry (as he called himself) had been privately tutored by John Dalton, whose biography he was later to write (rather badly) before studying medicine at the University of Edinburgh, where he graduated in 1827. Following a tour of Continental medical schools, he had settled in Manchester as an honorary physician at the Manchester Infirmary, only to interrupt this in May 1835 with a year at Berlin and Giessen. Henry's stay in Giessen for the winter semester of 1835 was followed in the winter session of 1836 by another Englishman, Thomas Richardson, a pupil of the highly regarded professor of chemistry at the University of Glasgow, Thomas Thomson (1773–1852), who sent his own son, another Thomas Thomson (1817–78), to study with Liebig during the summer semester of 1837. Thomson senior, who kept up with the overseas chemical literature, was aware of how behind the small community of British chemists was in organic chemistry, and he exploited the annual platform provided by meetings of the BAAS to make his colleagues and rivals in London take notice of this neglect.[2] It was Thomson who hit upon the idea of bringing Liebig to England to address the Liverpool meeting of the BAAS in September 1837. Liebig, who had learned some English as a student in Paris and who was perhaps a little envious of Wöhler's visit to Manchester and London in 1835, leaped at the opportunity.[3]

Thomson's son met Liebig at Hull docks and took him by stagecoach across the Pennines via York and Leeds to Manchester, where Charles Henry was his host for a week. As Liebig told his wife, this stay was an eyeopener on how the British middle classes lived and dined in opulence unknown in Hessen-Darmstadt:

> The district between Leeds and Manchester is one big smoking chimney. . . . I found Henry's house easily, and entered into a kind of palace. . . . I was rather taken aback by the massive elegance of a rich English household. . . . My room is provided with a number of things which only an Englishman is accustomed to use; four kinds of wash-basin, one for the head and face, one for the teeth, one for the hands, and a bidet. In the evening, Henry had friends for dinner, which was dreadfully boring for me [Liebig's English was not yet perfect]; the servants came in black tailcoats, knee breeches and stockings, white gloves, three slaves [i.e., negro servants] behind us; in short, it was princely, but for me very dreary. I will say nothing about the food, still less of the dozen or so wines, ices, Spanische Frische, black and white grapes, none of this was of much interest to me. Obviously I must get used to this way of life.[4]

Not surprisingly, Liebig was ill after this banquet, though he recovered sufficiently to admire the Henrys' scullery "shut off from the kitchen," the latter being magnificently equipped.

On 16 August Liebig crossed the Irish Channel from Liverpool to Dublin, where he met another two former pupils of 1835 and 1836, the Irishman Robert Kane (1809–90) and the Scotsman William Gregory, who was then working in Ireland. Both keen organic chemists, they showed him around Dublin before taking him to Belfast to meet Thomas Andrews, a distinguished chemist at Queen's College. From Belfast he was shipped to Glasgow, where he lodged with the Thomsons. Thomas

2. R. F. Bud and G. K. Roberts, *Science versus Practice* (Manchester, 1984), pp. 35–45.
3. J. B. Morrell and A. Thackray, *The Gentlemen of Science* (1981), pp. 378, 488–89.
4. W. V. and K. R. Farrar and E. R. Scott, "The Henrys of Manchester," *Ambix* 24(1977), 5–6.

Graham, professor of chemistry at the Andersonian Institution in Glasgow, whom Liebig had met in Germany the previous year, showed him local factories for the manufacture of silk, sulphuric acid, soda, bleach, soap, steel, and pottery.[5] Thomson also introduced Liebig to Walter Crum (1796–1867), who proudly showed him around his huge calico printing factory, which employed 1,500 workers. They became firm friends. There followed a brief visit to the famous Edinburgh chemistry teacher Thomas Charles Hope, and to a paper mill, and attendance as the honoured guest at a farewell banquet for Graham, who had just been appointed to the chair of chemistry at University College, London, in succession to Edward Turner.

On 7 September Liebig sailed back to Liverpool to participate in the BAAS meeting at which Michael Faraday, that year's president of the chemistry section, read Liebig's paper "On the decomposition products of uric acid,"[6] after which Liebig became "the lion of the evening." The eight days in Liverpool also included visits to James Muspratt's huge soda factory and salt extraction works, for which Liebig made his first train journey, on the Liverpool–Birmingham railway. After a further stay in Manchester, where he toured Charles Macintosh's factory for making waterproof clothing, he stage-coached to London via Birmingham, where he rested to buy presents for his family. In London, he further cultivated his growing friendship with Faraday, who introduced him to the pharmacists Thomas Morson and Robert Warington. He was able to view Trueman's brewery and Brunel's newly constructed Thames Tunnel before travelling to Portsmouth on 6 October, from where he slowly returned to Giessen via Le Havre, Rouen, and Paris. He spent a fortnight in Paris with his former French pupil Théophile Pelouze.[7]

This two months' absence from Giessen was of great importance in Liebig's intellectual development. He consolidated friendships with former British pupils who were themselves becoming well known, he forged important new friendships with, in particular, Graham, Crum, Muspratt, and Faraday; he increased Thomson's and Graham's desire to send their best pupils to Giessen for further training in organic analysis; and he had seen the advanced state of British industrialization and the high standard of living of its middle classes. On the other hand, his conversations with manufacturers and their workforce had not impressed him. As he wrote Berzelius 26 November 1837:

5. They had already struck up a correspondence before meeting; see Graham to Liebig, 22 July 1837, Dibner MSS 609A (Smithsonian Institution Libraries, Washington).

6. *BAAS Reports*, 1837, pp. 38–41.

7. For full details of the English journey, see Volhard, vol. i, 131–49.

> England is not the land of science, only widespread dilitantism presides here. Chemists are ashamed to call themselves chemists because the apothecaries (who are despised) have appropriated this name.[8]

Most important of all, he had so impressed the council of the BAAS at Liverpool that he was asked to prepare a detailed report on the current state of organic chemistry. As we shall see in Chapter 6, Liebig executed this commission rather differently, but it represented the start of his special relationship with Britain. For its part, the British scientific community recognized Liebig in June 1840 when the Royal Society elected him a Foreign Fellow and Copley medalist, the Society's highest honour.[9]

There is no doubt that Liebig had been somewhat awestruck by Britain, and from 1837 onwards he was to ensure through his pupils' help that everything he wrote would be made available in English. This arrangement worked so well that, following the very considerable impact of his *Agricultural Chemistry*[10] among landowners and farmers in 1840, his next visit to Britain in 1842 was like a Royal Progress – indeed, on showing his passport at the Aliens' Office in London (as foreign visitors were required to do), the official shook Liebig by the hand, exclaiming how pleased he was to meet the author of *Agricultural Chemistry,* which he had read.[11] On this occasion, after staying with Graham for a few days (Graham was by then at University College, London), his guide was Lyon Playfair, a pupil of 1840 whose English translation of *Agricultural Chemistry* had impressed both Henry De la Beche, director of the Geological Survey, and William Buckland professor of geology at Oxford, when Playfair (on Liebig's behalf) had presented an account of *Agricultural Chemistry* to the BAAS meeting in Glasgow in 1840. This had resulted in Playfair's being called into government service as an administrator of sanitary projects for improving public health and as scientific advisor on the threatened failure of the potato crop in Ireland. Playfair escorted Liebig to Oxford to meet the triple-chaired professor of chemistry, botany, and rural economy, Charles Daubeny (1795–1867), one of the "Gentlemen of Science" who, like Playfair, was a graduate of Edinburgh University. Prior to meeting Liebig, with whom he struck up a correspondence, Daubeny had already begun to revise his views on agriculture to incorporate Liebig's ideas, as he acknowledged when he published his Oxford lectures.[12] Daubeny introduced Liebig to Buckland, who by then was a

8. J. Carrière, ed., *Berzelius und Liebig ihre Briefe* (München, 1898), p. 134.
9. J. F. Daniell to Liebig, 15 July 1840 (Staatsbibliothek Preussische Kulturbesitz Berlin).
10. J. Liebig, *Organic Chemistry in Its Applications to Agriculture* (London, 1840). This will be referred to as *Agricultural Chemistry.*
11. Volhard, vol. i, 159–65; T. W. Reid, *Memoir of Playfair* (London, 1899), Chap. 4.
12. C. Daubeny, *Three Lectures on Agriculture* (Oxford and London, 1841).

close friend of the prime minister, Sir Robert Peel. In company with Daubeny and Playfair, Liebig travelled to Peel's country house, Drayton Manor, and on to Bath, Bristol, Clifton (to view Brunel's suspension bridge), and Bridgewater, where Buckland, a Devon man, took them to see several farms and estates where agricultural improvements were being made, as well as to see beds of fossil dung, or coprolites, which Liebig instantly recognised as of potential use as a fertilizer. During this "Cook's Tour," Liebig was introduced to leading agriculturists such as Lord Spencer (at Althorp in Northamptonshire), Lord Ducie (at Totworth Court in Gloucestershire), Lord Fitzwilliam (Wentworth Castle in Yorkshire), and Philip Pusey at Oxford. Not surprisingly, Liebig was quickly made a Fellow of the Royal Agricultural Society that had been formed in 1838.

In Bristol he met the young pathologist William Budd, who reported excitedly to his brother on 5 September 1842:

> Don't you envy me? he is a charming fellow. Quite a young man – not over 37 – like myself, a very lean creature . . . thoroughly impregnated with tobacco as all other Germans. I had him to myself for nearly four hours. In that time we went through the whole range of his own topics. You may fancy how intensely interesting and improving the conversation was for me. I left him having a higher opinion than ever of his profound talent.[13]

Liebig also had the opportunity to stay with his friend Charles Henry in his stately home at Hatfield in Hertfordshire, where Henry had been living in semiseclusion since shortly after Liebig's previous visit, having abandoned all pretension of being a practising chemist.

Although the famous English analytical chemist Edward Turner had died in February 1837 before Liebig's first visit to England, his brother Wilton George Turner (1811–55) attended Liebig's laboratory at about this time. As a prospective chemist himself – he later became an industrial chemist in Newcastle – and as heir to his brother's literary estate, Wilton was keen to see that his brother's textbook *Elements of Chemistry* remained in print and up-to-date.[14] Accordingly, while he was at Giessen (where Vogt seems to have confused him with his deceased brother[15]), Turner persuaded Liebig to extend the coverage of organic chemistry in the text. Liebig did so from his Giessen lecture notes, which William Gregory translated into English as a serial in *The Lancet*, 1844–45. Two

13. Margaret Pelling, *Cholera, Fever and English Medicine* (Oxford, 1978), pp. 128–9.
14. E. Turner, *Elements of Chemistry* (Edinburgh, 1827); 2nd ed. (London, 1828); 3rd, 1831; 4th, 1833; 5th, 1834; 6th, 1842 ; 7th 1842; 8th, 1847.
15. C. Vogt, *Aus meinen Leben* (Stuttgart, 1896); see Good (1936), p. 557.

printings, labelled as the sixth and seventh editions of Turner's *Elements,* were published by Taylor & Walton in 1842. Puzzingly the sixth edition was credited to Liebig, Gregory, and W. G. Turner, whereas the seventh edition dropped Turner's name. The problem seems to have been, as Faraday noted privately to Gregory on 24 October 1842, that the preliminary section on the general principles of organic chemistry had been badly done by W. G. Turner, "a gentleman who had no practical acquaintance with organic chemistry, and who in consequence committed some mistakes, occasionally even reversing the author's meaning."[16] Gregory's role, he claimed, had been purely to translate the factual material on organic compounds supplied by Liebig. Aware of the criticisms, Gregory subsequently took it upon himself to revise the textbook from scratch, and the revision appeared as a seventh edition in December of the same year. In actuality, as Gregory told Liebig, "Turner" was too expensive and too advanced for British students, and in consequence he planned to compile his own simpler textbook.[17] Gregory's highly successful *Outlines of Chemistry* was duly published in 1845.

Liebig's attendance at the British Association meeting at York in 1844 attracted many landowners who, as he told Faraday in December 1844, "afforded him rare proofs of recognition."[18] He was pleased to meet so many famous men, though he was dissatisfied by the way geologists seemed to run the whole show, so that other sciences merely served "to decorate the table." Liebig deplored the empirical approach of British geologists and the way they were hailed as great authorities without seemingly knowing any physics or chemistry. He had gone on to say, in a much quoted remark:

> What struck me most in England was the perception that only those works which have a practical tendency awake attention and command respect, while the purely scientific works which possess far greater merit are almost unknown. And yet the latter are the proper and true source from which the others flow. Practice alone can never lead to the discovery of a truth or a principle. In Germany it is quite the contrary. Here in the eyes of scientific men, no value, or at least but a trifling one is placed on the practical results. The enrichment of Science is alone considered worthy of attention. I do not mean to say

16. F. A. J. L. James, ed., *The Correspondence of Michael Faraday,* vol. 2 (London, 1993), letter 1319, pp. 699–70. Gregory misdated the letter 24 October 1824 (*sic*), which James misconstrued as 24 October 1840.
17. Gregory to Liebig, 17 December 1842, Staatsbibliothek, München.
18. L. P. William, ed., *The Selected Correspondence of Michael Faraday* (Cambridge, 1971), vol. 1, p. 428.

> that this is better, for both nations the golden medium would certainly be a real good fortune.[19]

From then on, especially in his contributions to agriculture and animal chemistry, Liebig was always to claim that mere empiricism could never by itself "lead to the discovery of a truth or a principle." With "the leave of both correspondents," Liebig's statement to Faraday was given considerable publicity by the geologist Charles Lyell when he compared English and American university and college education in his *Travels in North America* (1845).[20] For Lyell, the prevailing utilitarian spirit of the teaching system at Oxford and Cambridge that was aimed solely at the *Brodstudien* of classics for clerics and schoolmasters had failed to inculcate knowledge of the progressive (i.e., research-based) sciences. Lyell's and others' opinions were factors in the approach made to Liebig by the Royal Commission on Oxford University in 1851. In his written evidence to the commissioners, Liebig was uncompromisingly in favour of examinations as the only effective way to promote the study of science at Oxford.[21]

The undoubted highlight of Liebig's 1844 tour was his return to Glasgow, where a great dinner for 300 guests was held in his honour in the Trades' Hall on 13 October on the eve of his receiving the freedom of the city. The grand affair was chaired by the Earl of Eglinton, the Lord Lieutenant for Ayrshire, who was famed for holding a medieval tournament at Egmont in 1839, and it included aristocracy, the Church of Scotland (Dr. Norman McLeod), science (Thomas Thomson, William Thomson, Playfair, and Gregory), and politicians, landowners, and manufacturers. Judging by *The Lancet*'s six-page report of the twenty-three speeches and toasts that included those made by Liebig and by William Gregory (who, with strong support from Liebig, had just been appointed professor of chemistry at Edinburgh University[22]), and the numerous toasts by other members of the distinguished gathering, this must have been a memorable occasion.[23] A thoughtful touch were the toasts drunk in their absence to

19. *Ibid.* For discussion, see W. H. Brock, "Chemical geology or geological chemistry," in L. J. Jordanova and R. Porter, eds., *Images of the Earth* (Chalfont St. Giles, 1979), pp. 147–70.
20. C. Lyell, *Travels in North America,* 2 vols. (London, 1845), vol. i, p. 309.
21. *Report of the Commission on the State . . . of Oxford University,* Parliamentary Papers, 1852 (1482), vol. xxii, p. 124.
22. Liebig to Gregory, 20 September 1843 (National Library of Scotland MS 19335 ff230–1).
23. Volhard, vol. i, pp. 166–72; *The Lancet* 2 November 1844, pp. 170–77. *The Lancet* also printed a petition from students of University College, London, asking Liebig to a public dinner; lack of time prevented Liebig's acceptance, though he did attend one of Graham's lectures in the students' company.

the grand duke of Hessen-Darmstadt for his patronage of Liebig and the endowment of the Giessen laboratory, and to Frau Liebig. (Playfair spoke of the uniform kindness she had always shown to British students at Giessen.) Liebig was so proud of this Glaswegian experience that he had *The Lancet*'s report translated into German and published by Vieweg in 1845. He told Wöhler, "When one is showered with honours the results is a Falstaffian belly; I've been stuffed to bursting point. I enclose an English [Glasgow] newspaper account of the *public dinner* which you may find interesting."[24]

By 1844, like Reichenbach in Austria, Gregory was becoming more and more drawn to the investigation of psychological and paranormal phenomena. Long fascinated by and committed to phrenology, he arranged for Liebig to meet Britain's leading phrenologist, the Edinburgh lawyer George Combe (1788–1858), who had seen Liebig's books on agricultural and animal chemistry as fortifying the views he had promulgated in his best-selling *Constitution of Man* since 1828.[25] This extraordinary meeting was duly reported by Gregory in the *Phrenological Journal* and is interesting with respect to the reading of Liebig's character, supposedly from the shape of his skull.[26] Liebig's temperament was registered as "bilious-nervous, with a little sanguine," which Gregory interpreted as of "the first quality, both for activity and endurance." These, combined with his highly developed moral and intellectual powers, made it difficult to conceive of a better example "to preserve a place in the highest rank among men of science." As confirmation of the high state of development of Liebig's organs of Individuality, Form, Size, Weight, Colour, and Order, Gregory mentioned how in 1829 Liebig had identified a crystalline allanotoic acid in the allanotoic fluid of a foetal calf. Then almost ten years later, in his joint research with Wöhler on uric acid, he had been able to identify allanotoic acid from memory of its crystalline form. Large cranial bumps of Benevolence, Veneration, and Conscientiousness explained Liebig's intense love of truth, the striking illustrations of God's design to be found in his agricultural and physiological writings, and his kindness to pupils to whom "he systematically furnishes . . . interesting subjects of research, guides and assists them with his advice, and thus is enabled to point every year, to a new series of important papers, produced by his pupils."

24. A. W. Hofmann, ed., *Aus Justus Liebig's und Friedrich Wöhler's Briefwechsel,* 2 vols. (Braunschweig, 1888), vol. 1, p. 246.
25. Charles Gibbon, *The Life of George Combe,* 2 vols. (London, 1878; repr. 1980), vol. ii, p. 162.
26. W. Gregory, "The cerebral development of Dr Justus Liebig," *Phrenological Journal* 18(1845), 54–60.

Although most of Liebig's cranial organs were massively well developed, Gregory admitted that Self-Esteem and Love of Approbation were relatively less well developed – a conclusion that does seem askew, judging by Liebig's actual behaviour and motivation. Gregory was not, however, totally uncritical, admitting that the organs of Destructiveness and Combativeness were extremely large. This explained deterministically why Liebig found it necessary to criticise and enter into controversy. Liebig's weakest organs – Language, Number, and Tune – explained why his time at the Darmstadt Gymnasium had proved irksome and why he frequently made "errors in the details of his numerical calculations" – an interesting admission from one of his translators who probably had learned to rework Liebig's analyses.

Gregory's report undoubtedly tells us more about phrenology than about Liebig. It scarcely hints at the reality of a complex human being who was full of contradictions and inner conflicts, who might at any moment be genial and charming but at others emotional, annoyed, and eager to quarrel.

The phrenological lecturer, mesmerist, and poet Spencer Hall was present at this meeting, which also included a demonstration of phrenomagnetism for Liebig's benefit. (In phrenomagnetism the validity of the functions of the supposed phrenological faculties of the brain was demonstrated under hypnosis. On touching the relevant "organ," or part of the skull, the subject would demonstrate appropriate behaviour.) According to Hall:

> We certainly could not have been at it less than four hours, when Liebig, who had been most calm and close in his scrutiny, got up like one who had arrived at an important conclusion, and his face glowed with a peculiar glow as he suddenly and silently left the room. Gregory followed, while Combe staid (sic) talking with the patient, who was now recalled to his normal state. What immediately succeeded was most strikingly indicative of the character of the three savants. Liebig was some time before he reappeared; but Gregory, who had been in conversation with him apart, returned with great animation and something of enthusiasm in his manner, exclaiming "Liebig's *convinced*!" When Combe, in a tone and manner cool in proportion to the other's warmth – yet manifestly much interested – merely said, "Is he?" "Yes" (answered Gregory) Liebig's convinced of the validity of the phenomena – though of course he says nothing of any theory."[27]

27. Spencer T. Hall, *Biographical Sketches of Remarkable People* (London, 1873), pp. 174–83.

Liebig did not mention mesmeric phenomena in any of his later writings, including *Familiar Letters on Chemistry,* though he did, as we have seen, eventually dismiss Reichenbach's odylism. No doubt Liebig accepted that something odd occurred but thought that without a theoretical explanation, such phenomena did not belong to the field of scientific investigation.

The sole motivation of Liebig's visit to London and Liverpool in March 1845 was commercial: to negotiate the commercial sales of his projected fertilisers and the disposal of quinidine wastes (see Chapter 5).[28] By 1845, the publication of Liebig's agricultural and physiological theories, the appearance of his *Familiar Letters on Chemistry* in 1843, as well as the agricultural tour of 1844, had aroused widespread interest in Liebig among landowners, medical practitioners, industrialists, and educators. A Museum of Economic Geology had already been established in London in 1842, where Playfair was given a small analytical chemistry laboratory, and on the crest of the wave of interest in Liebig, two pharmaceutical entrepreneurs who had both visited Liebig at Giessen, Lloyd Bullock and John Gardner, launched the suggestion of creating a proprietory school of chemistry at the Royal Institution. In effect, they were transplanting the German idea of a private institute such as Trommsdorff and other pharmacists had found profitable and that Liebig himself had exploited during his first decade at Giessen. They proposed that the school should have both a pure and an applied department, the former being administered by the Royal Institution. The latter, to be managed by Gardner and Bullock as a commercial enterprise, would be physically located elsewhere in London and devoted to agricultural, medical, and industrial consultancy.[29] Although the suggestion did not meet with Faraday's or the Royal Institution Managers' approval, it did stimulate the queen's physician Sir James Clark, who had the ear of the queen's consort, Prince Albert, to suggest the formation of a Royal College of Chemistry in London in 1845. At this stage the plan was taken out of the hands of Gardner and Bullock, although Gardner was to act as the college's secretary until 1846, when he was sacked.

Naturally, Liebig was asked whether he would consider directing this academic teaching institution, but he declined, recommending instead three of his recent pupils: Will (who had already unsuccessfully negotiated for a post of chemist to the Highland Agricultural Society following a visit

28. Volhard, vol. i, pp. 172–3.

29. *Proposal for Establishing a College of Chemistry* (London, July, 1844). See G. K. Roberts, "The establishment of the Royal College of Chemistry," *Hist. Stud. Physical Sciences* 7(1976), 437–85.

to the BAAS in Glasgow in 1842), August Wilhelm Hofmann, and Remegius Fresenius (1818–97). Will declined, realizing that there were growing opportunities for his advancement at Giessen, as did Fresenius, who was to make a name for himself as an analyst and textbook writer at his own private analytical school at Wiesbaden. It was Hofmann, who had recently become a Privatdozent at Bonn, who was personally persuaded by Prince Albert to direct the London College, which he did successfully for twenty years.

In the same year, 1845, following the unexpected death of J. F. Daniell, the professor of chemistry at King's College, London, Daniell's colleague, the electrician Charles Wheatstone invited Liebig to apply for the chair.[30] Daniell's annual income from pupils alone, he stated, had been £700, to which he had added £400 from private pupils and another £400 from consultancy. Liebig could be guaranteed £1,000 to £1,500. Since Wheatstone was a close friend of Faraday's (as Daniell had been), it is possible that Faraday was behind the invitation. Liebig must have discussed the possibility with Benjamin Brodie, who was studying at Giessen. Brodie's father, also Benjamin, the surgeon and president of the Royal Society, told his son to warn Liebig that "engagements of societies, committees, and other people's business of all kinds and sorts . . . interfere vastly with the calm pursuits of science in this great metropolis."[31] Negotiations had hardly started, however, before they were quashed by Charles Blomfield, the Bishop of London, who had close connections with King's. He claimed that since Liebig was a Lutheran, he could not be appointed to an Anglican college. This argument, which has always seemed somewhat specious since Lutheranism was a primary source of Protestantism, must probably be seen in the light of the controversy over whether an Anglo-German Bishop of Jersualem should be ordained by a Lutheran or Anglican rite.[32] Blomfield was a Tractarian, or High Anglican, who had little sympathy for what he assumed were the "low church" aspirations of Lutheranism. The chair went instead to William Allen Miller (1817–70), one of Liebig's pupils of 1840. Interestingly, however, Liebig awarded Wheatstone an honorary Giessen degree in 1852 when Liebig was acting as dean of the philosophical faculty.

One other good reason Liebig could not have accepted a call to London in 1845 was his expectation of a baronetcy that year; it would have been ungenerous, to say the least, had he immediately left Hessen-Darmstadt

30. Wheatstone to Liebig, 19 March 1845 (Staatsbibliothek Preussische Kulturbesitz Berlin).
31. T. Holmes, *Sir Benjamin Collins Brodie* (London, 1898), pp. 232–4.
32. See Owen Chadwick, *The Victorian Church* (1966), pp. 189–93.

after his country had made him a member of the aristocracy. Liebig had already solicited a state decoration in 1836 and had been awarded first class membership in Grand Duke Ludwig II's Order on 24 July 1837.[33] That his baronetcy was similarly not unexpected is clear from Liebig's correspondence with the chancellor of Giessen University, the lawyer Justin von Linde (1797–1870). Linde had been appointed a professor of law at Giessen in the same year as Liebig, and the two men remained friends after Linde's appointment to the Ministry of Justice in Darmstadt in 1829. In 1833 Linde was made chancellor of the university, which also gave him a seat in the upper chamber of the government. Linde shared with Liebig a profound belief that "our *Land* [Hessen] is too small and not powerful or rich enough among the German states to achieve fame in anything else except its University."[34] Linde strove hard, therefore, to transform Giessen into a major university.

In July 1845 Liebig drew Linde's attention to the fact that a number of distinguished foreign chemists had been made barons by their governments. In England, Sir Humphry Davy had been created a baronet in 1818 (following a knighthood in 1812); in France Liebig's teachers Gay-Lussac and Thenard had been made barons in 1839 and 1825, respectively; and in Stockholm, Europe's greatest chemist Berzelius had been raised to the nobility on his marriage in 1835. A similar distinction, Liebig told Linde, would be his country's most enduring and worthiest recognition for his achievements at Giessen. A *Freiherrnstand* (a baron with the right to add *von* to the surname) would be proper, for any other lower order of nobility would be worse than nothing to him. Liebig then used the offer from the Royal College of Chemistry as his lever:

> Obviously the decision must be the personal and impartial decision of the Grand Duke; but perhaps you can gild the lily. In England, where no Orders are awarded, honours come from recognition by the state. It may be of some significance in this matter if you are aware that I have been offered a position at the Royal College of Chemistry in London with a salary of £2000–£3000 [24–36,000 gulden]. Prince Albert himself has made the proposal. This income is double what I currently earn here. If need be, I will send you the relevant letter.[35]

Liebig's blatant lobbying for honours worked, for while he was spending the New Year with Wöhler at Göttingen, the baronetcy was conferred on 29 December 1845. Liebig thanked Linde profusely for his intercession

33. Liebig to Linde, 2 November 1837. See E. M. Felschow and E. Heuser, eds., *Universität und Ministerium im Vormärz* (Giessen, 1992), p. 61.
34. *Ibid.*, p. 192.
35. Liebig to Linde, 14 July 1845, *ibid.*, p. 226. Prince Albert's letter does not seem to have survived in Liebigiana.

and sent the grand duke a box of Havanna cigars that he had brought back from London.

There is no reason to doubt that Ludwig II was delighted by what Liebig had achieved at Giessen and for the state of Hessen-Darmstadt, or that he did not feel Liebig worthy of this distinction. As Horsford reported to his father: "We are all gratified, since he is one of the noblemen of nature."[36] No doubt, too, the honour would have eventually come to Liebig anyway. But the baronetcy loses some of its gloss when it is realised that, like the Erlangen doctorate, it was actively solicited. Moreover, as we shall discover in Chapter 5, there must remain a suspicion that Linde was happy to intervene on Liebig's behalf in exchange for insider information concerning a valuable chemical commodity.

As if in anticipation of being made a baron, Liebig had already purchased an estate of ten acres at the end of 1844. Local people christened this Liebig's Heights. He had grand plans to develop this property as an English landscape garden and to use it for agricultural experiments as well as to satisfy his wife's delight in gardening. Just as he had proudly signed himself "Dr." after obtaining the Erlangen degree in 1821, he now made sure his *von* appeared on publications, though he told Linde "*von* doesn't come easily to me; I scrawl it to make it unreadable."[37] Probably it made little difference to his career at Giessen, though by 1848, as Europe rose up in revolution, it began to look like a liability. Recalling the fate of the French nobility, he wondered to Emanuel Merck whether he should renounce his title:

> I have agreed with my children to give up our title of nobility. The aristocracy belongs no longer to our time, it is a decayed institution that has outlived its purpose. From the first it was like a hot four-pronged iron to me, for it could not contribute the least either to my satisfaction or to my children's happiness. Since this unhappy promotion to the nobility, I know what hate and envy are. Before this I did not know these human weaknesses.[38]

Indeed, worried by the after events of March, he ordered Wöhler to leave *Freiherr* (baron) off his address because "the title is comical without a domain to go with it."[39] By 1850, when the political fervour had abated, Liebig and his children were again happy to be known as members of the *Adelstände* (nobility); ironically, like a fairy story Baron Broke, Liebig had

36. S. Rezneck, "The European education of an American chemist," *Technology and Culture* 11(1970), 371.
37. Felschow and Heuser, *Universität und Ministerium*, p. 256.
38. P. Munday, *Sturm und Dung. Justus von Liebig and the Chemistry of Agriculture* (PhD, Cornell University, 1990), p. 265.
39. Hofmann, *Liebig's und Wöhler's Briefwechsel*, vol. 1, p. 333.

been forced to sell his domain Liebig Heights after overcommitting himself financially.

Other honours poured upon Liebig after 1845. Friedrich Wilhelm IV of Prussia decorated him with the Prussian Order of Merit for Science and Art, the French made him a member of the Legion d'Honneur, the Royal Society awarded him the Copley Medal, the Society of Arts its Albert Medal, he was a Freeman of the cities of Glasgow (1844), London (1865), and Munich (1870), and at the end of his life in 1870, German agriculturists struck a Liebig Medal in his honour. In the Liebigiana catalogue of the Staatsbibliothek in München, Liebig's honorary degrees, orders, and medals occupy a page.[40]

In 1851 Liebig celebrated his twenty-fifth wedding anniversary and was persuaded by his British students that he ought to visit the forthcoming Great Exhibition in London. Since he was also keen to see Ireland again after its aweful famine and epidemic of potato disease, he gladly accepted an offer of hospitality from James and Edmund Muspratt. Disappointingly, Wöhler was unable to accompany him. The visit was complicated by the knowledge that Leopold Gmelin had fallen ill the previous year and retired from the chair of chemistry at Heidelberg. The Baden government, mindful of Liebig's great polemic of 1840 and of his great achievements at Giessen, accordingly invited him to succeed Gmelin.[41] It does not seem that Liebig had any great wish to move to Heidelberg, but the call was once again a useful bargaining point for improvements in conditions at Giessen. It is interesting that for once Liebig did not ask for improvements in his own working conditions but for general improvements in the facilities for science teaching, including higher salaries so that the very best professors would be attracted to Giessen. These were his conditions if he was to remain at Giessen.

Bureaucracies by definition are sluggish beasts, and the matter had not been resolved by August when Liebig sailed to Liverpool to stay with James Muspratt. Then, accompanied by his student Edmund Muspratt, as well as by Samuel Lover (1797–1868), an Irish poet and painter; Michael Whitty, the then editor of the *Liverpool Journal;* and a Mr. Fincham, a political economist who managed a glass factory at St. Helen's, Liebig sailed to postfamine Ireland.[42] The weather and travelling conditions were atrocious when they visted the Connemara region of Galway, where the worst poverty was to be found on the large estates, most of which were

40. See also *Almanach der k.b. Akademie der Wissenschaft* (München, 1867), pp. 53–4.
41. Volhard, vol. i, pp. 181–2.
42. The episode was seemingly unknown to Volhard; see E. K. Muspratt, *My Life and Work* (London, 1917), and M. and W. Schneider, eds., *Justus von Liebig Briefe an Vieweg* (Braunschweig, 1986), pp. 254, 257.

up for sale. He found that the reports he had read concerning Irish poverty and misery were only too true, and he was appalled by the children's stunted growth and by the fact that both girls and youths seemed old and lacking in energy. It was the Irish peasantry rather than his own Hessen countrymen who first prompted Liebig's Malthusian views on the relationship between agriculture, prosperity, and the circulation of the elements.

In Galway itself Liebig was reunited with his former pupil Edmund Ronalds (1819–89), who was professor of chemistry at the University College. It was there in the library that Liebig seized upon a copy of Webster's *Dictionary* to prove to Whitty that "theory" was not opposed to "practice" but meant "an explanation of facts"; hence it was not "inimical to the progress of industry," as Whitty had argued during their journey.

After returning to Liverpool through Dublin and County Wicklow, Liebig travelled alone to Balmoral House, where he was the personal guest of the queen and Prince Albert. Balmoral Castle had not yet been completed, so that the royal living conditions were spartan. Playfair, who had helped Albert with the planning of the Great Exhibition, and James Clark were also house guests. Intriguingly, Liebig was quite overcome by the extraordinary beauty of one of the queen's attendants, Lady Evelyn Blantyre (1825–69), the Marchioness of Douro. He was so struck that he subsequently sent her portrait to Wöhler.[43] The queen found Liebig "a pleasing and quiet man . . . the most distinguished professor of Chemistry in Germany, or anywhere."[44] No details of his subsequent visit to the Great Exhibition in Hyde Park appear to have survived, which is a great pity.

Meanwhile, in Liebig's absence abroad, there was stalemate in Giessen and Darmstadt over Liebig's ultimatum, despite the personal efforts of Frau Liebig to extract a decision from the Hessen government.[45] It was not until after Liebig's return in November 1851 that the government agreed to increase its commitment to science teaching at the university. In the long term this commitment fell short of what Liebig wanted, and he was undoubtedly within his rights to begin negotiations with the Bavarian government a few months later. As for Heidelberg, failing to catch Liebig, the Baden government was able to recruit another great teacher and researcher from Breslau, Robert Bunsen. An award of another ducal medal

43. F. Boase, *Modern English Biography,* vol. i, p. 310; Hofmann, *Liebig's und Wöhler's Briefwechsel,* vol. 1, p. 372.
44. Muspratt, *My Life,* p. 88.
45. Volhard, vol. i, pp. 184–5.

Sketch of Liebig's enlarged Chemical Institute as seen in 1845 by J. M. Bayrer. Liebig used an engraved version of this on his notepaper. (Courtesy Liebig-Museum Giessen)

in November 1851,[46] which was clearly motivated to keep Liebig's loyalty to Hessen-Darmstadt, and hints of further glory and honour to come, were, Liebig soon decided, worth nothing compared to freedom from teaching. We can well understand Liebig's feelings when we read Edmund Muspratt's memories of Liebig at this time:

> When I first arrived at Giessen [in May 1850], Professor Liebig, who was only about fifty or fifty-one years of age [sic, he was 47], appeared an old and broken-down man. When he entered the lecture room he could hardly walk firmly, but glided in and appeared exhausted with the effort. In a few minutes all was changed, when he became inspired by the subject of his lecture.[47]

Baron Liebig's sixth and final visit to Britain did not take place until ten years later in September 1855. Immediately following his daughter's wedding to Thiersch in August 1855 and accompanied by his favourite English pupil Edmund Muspratt, he travelled to Paris to see France's answer

46. Felschow and Heuser, *Universität und Ministerium*, p. 355.
47. Muspratt, *My Life*, p. 42.

to the Great Exhibition, the *Exposition Universelle.* Among the visitors were a royal party from England, Queen Victoria, Prince Albert, and their clever daughter Vicky, who was destined to become empress of Germany in 1858. Perhaps because Albert had enjoyed Liebig's company at Balmoral, he invited Liebig to see the newly completed royal palace at Osborne on the Isle of Wight and to give a talk on chemistry to the younger royal children. By then Liebig was an old hand at meeting and lecturing to royalty, having appeared before the king of Bavaria and taught his deluded son Ludwig. On one notorious occasion in April 1853, a violent explosion occurred when Liebig was lecturing before the Bavarian royal family; glass struck the faces of Queen Therese and her brother Prince Luitpold, as well as that of Liebig. Queen Marie had acted as nurse to all three.[48]

Accordingly, Liebig was able to see old friends such as Buckland, Graham, and Warington in London before travelling to Gosport in the company of Edmund Muspratt. From there he sailed on the royal yacht to the Isle of Wight and the house with its rooms "straight from the *Thousand and One Nights.*"[49] Apart from his audience with the queen, he spent some time with Prince Albert's private secretary Ernst Becker, who had been a student of physics with Buff at Giessen. Prince Albert personally showed Liebig around the Osborne farm and talked knowledgeably on agricultural chemistry. Liebig, who had made an admiring reference to Henry Palmerston (Peel's successor as prime minister) in the dedication of the third edition of his *Familiar Letters on Chemistry* to Sir James Clark in 1851, was delighted to find that Palmerston was also at Osborne. Palmerston, who was preoccupied by the Crimean War at this time, nevertheless engaged Liebig in a long discussion concerning cholera, the production of aluminium, and other matters of national concern to the British. Liebig returned to London via Southampton, from where he went to Oxford to see Daubeny and Brodie (another pupil who was now Waynflete Professor of Chemistry). From Oxford, he and Daubeny travelled to the Muspratts' home in Liverpool, from where they wandered together through the English Lake District, which Liebig had not seen before. This considerable holiday was completed by a visit to Scotland to stay with Sir James Clark in Aberdeen and with Walter Crum in Glasgow during the meeting of the British Association. After a day in Edinburgh, where he missed meeting George Combe, he returned to Germany in October.[50]

Albert's relationship with Liebig is intriguing. Both Hofmann and

48. Volhard, vol. ii, pp. 350–1.
49. *Ibid.,* vol. ii, p. 362.
50. Volhard, vol. ii, pp. 360–2; Combe to Liebig, 4 October 1855 (National Library of Scotland Acc. 8459).

Liebig were privy to Albert's original and modified plans for using the £186,000 profits and proceeds of the Great Exhibition for the creation of a polytechnic of teaching institutions, museums, and libraries on the Gore estate at South Kensington. Liebig thought Albert's plan was "so fine and in its direct results so blessed that everyone must want to make it possible." And he recalled:

> I once had a talk with the late Sir Robert Peel about a similar subject. This was before the erection of the Museum of Practical Geology. Like most English statesmen he was an opponent of the notion that the state should do something because, he said, everything goes quite splendidly with the manufacturers producing innovations, which science has done nothing for. This point of view is unacceptable to me, who understands the inner processes of progress; but this only caused the great man of whom I thought so highly to smile. I told him that the very manufacturers that he thought of so highly, instead of being able to exploit all knowledge and new procedures as a matter of course, would have to purchase this knowledge from younger people who had learned something. Manufacturing was like searching for gold. So long as it lay on the surface, anyone could find it; but once mining operations were required, then what was needed was knowledge learned and great dexterity.[51]

It is notable that the privately endowed Royal College of Chemistry, with Hofmann as its director, became part of the new School of Mines under the government's Department of Science and Art in 1853. But for Albert's early death in 1861, Hofmann would undoubtedly have stayed on in England. As he told the sorrowing queen in 1863, after receiving a gift of her husband's speeches:

> [Albert's] early kindness exercised so powerful an influence upon the destinies of my existence. Year by year do I feel more deeply the debt of gratitude which I owe to him. How many endowed with the highest powers and the noblest enthusiasm have I seen failing in the pursuit of science for want of that early advice and help which he generously accorded to me. So far as it has been within the humble scope of my ability to promote the science of my predilection, it is to him, I feel, that I owe my opportunities through life.[52]

51. Liebig to unknown correspondent at Court, 24 January 1853. The letter is unfortunately incomplete. Cited by gracious permission of Her Majesty The Queen. F25 Windsor Archive.
52. Hofmann to Sir James Clark, 9 February 1863. Cited by the gracious permission of Her Majesty The Queen, RA R4/59, Windsor Archives. The gift was *The Principal Speeches and Addresses of His Royal Highness The Prince Consort* (London, 1862).

Liebig, on receiving the same book from Her Majesty, expressed himself to Becker as follows:

> I have sincerely revered and was whole-heartedly devoted to him and no one valued his spirit and soul more than I did. In him I lost a friend that I'll never forget. I was working on my latest work, *The Natural Laws of Husbandry,* when I learned of his death, and I was robbed by this of the pleasure of being able to offer ([i.e., dedicate] the book to him as a strong demonstration of my great admiration; I know that, of all my works, it would have delighted him. The death of my dear daughter Agnes has depressed me so greatly that I can understand what the loss of such a man must mean to the Queen. Please thank the Queen, dear Becker, most sincerely for such a touching present.[53]

Although invited to London in the 1860s, Liebig did not travel across the Channel again. But from 1837 until his death in 1873, Liebig's wide-ranging theories and speculations on organic chemistry and its applications to agriculture, medicine, and industry exerted an impact on all levels of British society, from the queen and her consort through the aristocracy and so-called improving landlords, civic dignitaries, and government adminstrators, down to academics, industrialists, and manufacturers, doctors, chemists, and working men. The means by which this was achieved included his own charismatic presence at meetings of the BAAS; the numerous personal contacts he made in various journeys and through his voluminous correspondence; the influence of the translations of his writings by Playfair, Gregory, Hofmann, Blyth, and Gardner; the extensive literary exposure he received in important periodicals and newspapers such as *The Lancet* and *The Times;* the widespread influence of his pupils as his chemical missionaries, and the work of Hofmann at the Royal College of Chemistry in London. Overall, Liebig might well have regarded himself as an honorary Englishman.

How careful of Liebig's reputation his British disciples were can be seen from the curious episode concerning Dennis Cronin in 1847. Cronin, who had acquired one of the medical degrees that could notoriously be bought from the University of Giessen, had used it to raise his professional standing as a legally qualified London apothecary. A muddled prescription for bitter almond water led to the death of one of his female patients, but he was found not guilty of manslaughter. Liebig's involvement is unfortunately unclear from the surviving correspondence, but it seems likely that Cronin's defence lawyer had tried to obtain a guarantee of the validity of Cronin's Giessen degree, which Liebig then wished to expose as pur-

53. Liebig to Becker, 3 November 1863. Cited by gracious permission of Her Majesty The Queen. R4/108, Windsor Archives.

chased.[54] However, once Cronin had been acquited, Liebig was in danger of slander if he publicly questioned Cronin's medical competence. It was Graham who carefully smoothed away the legal difficulties.

As research has shown over the past decade, Liebig's relationship with Great Britain and with Englishmen was a symbiosis. Liebig looked to Britain for a receptive and sympathetic audience for his ideas on and discoveries in chemistry, agriculture, and biochemistry. Although these frequently provoked controversy in Britain, on the whole Liebig's views were accorded greater respect and attention there than in Germany. That this was so had much to do with the fact that young groups of British chemists, doctors, engineers, and educators who were keen to advance their own status within British society found Liebig an effective publicity agent and figurehead for their own campaigns and self-advancement. Together with Liebig's British pupils who held academic and industrial posts throughout the British Isles, Hofmann in particular ensured that Liebig's name and work were kept prominently before the public eye.

54. *The Lancet* (17 April 1847), vol. i, pp. 415–17.

5

Liebig and Commerce

> Wealth! wealth! wealth! Praise be to the god of the nineteenth century! The golden idol! the mighty Mammon! Such are the accents of the time, such the cry of the [British] nation. . . . There may be here and there an individual who does not spend his heart in labouring for riches; but there is nothing approaching to a class of persons actuated by other desire. [John Sterling, 1828]

As Walter Houghton observed, in commenting on Sterling's generalization, the commercial spirit, which emphasized money, respectability, leisure, and success, affected all the professions and aspiring professional groups in the nineteenth century.[1] In Germany as well as in Britain, this commercial spirit was a strong motive in pushing apothecaries and chemists into pharmaceutical and chemical manufacturing. Because apothecaries were often the sole source of scientific expertise in local communities, it is not really surprising that they often became technical advisors to chemical manufacturers or that they turned towards manufacturing on an industrial scale. And there was nothing unusual about university professors of chemistry and pharmacy becoming involved in manufacturing.[2]

Although the evidence has existed in Volhard's biography that Liebig cultivated close connections with the chemical and pharmaceutical industries throughout his life, until recently historians have mainly concentrated on his contributions to teaching and to the theoretical disputes that confronted organic chemists between 1820 and 1840. This chapter seeks to demonstrate that Liebig's conviction that chemistry was the central science that underwrote the modern economy and man's intellectual understanding of nature caused him to believe it could be applied to the relief of man's estate in the sense advocated by Francis Bacon in the seventeenth century.

As we have already seen, Liebig's ducal scholarship to Paris in 1822 was

1. W. Houghton, *The Victorian Frame of Mind* (New Haven, London, 1957), p. 183.
2. B. H. Gustin, *The Emergence of the German Chemical Profession 1790–1867* (PhD, University of Chicago, 1975), pp. 141–3.

Photograph of Liebig in 1852 taken by Hanfstaengl in Munich shortly after the move from Giessen. Note the inclusion of the potash bulbs. (Courtesy W. Lewicki)

probably awarded in return for information concerning French industries, such as sugar beet, that might be usefully introduced into Hessen-Darmstadt. Letters to his parents from Paris also show that Liebig was on the lookout for processes that his father might adopt for commercial purposes. For example, in 1823 he made a deal for his father to supply arsenic to a French dyer to whom Liebig had sold a recipe for a green dye that he had devised.[3] The dye was apparently marketed as "Paris Green," though it is unlikely to have been the same material as the copper arsenic acetate compound of that name later devised by Schweinfurth. One of the additional purposes of the private pharmaceutical institute Liebig had established at Giessen, with obvious state approval, was the training of industrial chemists (*Fabrikanten*). Moreover, as a professor in a state university, Liebig was always expected to advise Duke Ludwig and his officials. In

3. E. Berl, *Liebig und die Bittersalz- und Salz-Säurefabrik zu Salzhausen 1824–31* (Giessen, 1931), pp. 35–40.

this capacity, at the end of 1824 he was requested to analyze the salts and brines of Salzhausen, where the government and inhabitants of the nearest town, Kreuznach, were hoping to establish a spa.[4] As Liebig told his friend Schulz, "I am only too pleased to do this, since it means I become closely connected with the state and hope that other investigations of the same kind concerning mining and other matters will come my way."[5]

Unfortunately, as he confessed twelve years later, his initial sloppy analyses of these waters were the cause of his failure to discover an unknown element, bromine, whose presence he misinterpreted as a hitherto unrecognized chloride of iodine. Bromine was distinguished instead by the French apothecary Antoine J. Balard (1802–76), leading Liebig to coin the waspish aphorism, "Balard was discovered by bromine."[6] (Ironically, at a chemists' banquet held during the Universal Exposition in Paris in April 1867, it fell to Balard to propose the toast in Liebig's honour.[7])

Although like most Germans Liebig later frequented spas for health reasons, he had no interest in exploring further the therapeutic value of Salzhausen's waters. On the other hand, because they were rich in "halogens," as Berzelius called the family of chlorine, iodine, and bromine, as well as magnesium, Liebig saw their potential industrial significance. In a report to the government in 1825, he pointed out that if the brines were concentrated by evaporation and then treated with sulphuric acid, a factory for the production of hydrochloric acid and magnesium sulphate (the purgative, Epsom salts) could be operated. Accordingly, a plant directed by a man named Reuss, whom Liebig soon came to regard as incompetent, began operations in 1826. Liebig's reward was a lump sum of 500 Gulden (£40) and a 7.5 percent commission on the concern's profits in exchange for his acting as technical consultant.

Liebig seems to have earned his lump sum (the commission never materialised), for by 1828 the factory was daily producing two tons each of hydrochloric acid and Epsom salts. Profitability was put at risk, however, by the fact that no sulphuric acid was made in Hessen-Darmstadt at that time, and it was expensive to import. Operations became profitable only in 1827 when Liebig obtained supplies more cheaply from Kestner in Oppenheim. This experience may well have been the factor that led

4. *Ibid.*, p. 4.
5. P. Munday, *Sturm and Dung. Justus von Liebig and the Chemistry of Agriculture* (Ph.D., Cornell University, 1990), p. 81.
6. J. Liebig, "Chemische Untersuchungen der Salz zu Salzhausen," *Kastners Archiv für die gesamte Naturlehre* 5(1825), 454–62, and "Ueber Laurents Theorie der organischen Verbindungen," *Annalen* 25(1838), 29.
7. C. A. Browne, "The banquet des chimistes, Paris, April 22, 1867," *J. Chem. Educ.*, 15(1938), 254–59.

Liebig to state the much-quoted principle in his later *Familiar Letters on Chemistry:*

> It is no exaggeration to say, we may judge, with great accuracy, of the chemical prosperity of a country from the amount of sulphuric acid it consumes. In this point of view, there is no manufacture worthy of greater attention on the part of governments.[8]

Liebig was absolutely right, and until recently, when the touchstone of industrialism became nitrogen production, sulphuric acid was the fundamental feedstock of the chemical industry.

As Liebig's reports to the Hessen government and his abrasive letters to Reuss reveal, Liebig believed that he could run the Salzhausen manufactory much more competently and profitably himself, and he offered to take it over. No doubt he was thinking of his father and brother as partners. In 1828, Liebig drew up elaborate plans for the extension of the Salzhausen works into the manufacture of soda, calcium chloride (for bleaching paper and for disinfection), bone glue, tin salt (stannous chloride for dyeing), magnesium carbonate, and a green dyestuff that was probably the Paris Green he had developed five years earlier; but the government was not willing to add to the 12,000 gulden (£1,000) it had already invested in the works. By 1831, when Liebig's interests had shifted completely to organic chemistry, he passed the consultancy to others. The Salzhausen plant was finally abandoned in 1853.

Because of recent rural unrest and bad harvests, the Hessen-Darmstadt government was much more interested in agricultural investments at that time. Sugar had been isolated from the beet plant by Andreas Marggraf (1709–82) in 1747. Wise investment by the French government had released France from its dependence on West Indian sugar imports during the Napoleonic Wars. This successful sidestepping of possible embargos stimulated several landlocked German states, including Hessen, to look with interest at French beet farming, particularly because in a disunited Germany sugar prices were made artificially high from incremental border customs taxes. Accordingly, in 1828 Liebig was dispatched back to France to report on beet sugar refining. Surprisingly, in view of his later commitment to agricultural reforms, he did not find the experience congenial, though it provided him with the welcome opportunity to meet old Parisian friends, including Humboldt.

On the basis of Liebig's findings, a state sugar refinery was established near Darmstadt. Unfortunately, other neighbouring states had a head

8. J. Liebig, *Familiar Letters on Chemistry,* 1st ed. (London, 1843), p. 30; 3rd ed.(1851), p. 146. The saying is often ascribed to Disraeli, who must have used it during a parliamentary speech at some stage in his career.

start, and, unable to compete with their prices, the refinery soon went bankrupt.[9] Because Liebig was a member of the government investment group (*Aktiengesellschaft*), it seems likely that he lost some money in the venture. For this reason, in 1835 he strongly advised Vieweg against investing in the sugar industry, even though it was to provide positions for some of his students.[10]

Because both these state ventures in which Liebig was involved proved personally unprofitable, they must have taught him several valuable lessons. As Pat Munday has observed with particular reference to the Salzhausen experience:

> First of all, Liebig's reputation was by no means established as of 1830. He had published few significant articles, no books, and had few students. Furthermore, from Salzhausen he learned some important and useful lessons about the importance of chemical industry to the state, the difficulties of translating true chemical knowledge into successful technical operation, and the need for chemical education among apothecaries and manufacturers.[11]

There is no doubt that Liebig was continually on the alert to the industrial and commercial implications of his current investigations. For example, in 1838, in the epoch-making collaborative paper that he and Wöhler wrote on the degradation products of animal metabolism to be found in urine, they re-identified Prout's murexide, or ammonium purpurate, as alloxan.[12] Both Wöhler and Liebig tried this beautiful purple dye on silk, but in the absence of a suitable mordant, they were unable to fix it to cloth. After 1840, with the easy availability of uric acid in the form of imported guano, Liebig's pupil Adolf Schlieper (1825–87) developed a murexide dyeing process at his father's calico print works at Elberfeld. This was in effect the first artificial dyestuff. Calico printers in Manchester also exploited the process, probably independently. It is interesting that William Perkin's synthetic mauve dye was a striking commercial success because of the European popularity of this earlier colour.[13] Wöhler became part owner of a nickel factory in 1832; it was a wise investment that gave him a comfortable extra income throughout his life. Liebig was not so fortunate until middle age.

9. Volhard, vol. i, 122–8, for details. Liebig, *Familiar Letters*, 3rd ed., 1851, pp. 154–7.
10. Notably Karl Weinrich at Prague. See "Moritz Carrierè's Lebenserinnerungen," *Archiv für hessische Geschichte* 10(1914), 133–301, at p. 165.
11. Munday, *Sturm and Dung*, p. 89.
12. J. Liebig and F. Wöhler, "Untersuchungen über die Natur der Harnsäure," *Ann. Pharm.* 26 (1838), 241–340.
13. C. M. Mellor and D. S. L. Cardwell, "Dyes and Dyeing 1775–1860," *British Journal History of Science* 1(1963), 265–79.

Given Liebig's close involvement in the training of pharmacists for Hessen-Darmstadt, for which he was also a state inspector for the Oberhessen region in the 1830s, it is not surprising that several of his pharmaceutical pupils became the founders of chemical and pharmaceutical industries. Perhaps the best known of these pupils was Georg Merck (1825–73), the son of the apothecary Emanuel Merck (1794–1855) of Darmstadt, who had redirected the old family pharmacy into the refinement of chemicals for sale to chemists and pharmacists all over Europe. As Ulricke Thomas has shown, one of the main purposes of bringing Liebig onto the editorial board of the *Annalen der Pharmacie* in 1831 was so that he could contribute to the standardization of drugs that Geiger and Emanuel Merck were promoting.[14] During the next few years Liebig collaborated with Merck on the purification of potentially useful plant extracts. In particular, under Liebig's guidance, Georg Merck became expert in applying organic analysis to alkaloids. After further training with Hofmann in London, Georg rejoined his father's firm, which by the 1860s had become the largest pharmaceutical company in Europe, specialising in particular in the supply of quinine, other alkaloids, and fine chemicals.

As far as is known, though, Liebig never ventured his capital in such private chemical or pharmaceutical industrial developments. He simply did not have the money to do so while he built up the reputation and success of his Giessen laboratory. In the mid-1840s, however, with extra income rolling in from his literary activities, including the editing of *Annalen der Pharmacie,* he was in a position to take a gamble, only to have his fingers badly scorched.

Fertilizers

The first two editions of *Agricultural Chemistry* (1840 and 1842) contain plenty of textual evidence that Liebig saw the theoretical and practical value of nitrogeneous fertilizers laced with minerals. His tour of English agricultural regions and estates in 1842 convinced him that farmers were using their land inefficiently because of their regular use of the crop-rotation–fallow system. With an insider's nod towards the repeal of the Corn Laws, he told Robert Peel, Britain's prime minister in 1843:

> I have arrived at the conclusion that the most indispensable nourishment taken up from the soil is the phosphate of lime. In the course of a

14. U. Thomas, "Philipp Geiger and Justus Liebig," *Ambix* 35(1988), 77–90; Anon, *E. Merck–Darmstadt. Die Chemische Fabrik* (Darmstadt, 1952); A. W. Hofmann, "Georg Merck (1825–73)," *Berichte* 6(1873),1582–5.

> Journey through England last year [1842] I have become convinced that all the fields of that country are in such a state [of soil exhaustion. However,] there is no Question that in a very short time England could entirely [dispense] with the importation of foreign grains if a rich and cheap source was opened to her for the supply of phosphates. . . . The reasons of my writing this letter is to acquaint you with the fact that there exist in England layers of fossil guano in a quantity sufficient to provide her with phosphates for centuries to come. . . . As this material must acquire great value I was led to consider that I might draw from my discovery the advantage for myself. . . . [But] as a man of science and not of Commerce I have thought it therefore preferable to communicate my discovery to her Majesty's government.[15]

Two years later, Liebig had no such commercial inhibitions; though ironically it was his English rival John Lawes who captured the coprolites phosphate market.

Yields of crops could be maximized annually, Liebig reasoned, simply by returning to the soil the mineral agents found in the ash analysis of specific crops. He was also convinced that there was more than sufficient nitrogen reaching the soil through the precipitation of ammonia from the atmosphere. Theory implied, therefore, that crops could be grown continuously on the same land (without fallowing) simply by devising an appropriate chemical dressing for each soil type and particular crop.[16]

There were already English commercial enterprises that offered Liebig models. In 1841 the Irish doctor and agriculturist, Sir James Murray, offered for sale three types of oxygenated fertilizers for various types of crops. These fertilisers were made by mixing liquid superphosphate (i.e., bones dissolved in sulphuric acid) with organic materials such as bran, sawdust, and rags.[17] Two years later, the English landowner Sir John Bennet Lawes opened a superphosphate factory at Deptford Creek on the River Thames. His patent manure "composed of Super-Phosphate of Lime, Phosphate of Ammonia, Silicate of Potass, etc." was first advertised in the *Gardeners' Chronicle* in July 1843 and regularly thereafter.

Murray's enterprise seems to have been a failure, and he was bought out by Lawes. Murray's mistake, which was to be repeated by Liebig, was not to test his dressings in the field. He was also undercapitalised. Lawes, on

15. Robert Kargon, *Science in Victorian Manchester: Enterprise and Expertise* (Baltimore, 1977), pp. 104–5
16. Vance Hall, *A History of the Yorkshire Agricultural Society 1837–1987* (London, 1987), p. 210; L. F. Haber, *The Chemical Industry* (Oxford, 1958), p. 60; Munday, *Sturm and Dung*, Chap. 7.
17. J. Murray, *Advice to Farmers* (London, 1841).

the other hand, had capital, he had been testing his manures for five years, first on pot plants and then on fields, and apart from good marketing skills, he also captured the interest and confidence of the agricultural press.

In 1837 Liebig had met the successful Liverpool-Irish alkali manufacturer James Muspratt (1793–1886), and a close friendship developed over the subsequent years. At the beginning of the summer semester of 1843, Muspratt sent two of his sons, James Sheridan Muspratt (1821–71) and Frederick Muspratt (1825–72), to study with Liebig at Giessen. Another son Richard (1822–85) had been the first to study with Liebig in 1840. James Sheridan, the most maverick and entrepreneurial of the brothers, returned to Giessen for a second spell in 1845 and arranged with Liebig for himself and Richard to manufacture Liebig's artificial manures at their father's factory at Newton, near Liverpool. Liebig also did a deal with the Muspratt family (and seemingly with his Glaswegian friend Walter Crum) in which Liebig's brother Carl, who had taken over their ageing father's materials business in Darmstadt, would export guano to Hessen farmers from Liverpool and Glasgow.[18] Meanwhile, Sheridan Muspratt took out a British patent (No. 10,616) on Liebig's behalf, for foreigners were not allowed patent protection at that time even if they intended to work a process in Great Britain. To capitalize the project, the Muspratt brothers succeeded in interesting the aristocratic landowner Sir John Wolseley (1803–90), who presumably was to have a major share in any profits. Another of Liebig's former German students Ernst Dieffenbach (1811–55), who had recently returned from a pioneering exploration of New Zealand, was appointed the business and advertising manager (see Chapter 10).

Liebig shared his own excitement and confidence in the likely success of the enterprise with his confidante Justin von Linde, the chancellor-rector of his university, explaining how, if all went well, he would start a fertilizer factory on the Mosel. As we learn from Liebig's correspondence with his publisher Vieweg, this Hessen factory would have been a partnership between Vieweg and Liebig.[19] Liebig also told Linde how he had sent his former pupil and brother-in-law Friedrich Knapp to Liverpool to help bring the fertilizers to market. Aware that Knapp was taking extra time out from his teaching duties at Giessen to do this, Liebig carefully emphasized that it would be a most useful learning experience for a future professor of chemical technology to study English manufacturing

18. Munday, *Sturn and Dung*, p. 250.
19. Munday, *Sturm and Dung*, pp. 252–58.

methods – as indeed it was.[20] In September 1845, Liebig visited the Muspratt works personally to make sure that everything was proceeding satisfactorily, but he had to cut short his visit to England when his daughter Agnes fell ill.

Six types of dressing were prepared under Knapp's supervision and costing of the ingredients. These, labelled Nos. 1 to 6, respectively, were for cereals, root crops, grasses, legumes, tobacco, and flax. Each of these was bagged separately for sale to accord with a farmer's cash crop and type of soil. The tobacco fertilizer was presumably intended for sale to American farmers, though there is no evidence that it was ever used in the United States. Liebig's formulas were mixtures of plant ashes, gypsum, calcined bones, potassium silicate, and magnesium sulphate, and significantly, some ammonium phosphates. The mixes were fused in a furnace with the deliberate intention of decreasing their solubility, hence ensuring that the ingredients were only slowly diffused into the soil. Liebig had been most concerned that in heavy rainfall the not-inexpensive dressings (they were to be £10 a ton spread at the rate of a ton per acre) would be washed away or soak too far and too rapidly into the subsoil to be absorbed by growing plants.

No field trials were made before the materials were launched for marketing in the autumn of 1845 – a fatal mistake. When, following the bad harvest of 1844 and the onset of the potato famine in 1845, enthusiastic British farmers tried the products they rapidly discovered that the materials remained on the surface like a grass dressing unless at further expense they were ploughed in. The manures lacked the properties extolled by the manufacturer. Conceivably today, Liebig or, rather, the Muspratts would have been prosecuted for deception.

As mentioned in an earlier chapter, towards the end of 1844, Liebig bought some ten acres of land (Liebig Heights) outside Giessen from a city alderman. This purchase had little or nothing to do with practical agriculture, for it is clear that his plan was to turn it into an English garden, to which end a Frankfurt architect was engaged to draw up expensive plans. The park was also eventually to include a summer home for his family. It was only in later publications that he asserted that its purpose had been to test his artificial manures. From the comparatively short time that he owned the land, these experiments cannot have been as extensive and systematic as he claimed.[21]

20. E. M. Felschow and E. Heuser, eds., *Universität und Ministerium im Vormärz* (Giessen, 1992), pp. 215, 217.
21. The matter is discussed by Munday, *Sturm and Dung*, pp. 262–8.

Even when the Muspratts, realizing the problem caused by the formulas's insolubility, revised the manufacturing process to make the dressings more soluble, they did not work as well as Lawes's patent superphosphate manure and its competitors, or the expensive guano from South America that had been available since 1840. The reason for this problem, as we shall see in the next chapter, is that after 1843 Liebig had decided that crops received sufficient nitrogen from the atmosphere and that what they most lacked (and the dressings were designed to supply) were inorganic minerals. So, although the dressings contained some ammonium phosphate, as Lawes noted in 1847, they contained insufficient nitrogen to be effective as fertilizers.

Not surprisingly, British farmers, the agricultural press, and the agricultural societies never forgave Liebig, and the affair seriously interfered with the reception of his message. In a letter to Hofmann (12 February 1851),[22] he cursed the whole business of "the stupid patent manures." At a personal level, both he and the elder James Muspratt lost money in the venture. The episode also nearly wrecked the friendship between Liebig and Sheridan Muspratt.

The patent manures were launched simultaneously with an accompanying English and German pamphlet to explain and puff the products' functions and virtues: *An Address to the Agriculturists of Great Britain Explaining the Principles and Use of Artificial Manures* (1845).[23] The pamphlet was translated into German by Alexander Petzholdt (1810–89), a Saxon apothecary turned geologist and agriculturist, who had come to Liebig's attention by enthusiastically endorsing and popularising Liebig's *Agricultural Chemistry.*[24] As a reward for services rendered, Petzholdt was offered the position of agent for German sales of the dressings imported from Liverpool. To Liebig's consternation, however, Sheridan Muspratt privately revealed the ingredients and details of the manufacturing process to Petzholdt, who – or so Liebig thought – might then have set up a German company before he and Vieweg had made their own arrangements with the Kassel firm of Pfeffer & Schwarzenberg. There are some choice remarks about Muspratt's stupidity and criminal indiscretion in

22. W. H. Brock, ed., *Justus von Liebig und August Wilhelm Hofmann in ihren Briefen* (Weinheim, 1984), p. 104.
23. French and Italian versions appeared in 1846. These do not include the "Preface by the Manufacturers of Professor Liebig's Manures."
24. A. Petzholdt, *Populäre Vorlesungen über die Agriculturchemie* (Leipzig, 1844). The second edition (1846) included Liebig's "Der neu efundene Patent-Dunger," which was also offprinted.

Liebig's letter to Sheridan's father.[25] For a time Sheridan Muspratt became a convenient fall guy for the debacle.

Not that Sheridan Muspratt deserves much sympathy. Astonishingly vain – he dyed his great beard red, and a local phrenologist reported astutely that Sheridan loved fame and approbation to a very high degree – he frequently caused Liebig anxiety and public embarrassment. In 1838 his father had sent him to Philadelphia to sell alkali to the Americans, but Sheridan proved to be a flop as a businessman and lost James Muspratt thousands of pounds. Following his studies with Liebig (1843–44), when Sheridan gained a doctorate for studies of sulphites and indigo, he joined Hofmann in London at the Royal College of Chemistry between 1845 and 1848. With Hofmann he identified and separated nitrotoluene and toluidine from coal tar. This was good work, for as Liebig noted to Hofmann, "so far he has not yet worn holes in his trousers with studying" (2 February 1848).[26] He never had to.

Intensely ambitious and with the Giessen and London laboratories as his models, Sheridan returned to Liverpool in 1848 to create his own college of chemistry in a stable at the rear of his house. Thomas Spencer, who got into an analytical dispute with Muspratt, was mercilessly rude concerning Muspratt's pretentiousness:

> When put into a house in a respectable street, I could see no harm in the intention to metamorphose the one-horse stable behind into a sort of school, to teach a few pupils; judging that it would turn out to be something in the nature of a mutual instruction class, or under any circumstances that if Master James [Sheridan] kept himself – by having a book constantly within reach – perhaps half an hour in advance of his pupils – and they not too old, why, all might go on smooth enough. But the plot began to thicken faster than I expected; for no sooner was the manger of the stable pulled down, and the hole in the floor above, by which the hay used to descend – stopped up, than he took it into his head to entitle this dilapidated little stable, THE LIVERPOOL COLLEGE OF CHEMISTRY. Moreover, when doubtless, as elated as Robinson Crusoe, being lord of all he surveyed – so long as he managed to pay the rent, – like the hero of Cervantes, he at once *dubbed himself* – PRINCIPAL and PROFESSOR . . . and forthwith issued out prospectuses to announce the fact. . . . People at a distance are apt to suppose, indeed, to take it for granted, that this apparent institution, has been got up by the public spirit of the people of

25. M. D. Stephens and G. W. Roderick, "The Muspratts of Liverpool," *Annals of Science* 29(1972), 287–311, at p. 295.
26. Brock, *Liebig und Hofmann in ihren Briefen*, p. 77.

> Liverpool, for the purpose of teaching this all important science, on an extended scale, and, that its principal and professor, has been selected from other competitors for his superior talents and competency.[27]

Invective aside, Muspratt's enterprise was harmless enough, since it was precisely what Trommsdorff and Liebig had done with private enterprise institutes for the teaching of pharmacy. The harm came in the puffery that Muspratt used in promoting his institution that included extensive references to the fact that he was Liebig's favoured, indeed favourite, pupil. In 1850, aware perhaps that Liebig was already thinking of leaving Giessen, he announced that by a tacit agreement Liebig intended to cede the chair to "his great pupil and friend, the celebrated [sic] Muspratt."[28] When Liebig heard about this puffery from Hofmann and other English pupils, he was appalled, especially when he also learned during his visit to the Great Exhibition in 1851 that Muspratt's analytical competence was being continually challenged in the Liverpool press and that Muspratt sometimes traded on Liebig's name when arranging his bills.[29]

This behaviour convinced Liebig that Sheridan was mad. Liebig reprimanded Sheridan, saying:

> Although I rendered him friendship, this did not give him the right to put himself on a par with Wöhler, Magnus, &c. I asked him to stop using my name in publications as a means of achieving his ends, otherwise I would protest publicly; and I told him that his way of gaining recognition was not the right way and that it should instead be done through good work and not by irrational attention-seeking. I did not think him skilled and talented enough to maintain the position which he publicly awarded himself; his most recent work on carmfuleic acid was the work of a tyro. (Liebig to Hofmann, 15 November 1851)[30]

To this rebuke Muspratt replied unashamedly and preposterously:

> I must say that a great portion of the celebrity of your [Liebig's] school rests on me. I have published as many important investigations as any chemist in the present century. Liebig and Wöhler worked on uric acid, Muspratt and Hofmann on nitroaniline and toluidine. Your opinion on my paper on carmafuleic acid is only an opinion. I know the worth of it and so do others. You say "You are aware that I

27. Letter of Thomas Spencer printed in *Liverpool Mercury,* 12 April 1850, Rothamsted archives.
28. Frankland to Liebig, 22 November 1851, Frankland Archive, Open University.
29. Knapp to Warington, 22 April 1852, Rothamsted archives.
30. Brock, *Liebig und Hofmann in ihren Briefen,* p. 118.

> [Muspratt] am not a leading chemist." I say I am not aware of any such thing. *Punch* says "Dr Muspratt's reputation is more than European" and Mr Dickens writes "I hope Liverpool will one day endeavour to find some expression for the obligations she owes you."[31]

And in the same breath, he asked Liebig to get him elected to the Irish Academy in Dublin. "It really does one no honour to be connected with such an impudent man," Liebig concluded.[32]

By 1850, pulling strings through Hofmann's friendship with Sir James Clark paid off. The Liverpool Royal College of Chemistry received the patronage of Prince Albert and claimed to have twenty-two pupils. The college survived until about 1911, but apart from the pioneer of technical education Norman Tate, it never trained any chemists of distinction. The venture is nevertheless properly seen as paving the way for a university in Liverpool, which Sheridan's youngest brother Edmund Knowles Muspratt (1833–1923), who was also Liebig-trained, was to help finance in 1903. Sheridan Muspratt was a sham but a very successful one. He married the most beautiful actress of his day, and by getting others to do his analyses and even to produce his *magnum opus*, the book *Chemistry, Theoretical, Practical, Industrial,* he fooled some of the world.[33]

Because he was never one to bear a grudge for long and because of his close friendship with the rest of the Muspratt dynasty, Liebig resumed relations with Sheridan in the mid-1850s. By then Liebig was only too well aware that the patent manure fiasco had been entirely his own fault. Lawes had seen his advantage in 1846 when he stated in his own advertising brochure:

> The results of [Liebig's manures] have not, in general, been equal to the advantages offered by the manufacturer [and] public opinion is not, at the present day, altogether favourable to the science of agricultural chemistry. . . . The repeated declaration of chemists that farmers will be able to grow as fine crops by the aid of a few pounds of some chemical substance as by the same number of tons of farmyard dung is never likely to be realised. Neither will the farmer ever be likely to write prescriptions to be made up by the manufacturing chemist, as had lately been suggested [by Liebig]. . . . Until chemistry has solved the great problem of uniting hydrogen with nitrogen and can furnish us with a cheap and abundant supply of ammonia, the

31. *Ibid.*, p. 119.
32. *Ibid.*, p. 119.
33. S. Muspratt, *Chemistry, Theoretical, Practical & Analytical,* 2 vols. (Glasgow, 1854–60). See the chapter on Muspratt in Brock, *Fontana History of Chemistry* (London, 1992).

> employment of artificial manures must be confined to one or two crops.[34]

Lawes's prescient pessimism was a necessary antidote to Liebig's precipitate optimism: Agricultural science was a good deal more complicated than Liebig had initially thought.

Although the Muspratts appear to have continued marketing Liebig's manures until about 1848, they were then quietly withdrawn. Bad publicity, coupled with the perceived success of ammoniated superphosphates and imported guano, had spoiled Liebig's commercial venture. Significantly, when he did return to fertilizer speculation again in 1857, it was to invest in a superphosphate factory at Aibling, Heufeld, in Upper Bavaria. A few years later, Emil Güssefeld (1820–97), a Hamburg importer of guano and potash, began to consult Liebig over analyses of his imports in exchange for consignments of Liebig's favourite cigars. Liebig was able to show that the Baker Island guano imported by Güssefeld made a particularly good superphosphate. Royal support for these ventures came easily from Maximilian II, for by then there were at least twenty-five such factories in Britain and the Germanies, let alone in other parts of Europe and America. Most of these were created by Liebig's students: Ludwig Baist (1825–99) at Griesheim,[35] Carl Clemm-Lennig (1818–87) at Mannheim, Heinrich Albert in Amöneburg, Hermann Grüneberg(1827–94) in Köln, and Varrentrapp at Braunschweig. Liebig's son Hermann also patented an artificial fertilizer, though whether it was a success is unknown.[36]

In 1856, in the paper he submitted to the Royal Agricultural Society, Liebig made his confession:

> I most readily admit that the idea of these manures could only have arisen and taken root in the brain of a man of science, penetrated and inspired by the truth of his doctrines, but who had before his mind's eye not the actual state of things, but an ideal agriculture. It was folly to believe it even possible for a practical agriculturist to enter into these ideas, or to fancy that he could, or would convert all his fields into experimental fields for the sake of establishing fixed laws for agriculture; it was folly to call on others to enter on such an undertak-

34. J. B. Lawes, *A Few Words of Advice on Artificial Manures* (London, 1846), p. 4. Copy in Rothamsted archive.
35. Baist founded the Frankfurter Akteingesellschaft für landwirtenschaftlich-chemische Preparate in 1856. This was renamed the Chemische Fabrik Griesheim in 1863. See Haber, *The Chemical Industry*, p. 124.
36. Correspondence Hermann von Liebig and Gilbert, Rothamsted archives. For fertilizer enterprises, see P. Borscheid, *Naturwissenschaft, Staat und Industrie in Baden 1848–1914* (Stuttgart, 1976), pp. 92–3.

> ing without any prospect of being able to devote to it the time and energy which were indispensable to its prosecution.[37]

No such public admission concluded the quinidine–quinine affair of 1845 to 1847.

The Quinidine–Quinine Conspiracy

Quinidine is a *dextro*-stereoisomer of quinine that can be extracted from the resinous, amorphous substance known as quinidine, which is left in the mother liquor from which quinine has been extracted. It was first precisely defined in 1848 by an apothecary at Zwingenberg, Ferdinand Winckler (1801–?). A year later, the Dutch pharmacist J. van Heijningen showed by quantitative analysis that it was an isomer of quinine, which he referred to accordingly as β-quinine.

In a period of colonial aggression and expansion, quinine was especially in demand amongst the armies of nations whose imperial interests lay in tropical zones where malaria was endemic. In the 1840s, the East India Company in London was buying about half a ton of quinine per annum.[38] Both quinine and quinidine had been extracted and identified by P. S. Pelletier in the 1820s from the traditional antimalarial remedy known as cinchona, or Peruvian bark, stimulating a number of traditional apothecaries, such as Merck in Darmstadt and Zimmer in Stuttgart, to expand their activities into the production of pure quinine. Sertürner also claimed that quinidine, despite its noncrystalline nature, was as good a reducer of fever as quinine itself. As Liebig and other pharmaceutical chemists concluded, the "medical efficacy of quinoidine [probably varied] according to the greater or less proportion of quinine it may happen to contain."[39]

In his work on coal tar derivatives, Hofmann also extracted the base, quinoline, which Gerhardt had prepared from the cinchonine that was extractable from Peruvian bark. Hofmann's discovery appears to have given Liebig the idea of developing a method for the pharmaceutical assay of the quinine content of the quinoidine that was marketed commercially as a cheaper antipyretic remedy, by transforming it into quinoline. Recall that one of the purposes of Liebig's original editorial association with Geiger on the *Annalen der Pharmacie* had been to test the quality of commercial drugs. When Liebig tried the procedure, however, he found

37. *J. Roy. Agricultural Soc.*, 17(1856), 284–326, at p. 315.
38. Felschow and Heuser, *Universität und Ministerium*, pp. 246–7.
39. J. Liebig, "On amorphous quinine as it exists in the substance known in commerce as quinoidine [quinidine]," *The Lancet* 1846, i, 585; *Ann. Chem.* 58(1846), 348–56.

that he obtained quinidine, not quinoline. He now realized that analytically quinine and quinidine were identical, though it remained for Winckler and van Heiningen to show precisely how they were related. In early 1846 Hofmann showed that quinidine could be transformed into quinine, but this fact was not published in order to preserve the surprise value of Liebig's identification. In Liebig's process, waste quinidine was dissolved in dilute sulphuric acid, and metal impurities were removed by precipitation with hydrogen sulphide. After boiling, pure quinidine was precipitated with sodium carbonate and then dried and bottled. The so-called amorphous quinine involved the further processing of the purified quinidine by mixing it with sand and dissolving it in sulphuric ether. The ether was then distilled off and the product decanted and bottled.

But if the two substances were identical, huge amounts of commercially valuable quinine (in the form of pure quinidine) were available in the mother-liquor waste products of quinine sulphate refinement. If such wastes could be bought up cheaply from chemists and druggists all over Europe (perhaps ostensibly saying that they were to be used for research purposes) and processed centrally by a friendly manufacturing chemist, then when the announcement of the identification was made public, a handsome profit might be made as dealers rushed to buy quinidine and the price rose in consequence. Moreover, by patenting the process, a monopoly would provide a steady income for at least fifteen years. A decade later, a similar motive was to inspire W. H. Perkin, the eighteen-year-old pupil of Hofmann's, to attempt the synthesis of quinine. Failing in this, Perkin succeeded instead in preparing the dye, mauve.

As the now-published correspondence between Liebig, Hofmann, and Linde reveals, a consortium composed of Liebig, Hofmann, Lloyd Bullock, and John Gardner raised some 50,000 Gulden (about £4,000) from a London capitalist Robert Barclay, a dealer in patent medicines, to buy up large quantities of cheap quinidine from all over Europe for shipment via Liverpool to Bullock's operative chemists' workshop in Conduit Street, London.[40] Although Liebig took the advice of Merck as to the druggists from whom quinidine could be bought, and F. W. Roller (Merck's representative in London) was brought in to handle the complexities of shipping and customs duties, neither Merck nor Roller was a party to the conspiracy. The materials were seemingly "smuggled" through the Liver-

40. Brock, *Liebig und Hofmann in ihren Briefen* (1988); E.Heuser and R. Zott, eds., *Justus von Liebig und August Wilhelm Hofmann in ihren Briefen. Nachträge 1845–1869* (Mannheim, 1988); Felschow and Heuser, *Universität und Ministerium;* and W. Hornix, "Tales of Hofmann," *Annals of Science* 44(1987), 519–24.

pool customs disguised as "resin," or colophony, and through Hull as "asphalt."

The original plan was to allow Bullock and Gardner to sell and market the purified amorphous quinine on Liebig's and Hofmann's behalf once Bullock gave Liebig the signal to publish his authoritative but deliberately delayed announcement of the identification of quinidine with quinine. Thus, although Liebig's article was ready in June 1845, publication was delayed a year until June 1846. Bullock seems to have secretly tested the product in London hospitals to which he supplied quinine – presumably by substituting quinidine – and assuring Liebig that it worked equally well. Profits were to be shared three ways: a third jointly for Liebig and Hofmann, a third jointly for Bullock and Gardner, and a third for Robert Barclay. As with the patent fertilizer scheme that was playing out its course concurrently, roles were also played by Sheridan Muspratt, whose father's alkali works were used to store bulk supplies of quinidine, and Friedrich Knapp. However, as with Merck and Roller, neither Muspratt nor Knapp initially knew what was involved in the scheme. As Gerrylynn Roberts has noted, all the principal conspirators had connections with the Royal College of Chemistry.[41]

Possibly all would have gone according to plan had not Liebig decided to ingratiate himself further politically with Justin von Linde. Partly out of genuine friendship (for as their correspondence makes clear, the relationship was close and extended to their children), partly out of gratitude to Linde for promoting Heinrich Will to a second chemistry chair in 1845, but partly undoubtedly to sweeten further negotiations over university posts, Liebig decided in June 1845 to reveal the consortium's plans without consulting Bullock. Linde was therefore in the position to buy up quinidine of his own and be ready to sell it as soon as Liebig's announcement was made. Liebig promised Linde that he should make a profit of 35 gulden (about £2) on every pound of quinidine he bought. To make things easy for Linde, Liebig made him a capital loan of 1,000 gulden (£55) and further committed himself to repurchasing Linde's stock of the material if the enterprise should fail.[42]

Things now went badly wrong. Although Liebig had not thought of patenting the discovery initially – possibly because he thought Sertürner's

41. G. K. Roberts, *The Royal College of Chemistry (1845–1854); A Social History* (Johns Hopkins University, PhD thesis 1973); and her "The establishment of the RCC," *Hist. Studies Phys. Sci.* 7(1976), 437–85.

42. Felschow and Heuser, *Universität und Ministerium*, pp. 223–7. At this period there were 18 gulden (florins) to the £1, or 12 thaler (gulden) to the £1, where 1.5 gulden equalled 1 thaler.

original work would block it – when Bullock suggested patenting during a personal visit to Giessen on 4 February 1845, Liebig saw that it would make sense, even though it would give Bullock full control of the discovery. As a foreigner, Liebig could not take out a patent in his own name. However, if the consortium drew up a legally binding document and Hofmann was in London to see that there was fair play, Liebig foresaw no problem. Even though he now felt obliged to tell Bullock that he had given Linde "inside" information, Bullock agreed to purchase the quinidine from Linde if Linde would ship it to Liverpool. A legal agreement was accordingly signed by Hofmann on Liebig's behalf on 26 March 1846.

Liebig realized that patenting a medical discovery could be criticised:

> People will gossip about me for a couple of weeks and then everyone will say to themselves that they would have done exactly the same in my position.[43]

Bullock had intended to keep the patent application secret, but this proved impossible. Consequently, considerable unwelcome discussion developed in the English medical and pharmaceutical press when it was noticed that Bullock's patent specification for "Improvements in the Manufacture of Quinine" (BP 1846, No. 11,204), which he filed on 12 May 1846, bore an astonishing resemblance to Liebig's description "On amorphous quinine as it exists in the substance known in commerce as quinidine," which appeared over a week later in *The Lancet* on 23 May. Why had Liebig brought forward the announcement of his discovery and not waited for the patent to be granted on 12 November 1846?

The answer is that the conspirators had fallen out over Linde's stock of quinidine.

It must have been impossible for the quinine market and customs officials in the various ports of Europe not to have been suspicious. Consequently, quinidine prices were already rising during the early part of 1846. Having committed a great deal of capital to the speculation, it seems that Gardner and Bullock began to have cash flow problems by the summer. Bullock and Gardner also began to suspect, without warrant, that because of the way supplies of quinidine (including Linde's stock) were flooding into Liverpool and Hull, Liebig and Linde, perhaps in conjunction with Merck, were playing the market independently of the agreement. Even Hofmann became suspicious until Liebig was able to assure him categorically and with spirit that the charge was false. He backed up his claim of innocence with a letter from Linde showing that all his quinidine purchases had been made and shipped to Liverpool before January 1846, two

43. Brock, *Liebig und Hofmann in ihren Briefen*, pp. 48–9.

months before the agreement was signed. Even Liebig became suspicious and carefully checked with Knapp in Liverpool that no extra consignments had been received from Linde or Merck.[44] The accusations dismissed, he angrily turned his full weight against Gardner and Bullock. Both, however, now refused to buy Linde's consignment, even when it was offered to them at cost price. It was at this point that Liebig decided to publish, even though Gardner and Bullock had not given the signal.

Thoroughly exasperated and aware that his hope of seeing Karl Sell called from Bonn to Giessen as professor of law following the sudden death of his brother Georg Sell in 1846 depended upon Linde's good will, Liebig bought back Linde's quinidine himself. As he explained to Hofmann, "Sell's nomination was under discussion . . . and in those cases I know how to make sacrifices."[45] In the end, Hofmann agreed to share the additional loss with Liebig.[46] The exact outcome is not known, but it seems unlikely that Liebig recovered more than 1,080 thaler (£90), having paid out 3,427 thaler in the first place – a loss of nearly £200. Fortunately, by selling off Linde's quinidine in 1847, he reduced his loss to about 950 thaler (£80).[47] But there were also potential social problems to be faced.

Liebig had emphasized in his paper the commercial significance of his discovery because of (1) the high price of importing and processing Peruvian bark, (2) the possible hindrance to supplies from South America because of political events there, and (3) the significant amounts of quinidine contained in the waste heaps of quinine sulphate manufacture. From this, any pharmaceutical manufacturer would have concluded that it would be sensible to buy up supplies of waste quinidine. Not surprisingly, therefore, when on 6 June 1846, in a lecture given publicly at the Royal College of Chemistry (RCC) by its secretary John Gardner, the virtues of Liebig's discovery were extolled, suspicions were fuelled that Liebig's name was being exploited by Gardner and his known associate Bullock for commercial ends. So far, Liebig remained the innocent party, and only Hofmann knew otherwise.

In a letter to *The Lancet* written in July but not published until 15 August (the delay suggests embarrassment by the journal's Liebig-admiring editor Thomas Wakley), Theophilus Redwood professor of pharmacy at the Pharmaceutical Society, noted that Bullock's patented amorphous quinine was being advertised as "the pure alkaloid of the cinchona bark, in the form under which it has been described by Professor Liebig."[48] He

44. Felschow and Heuser, *Universität und Ministerium im Vormärz*, pp. 265–6.
45. Brock, *Liebig und Hofmann in ihren Briefen*, p. 71.
46. *Ibid.*, p. 68.
47. *Ibid.*, p. 73.
48. *The Lancet* 1846, ii, 180–81.

went on to object that Bullock's product was impure and not exactly what Liebig had described. Although Bullock in his reply blustered that Redwood's analyses demonstrated the backwardness of English pharmacy, Redwood (no doubt to Liebig's acute embarrassment) stuck to his guns, leading John Grove, a Wandsworth surgeon who was innocent of Liebig's role, to portray Bullock as the amoral exploiter of the English patent laws:

> The discovery is Liebig's; he threw it open for the public to use, by means of the pages of *The Lancet* and Mr Bullock has appropriated that to himself which Liebig has publicly distributed.[49]

Winckler reached the same conclusion in Germany, pointing out also that by charging just over 3 gulden per ounce, Bullock was exploiting Liebig's discovery and profiteering, as a pound of quinidine could be bought for 10 gulden.[50]

Forced to exculpate himself by pointing out that he had registered the patent ten days *before* Liebig's paper on the subject was published, Bullock undoubtedly sowed the seed of suspicion that Liebig was somehow personally involved in commercial puffery.[51] Jacob Bell's *Pharmaceutical Journal* concluded in an editorial:

> In cases of this kind, a departure from the usual, straightforward course naturally gives rise to various conjectures, and in this instance it was correctly reported, and generally believed that several parties, beside the patentee, had an interest in the speculation; and that the plan [to use the Royal College of Chemistry to promote pharmaceutical remedies] had been resorted to for the purpose of ensuring its success by bringing the new remedy before the profession as a scientific discovery, eulogized in a widely circulated and influential journal, and in other quarters in a manner which would have been indelicate had it been known to be a merely commercial transaction.[52]

By the autumn of 1846, the speculation had not only threatened to tarnish Liebig's British reputation but also the images of both Hofmann and the new chemical institution he directed.[53] Although one of Hofmann's letters makes it clear that he was able to justify his innocence insofar as he had joined the quinidine speculation before he had signed his contract with the RCC, his role was no doubt kept quiet by the managers in order to safeguard the college's still shaky foundation. The need for

49. *Ibid.*, ii, 400.
50. *Jahrbuch für praktische Pharmacie* 13(1846), 361–89, quoted by Felschow and Heuser, *Universität und Ministerium*, pp. xxx–xxxi.
51. *The Lancet* (1846), ii, 436 (17 October).
52. *Pharmaceutical Journal* 6(1846–7), 160–72.
53. Roberts, *Royal College of Chemistry*, p. 147.

such secrecy meant, moreover, that Liebig could not force Gardner and Bullock "to exonerate him of using the scientific press for commercial purposes."[54]

The whole affair does raise the question whether Hofmann's involvement in the speculation was in any way connected with his appointment as director of the RCC, for, as Roberts has shown, both Gardner and Bullock had intended the college to have a commercial as well as an academic function. Gardner certainly saw the college's evening lectures as opportunities to publicise commercial ventures, as well as to review chemical developments. Moreover, a letter of Hofmann's of 24 June 1845, while he was still at Bonn, shows him calculating the profits of the quinidine venture at a time when negotiations over his London appointment were under way. And there is evidence that two of Liebig's other pupils, Will and Fresenius, both of whom were offered the London post before Hofmann, had business dealings with Bullock and Gardner.[55]

Although public discussion and speculation abated after Gardner was forced to resign as secretary of the RCC in August 1846 – probably through the influence of Liebig – the affair had proved disastrous for Liebig and Hofmann. When the latter took stock in February 1847, he reckoned that his own share of the profits amounted to only about £30 compared with his previous estimate of £1,000 in June 1845. No doubt Bullock and Gardner did rather better financially in the short term because they must have held a virtual monopoly of quinidine stocks for several months. In the long term, Robert Warington's patent for purifying quinine using animal charcoal reduced the significance of Liebig's process, while on a social and professional level both Bullock and Gardner lost the confidence of Liebig and the possibility of translating his many future writings.

Liebig must have been in a financial pickle by 1847. He had spent money on land that needed improvement and development; he had improved his living quarters, the costs of which he tried unsuccessfully to reclaim from the government; and he had invested unwisely in manures and quinidine. He had a wife and four children to clothe and feed. It is little wonder, therefore, that he was forced to sell Liebig Heights in 1849,

54. *Ibid.*, p. 300; Liebig to Hofmann, 7 July 1846, in Brock, *Liebig und Hofmann in ihren Briefen*, p. 57.
55. *Elementary Instruction in Chemical Analysis* (London, 1843), C. R. Fresenius, trans. W. Bullock, with a preface by Liebig; and C. R. Fresenius and H. Will, *New Methods of Alkalimetry* (London,1843). Note also the photograph of Bullock, Will, Hofmann, Fresenius, and Gardner together at Giessen in 1845, reproduced in B. Lepsius, *Festschrift zur Feier des 50-jährigen Bestehen der Deutschen Chemischen Gesellschaft* (Berlin, 1918).

though, curiously, in later publications, he created the impression that his agricultural field work on this land had lasted for a longer time.[56] When Pettenkofer, on behalf of King Maximilian II, was trying to persuade Liebig to leave Giessen for Munich, where he would be expected to encourage the industrialization of Bavaria, Liebig freely confessed that he was a complete failure as a businessman.[57] As we shall see, on balance, his Bavarian ventures, apart from the silvering of mirrors, were a good deal more successful.

Mirrors

The traditional method of making mirrors, which had been devised by Venetian glassmakers and craftsmen in the sixteenth century, used an amalgam of mercury and tin to produce a reflecting surface. A specially constructed, cased marble table, with grooves around the edges to collect the mercury runoff, was lined with tinfoil and covered with mercury. By working the mercury into the tin with the hands or a brush, a malleable amalgam of tin and mercury paste accumulated on the surface of the tin. This amalgam was then smoothed onto glass sheets and allowed to dry. Because large amounts of mercury were used, as in traditional hatmaking, mirrormakers suffered notoriously ill health and early death. But there was no alternative technology available until the 1840s.

Liebig himself had provided the alternative in 1835 when he noticed that aldehydes reduce silver salts to metallic silver and recommended this as a simple test for the presence of aldehydes in organic materials.[58] In 1843, an English operative chemist Thomas Drayton patented a process for silvering glass. Silver was precipitated by adding an alcoholic solution of oil of cassia to ammonia and silver nitrate.[59] Although the patent (BP 9968, 25 November 1843) drew attention to its possible use for mirrors, its sole commercial use until the late 1850s was in the artistic ornamentation of goblets and other glass vessels with silver braids. Such objects came to be widely admired at the Great Exhibition in 1851, but they clearly belonged to the luxury trade.

The stimulus for Liebig to take up his aldehyde reaction for mirrormaking came from his friendship with the Munich physicist, astronomer, and

56. This question is discussed revealingly in Munday, *Sturm and Dung*, pp. 262–8.

57. *Ibid.*, p. 284.

58. J. Liebig, "Über die Produkte der Oxydation des Alkohols," *Ann. Pharm.*, 14(1835), 133–67, at p. 140.

59. Drayton's method was used by Foucault. See W. Tobin, "Foucault's invention of the silvered-glass reflecting telescope," *Vistas in Astronomy* 30(1987), 153–84.

instrument-maker Carl von Steinheil (1801–70), who had been attempting to improve the mirror flats of reflecting telescopes since 1847. Steinheil hoped that glass mirrors were a solution to the inherent technical problems of using flat metal surfaces as telescope reflectors, but he had found that Drayton's deposition method was inadequate for his specialised purposes. Before meeting Liebig, Steinheil had collaborated with a Frankfurt technical chemist Rudolph Boettger (1806–81), who had developed a number of interesting ways of depositing thin veneers of metal onto other metal surfaces, as well as glass and ceramics, by both chemical and electrical means. Tied to Steinheil's specific objective of an optical mirror, however, it seems that Boettger never realized that he had in fact developed processes that would have been suitable for the manufacture of domestic mirrors.

In 1856 Steinheil asked Liebig whether he knew of a solution to his problem. Following a burst of empirical activity, Liebig found that by adding small amounts of copper to the mixture of ammoniacal silver nitrate and a sugar solution, a blemish-free surface of silver was deposited.[60] In further refinements and obviously exploiting Steinheil's knowledge of the advances that Boettger had made, Liebig deposited electrically a surface layer of copper on the glass flat before it was bathed in ammoniated silver nitrate and sugar.

Another of Steinheil's friends was Johann Beeg (1809–67), who since 1844 had been the rector of the trade school (*Gewerbeschule*) at Fürth, near Nuremberg. Beeg used this important position in the community to promote new industries in the town, which was also the home of several Jewish glassmakers. Beeg, who learned of Liebig's unpublished experiments from Steinheil, took the initiative in February 1856 in suggesting to Liebig the merits his process would have for replacing the current, dangerously unhealthy mirrormaking industry:

> This matter could be of the utmost importance for our town, not least from health considerations, but because industrial interests make it a duty of ours to endeavour to make sure that we are the first to exploit new technical procedures.[61]

Liebig seized his chance, and as the director of the Bavarian Academy he ordered its technical commission to finance Beeg's investigations of how

60. J. Liebig, "Über die Versilberung und Vergoldung von Glas," *Ann. Chem.* 98(1856), 132–9; translated *Chemical Gazette* 14(1856), 213–16; and "Versilberung von Glas," *Ann. Chem.* 1867, Supplement 5, pp. 257–60; translated *Phil. Mag.*, 35(1868), 146–7.

61. E. C. Vaupel, "Justus von Liebig (1803–73) und die Anfänge der Silberspiegelfabrikation," in Deutsches Museum, *Wissenschaftliches Jahrbuch 1989*, pp. 189–226 at p. 204.

the laboratory procedure could be translated to the factory. It was not until the middle of 1858 that Beeg had solved the many technical problems of silver concentration, temperature control, the design of benches, and the prevention of surface blotching. Several test mirrors were produced successfully; Liebig proudly presented them to the wives of his friends.

Knowing from Beeg that the process worked, Liebig then prepared the ground internationally for the replacement of the amalgam method by his process. There was a fortune to be made. He wrote to several of his friends, including the Berlin chemist and physicist Heinrich Magnus (1802–70), justifying his intention of moving into commerce.

> I want to drive out mercury mirror-making and its injurious influence on workers' health and will offer replacement mirrors for them in the home which are more durable, cleaner and clearer than the amalgam ones and at the same time cheaper. If this should reward me with [financial] independence, my every wish would be fulfilled. My position here [in Munich], compared with that of others, is a most splendid one, but, dear Magnus, when one has reached his 72nd semester [i.e., thirty-sixth year of teaching; Liebig was 55] the perpetual schoolmaster has become exhausted. If, as Kuhlmann has done, we had turned to the applied side of our science, and had only half the successes he has had in this field, we would like him have made millions. In me there is no spark of envy, but when one has grown children whose paths through life one should smooth, one inevitably makes such comparisons. In France and England it is different; people don't lecture there until their teeth fall out. This mirror business perhaps offers the means to make me free and I want to put it to the test. I want to take out a patent in Prussia and in other countries and to acquire a firm basis for the manufacturing. If this eludes me, I am in danger of losing everything that I spent on it in time and unspeakable work.[62]

The said Bavarian patent, which related only to the electrical deposition of copper (because the basic silvering process had been in the public domain for years) was acquired in April 1858, and Liebig spent a great deal of money taking out similar patents in England, France, Russia, and the United States. In November of the same year he licensed the Bavarian rights for ten years to a Nuremberg business – Crämer, Vetter & Co. As

62. *Ibid.*, pp. 209–10. Charles Frédéric Kuhlmann (1803–81), a student of Dumas, had opened a sulphuric acid factory in Lille in 1825, from which he diversified into alkali and fertilizer. Liebig dedicated *Ueber Theorie und Praxis in der Landwirtschaft* (Braunschweig, 1856) to Kuhlmann.

the owners of a tinfoil factory, Crämer & Vetter had a vested interest in gaining entry to a technology that sought to replace one to which they supplied a primary raw material, namely tin. Liebig's financial terms are not known; but he undoubtedly drove a hard bargain that involved his receiving a fixed proportion of the price of every box of mirrors sold.

The site chosen for the factory was to Beeg's delight Doos, a small village just outside Fürth; it was probably chosen because it was near the town's glassmakers. Production began in July 1860 and was under the technical guidance of Liebig's two Munich assistants, Friedrich Geiger (1833–89), the son of Liebig's *Annalen* partner of the 1830s, and Carl Schindling, both of whom immediately faced severe technical and commercial problems. Unfortunately, as experience rapidly showed, Liebig's process could not compete commercially with the less expensive and well-established silver-mercury amalgam mirror industry. Also, by not owning and controlling their own glassmaking, the factory faced daily problems of supply and quality. It was duly closed in December 1862 after only two years in operation. Ironically, after Liebig's death and when he could not benefit financially, safety legislation concerning the industrial use of mercury in most countries forced the reconstruction of the entire industry and promoted the adoption of versions of his process. All modern mirrormaking relies on developments of Liebig's basic process.[63]

Writing and Consultancy

In other ways, and particularly because of Hofmann's central position in London, Liebig was able to earn a fair income from literary and other activities in England. As Hofmann advised Liebig cynically, "Whatever you do, don't lift a finger unless money is offered."[64] As a result of the widespread interest in Liebig's *Animal Chemistry,* among doctors Gardner had been able to sell Liebig's Giessen lectures on organic chemistry to *The Lancet.* These appeared as a serial between 1844 and 1845 amid a great deal of trumpeting from the editor Thomas Wakely. They were subsequently reworked by William Gregory into a revised edition of Turner's *Elements of Chemistry* (see Chapter 4). Each successive edition of *Agricultural Chemistry, Animal Chemistry,* and *Chemical Letters* (which first appeared in 1843) was separately negotiated with Liebig's London pub-

63. According to Sir Henry Bessemer, *Autobiography* (1905), Liebig's process was developed in the UK by Henry Draper.
64. Brock, *Liebig und Hofmann in ihren Briefen,* p. 159.

lisher Thomas Walton of Gower Street.[65] When, however, in 1851 Liebig tried to get Wakely to serialise the four new letters inserted in the third English edition of *Chemical Letters* for £150, in addition to the £150 he was getting from Walton, he was turned down.

More revealing still is the curious episode of beer strychnine, which occurred in 1852 when England's two leading chemists Thomas Graham and Hofmann investigated the damaging rumours that had emanated from a French chemist named Payen that the Burton-brewed pale ales of Allsopp and of Bass were deliberately contaminated with strychnine to increase their bitterness. Astutely, Hofmann persuaded Allsopp that, although his and Graham's official report would exonerate the ales, the public's mind would be put more securely at rest if the beers were tested by Baron Liebig. Accordingly, Liebig earned £100 for writing a gushingly favourable open letter in praise of the English brew. As he confided to Hofmann, "The main test consisted in my drinking a bottle with great enjoyment." This sounds less shocking when it is remembered that Liebig would have had complete confidence in Graham's and Hofmann's assurances that the beers were innocent.[66] What is shocking is the fact that the wording of Liebig's testimonial closely reflected Allsopp's wishes, as relayed by Hofmann. Liebig must have known the consequence, namely, that Allsopp would use his citation in an extensive advertising campaign on billboards and in the press.[67] Allsopp's great rival beermaker in Burton was Bass, who, not to be outsmarted, also paid Liebig an unknown sum to guarantee the worthiness of his beers. One concludes from this episode that after leaving Giessen, Liebig was quite happy to associate his authority with commercial interests, the epitome of which was to be the Liebig's Extract of Meat in 1865.

One final example of Liebig's openness in looking for money from English sources concerns the testimonial that was offered to him by his British friends in August 1854, ostensibly to mark his retirement from the chair at Giessen. Following the debacle with Muspratt over the patent manures, Liebig had hoped vainly that the Royal Agricultural Society would buy the patent from him. Also, recalling his services to the British government in revealing the importance of coprolites, he began to hope

65. J. Liebig, *Die organische Chemie in ihrer Anwendung auf Physiologie und Pathologie* (Braunschweig, 1962), translated by W. Gregory as *Animal Chemistry* (London, 1862); *Die organische Chemie in ihrer Anwendung auf Agricultur und Physiologie* (Braunschweig, 1860), translated by L. Playfair as *Organic Chemistry in its Applications to Agriculture and Physiology* (London, 1840).

66. A. W. Hofmann, "Zur Erinnerungen an Peter Griess," *Berichte* 24(1891), 1000–7; E. R. Ward, "Peter Griess and the Burton breweries," *J. Roy. Inst. Chemistry* (1958), 383–9.

67. See, for example, *Illustrated London News*, 10 December 1853.

Baron Liebig

Your Retirement

Signed on behalf of the Subscribers

Tho. Graham

Chairman of the General Committee

London, July, 1854.

Diploma presented to Liebig by Thomas Graham on behalf of British chemists and agriculturalists July 1854. The scroll was accompanied by a silver tea service. (Courtesy Staatsbibliothek München)

that the society would reward him financially, perhaps by a premium prize, for his overall service to British agriculture. But Buckland, who was close to Peel, and by this date Dean of Westminster, quickly quashed such hopes.[68] Nevertheless, in 1851 the Royal Agricultural Society did seriously discuss the possibility of awarding Liebig £1,000, only to decide against it on the advice of Phillip Pusey. When Liebig got wind of this from Hofmann, he felt it was a public slap in the face, which it was, and he became determined to see the decision reversed.

To a greater extent than any of Liebig's pupils who held industrial or academic posts throughout the British Isles, Hofmann ensured that Liebig's name and work were kept prominently and favourably before the public eye. Not surprisingly, therefore, as a result of growing criticism of Liebig's agricultural work in the 1850s by the Royal Agricultural Society and its spokesmen, Philip Pusey, Sir John Lawes, and J. H. Gilbert, Hofmann became a leading member of Liebig's defence team. Although Hofmann's attempt to defend and publicise Liebig's position through an essay written by William Gregory came to nothing, as a palliative to Liebig's fury at the attack on him by James Johnston (see Chapter 6), Hofmann, together with Thomas Graham and Benjamin Brodie, had the idea of offering Liebig a public testimonial of thanks for his services to British agriculture.

The organising committee was chaired by Thomas Graham. Among the seventy members of the committee, twenty were former pupils of Liebig; many others were connected with the Royal College of Chemistry. Subscribers included Sir James Clark, Walter Crum, Charles Daubeny, Michael Faraday, William Vernon Harcourt, Sir Charles Lyell, John Mechi, William Sharpey, H. S. Thompson, and even Liebig's opponents Philip Pusey and J. W. Johnston.[69] As correspondence between Liebig and Hofmann shows, the money collected (about £1,060) was mainly subscribed by chemists, not agriculturists – and it was largely spent upon an elaborate and exceedingly ugly collection of silverplate designed by Messrs Hunt & Roskell. At about the same time, in rivalry and as a deliberate snub to Liebig, the supporters of Lawes raised a testimonial to him in December 1853 that was sufficient to pay for a new agricultural research laboratory at Rothamsted. Because of the extra time needed to design and make Liebig's tea service, Liebig's testimonial was not ready until the summer of 1854 – just in time to be exhibited at the international

68. Buckland to Liebig, 23 May 1846. Liebigiana, Munich.
69. See printed pamphlet (26 March 1855) containing details of the testimonial. Copy among Frankland papers, Open University.

exhibition in Munich. Rather ungratefully, in view of Hofmann's efforts, Liebig made it clear that he would have preferred a pension from the British crown for his services. For a guaranteed annual sum of £300 to £400, he would have been willing to visit London or Edinburgh each summer vacation to help run a state school of practical farming and to tour farms and offer advice.[70] Amidst all this agricultural partisanship, it is morally uplifting to note that both Lawes and Gilbert contributed to Liebig's testimonial.

This chapter has given a rather negative image of Liebig as unsuccessful in his own applications of chemistry, the very science that he was advocating as the most important discipline for any modern industrial nation. Just because his own schemes were unsuccessful before the 1860s, it would be quite wrong not to emphasize that many of his pupils were very successful indeed. Some of them intermarried into others' families or formed partnerships. Ernst Sell (1808–54) began distilling tar at Offenbach in 1842 with Liebig's former Parisian roommate, Karl Oehler. Sell's wife, who was from Darmstadt, knew the Mercks and brought the two firms together. Meanwhile, Sell's former partner Conrad Zimmer, who had founded a quinine refinery in Frankfurt, joined up with Liebig's pupil Friedrich Jobst (1816–96) to purify and sell alkaloids in Stuttgart. To complete the knot, Jobst had married a sister of Liebig's pupils Gustav (1814–66) and Carl (1817–87) Clemm. Carl Clemm-Lennig (as he was known following his American marriage) established an important sulphuric acid and soda works at Mannheim, while Gustav Clemm, together with the pharmaceutical businessman Christian Boehringen, began a chemical works in Stuttgart. The two Clemms' activities were brought together at Mannheim in 1854 as the Verein Chemischer Fabriken, from which the great Badische Anilin Soda Fabrik was created at Ludwigshafen in 1865. As Borscheid observes, such businesses grew from apothecaries' workshops, as pharmacy under the guidance of Liebig's chemistry diversified and increased its output.[71] Liebig's teaching of analysis and quality control played an essential role in the lives of these future businessmen. Finally, but not least, these pupils carried Liebig's message back to other German states, which, apart from Prussia, were alerted to the potentialities of

70. Liebig did not abandon this idea. See his letter to Faraday, 12 September 1856, and Playfair to Liebig, 7 November 1856, Liebigiana, Munich. See also Chapter 9, this book.

71. P. Borscheid, *Naturwissenschaften*. Fruton has also drawn attention to Liebig's pupils who were active in the Portland cement industry, dyes, inorganic chemicals and soap manufacture, as well as pupils who became directors of chemical factories. See Joseph S. Fruton, *Contrasts in Scientific Style* (Philadelphia, 1990), pp. 64–7.

support for chemical education by Liebig's polemical essay of 1840.[72] If Liebig himself was not a model for making a living from commercial chemistry – at least until the mid-1860s – he was an inspiration to others to industrialize Germany.

72. See Borscheid, *Naturwissenschaft*, for elaboration. Also K. Ohlendorf and E. G. Franz, "Gustav Clemm. Vor demokratischen Verschwörer zum Wegbereiter der deutschen Kaliindustrie," *Archiv für hessische Geschichte* 45(1987), 249–68.

6

Liebig and the Farmers: Agricultural Chemistry

> A time will come when fields will be manured with a solution of glass (silicate of potash), with the ashes of burnt straw, and with the salts of phosphoric acid prepared in chemical manufactories, exactly as at present medicines are given for fever and goitre.[1]

With the publication of *Chemistry in Its Applications to Agriculture* in 1840, Liebig ceased to be a German chemist known only to those interested in finding a way through the jungle of organic chemistry and became an international public figure. This prominence was achieved not merely through the fact that the book was rapidly translated into French, English, Italian, Russian, Polish, Dutch, Danish, and Swedish, but because its optimistic message was that by scientific management food yields could be increased and "the hungry forties" be forever eradicated.

As in most fields for which he became famous, Liebig was not the first chemist to have attempted to understand the principles of empirical husbandry from a scientific, or theoretical, perspective. In the early 1800s, for example, the English Board of Agriculture, concerned with farming as a source of social stability during a period of European war and blockades, had encouraged the young Humphry Davy at the Royal Institution in London to give a series of annual public lectures (1802–12) on agriculture that would help improve practise and expand output. Davy took his brief seriously, spending some time touring estates in England and Ireland to discover and analyse the chemical basis of good farming practise.[2]

1. J. Liebig, *Die organische Chemie in ihrer Anwendung auf Agricultur und Physiologie* (Braunschweig, 1840); *Chemistry in Its Application to Agriculture and Physiology*, trans. Lyon Playfair (London, 1840); hereafter Liebig, *Agricultural Chemistry*. There were ten German editions: 1840, 1841 (two), 1842, 1843, 1846, 1862, 1865, and posthumously 1876 and 1877; six English, 1840, 1841, 1842, 1843, 1847, and 1862 (Part 2 only); and nineteen American editions. For details, see C. Paoloni, *Justus von Liebig. Eine Bibliographie* (Heidelberg, 1968). Although the first German edition was only 195 pages long, the definitive seventh edition (1862) comprised 1,130 pages – vol. 1 (650 pp.), vol. 2 (480 pp.).
2. D. M. Knight, "Agriculture and chemistry in Britain around 1800," *Annals of Science* 33(1976), 187–96; and his *Humphry Davy* (Oxford, 1992), Chap. 4.

The eighteenth century had been a period of agricultural improvement and innovation: Tillage was transformed by seed drills and horse hoes, new crops such as turnips and clover were introduced into the rotation pattern, and experiments were made with inorganic fertilizers such as nitre, marl, and common salt. The pneumatic chemistry of Joseph Priestley and others, and its dramatic catalysis through Lavoisier of a new chemistry based upon oxygen, acids, bases, and salts, had also led Jan Ingenhausz, Jean Senebier, and Theodore de Saussure to the concept of photosynthesis. Photosynthesis postulated that plants mainly obtained their nourishment through air and water, whereas their mineral content (as revealed by analysis) was obtained directly through the soil – explaining why some soils were fertile and others sterile. All this was noted by Davy, whose *Elements of Agricultural Chemistry,* published in 1813, represented the zenith of the subject before organic chemistry or, rather, organic analysis, transformed the subject. Not surprisingly, therefore, in Britain and the United States Davy's book remained in demand well into the 1840s when Liebig's more theoretical treatise rapidly replaced it.

Because Davy's work represented good practice, it is important to note that he recommended the use of fresh (i.e., soluble) dungs as manure, reinforced when necessary – though he laid no great stress on this – by soluble inorganic salts of the alkalies and alkaline earths. Liebig's approach in the 1840s was to lay stress on the crucial significance of inorganic salts and on their slow, controlled diffusion into the soil when distributed as fertilizers. Davy's advice was uncontroversial. No one reviewing his book suggested that he did not know what he was talking about, as they were to accuse Liebig.

Similar chemical summaries of contemporary good practice were also to be found on the Continent in the period before 1840. Here, however, advice was accompanied by a model of soil fertility referred to by historians as "the humus theory." No one agricultural scientist, botanist, or chemist can be said to have devised this theory, though it was given publicity by Daniel Thaer (1752–1828), a professor of agriculture at the new Prussian University of Berlin from 1810 until 1818 when, believing agricultural training would best improve practice, he founded the first of Germany's agricultural institutes at Möglin near Berlin. According to the humus theory, promulgated in Thaer's *Grundsätze der rationallen Landwirtschaft,*[3] plants were sustained by water and humus, a black, treacly insoluble organic decay product of vegetation that was absorbed through

3. A.D. Thaer, *Grundsätze der rationallen Landwirtschaft,* 4 vols., Berlin 1809–12. The book was eventually translated into English by W. Shaw and C. W. Johnson in 1844. See D. Thaer, *The Principles of Agriculture* (2 vols., London, 1844).

the roots of new plants. There had been various attempts to extract an active principle from humus at the end of the eighteenth century; by the 1820s vegetable chemists commonly supposed it to contain ulmic and humic acids, both of which were believed to be complex degradation products of vegetation. Embedded in the theory was the notion that, as living organisms, plants possessed a vital force that enabled them to transmute humus into lime and other inorganic materials found in plants after ash analysis. Despite the discovery of photosynthetic fixation of carbon by carbon dioxide assimilation, humus was still considered to be the principal source of plant carbon uptake. There seems little reason to doubt that Liebig adopted a version of the humus theory – probably that of Berzelius – in the lectures on agricultural chemistry and forestry that he gave at Giessen in the 1820s.[4]

As early as 1804, however, in his *Recherches chimiques sur la vegetation,* the Swiss chemist Saussure had cast doubt upon the humus transmutation model when, in further experiments on photosynthesis, he had shown that plants could be grown on sand and that rain (or even distilled water) contained the inorganic salts that the plants absorbed. The idea, dating back to the time of Paracelsus (1493–1541) and van Helmont (1579–1644), that air or water was transmuted into the "earth" of plants, was definitely outmoded. It was on this basis that Karl Sprengel (1787–1859), a student of Thaer's, and Germany's most influential agricultural chemist before Liebig, rejected the humus theory in 1837. In a careful series of analyses, using Liebig's new method, Sprengel concluded that the value of humus, or any manure, lay in the minerals it contained, including carbon (which he found to be nearly 60 percent of humus). He further concluded that, where any vital inorganic nutrient was absent or in short supply, plant growth was seriously affected.

Despite these earlier demonstrations and criticisms, which were in any case less well known in Britain, it was Liebig who most prominently succeeded in rejecting the humus model and in promulgating the law of minima and the mineral theory of plant nutrition.[5] (The law of minima asserts that plants' growth cannot be stronger than is permitted by the element or compound present in the soil in the least adequate amount.) Although Liebig was clearly indebted to Saussure and Sprengel for the mineral theory, as he acknowledged initially, Liebig's originality lay in his

4. See lecture lists in W. Conrad, *Justus von Liebig* (PhD, Technischen Hochschule, Darmstadt), pp. 65–72.
5. J. R. Partington, *A History of Chemistry,* vol. 4 (London, 1964), pp. 310–13; A. J. Ihde, *The Development of Modern Chemistry* (New York, Evanston, London, 1964), Chap. 16; Ursula Schling-Bodersen, *Entwicklung und Institutionalisierung der Agrikulturchemie in 19.Jahrhundert* (Stuttgart, 1989).

suggestion that inorganic fertilizers might be used instead of farmyard manure. Sprengel, on the other hand, stated explicitly that minerals already present within the organic molecules of humus made the best fertilizers. Ironically, Saussure, whose photosynthetic experiments had convinced Liebig that plants synthesized their entire fabric from inorganic compounds, still retained a vitalistic belief that humus played some essential role in plant nutrition. In 1862, following the publication of Liebig's *Agricultural Chemistry*, Saussure expressed his doubts in the *Annalen der Chemie* that the process was entirely inorganic, only to be dismissed by Liebig in an editorial comment.[6]

Another agriculturist by whom Liebig was strongly influenced was the Frenchman Jean-Baptiste Boussingault (1802–87). Following a career as a mining engineer in South America, Boussingault had married an heiress and begun to farm on scientific principles a large estate in the Alsace at Bechelbronn.[7] In a series of papers beginning in 1836 in the *Annales de chimie* and later summarized in his *Economie rurale* (1843–44), he made fundamental inquiries into crop rotation and into the source of plant nitrogen but without reaching any firm conclusion.[8] For example, he was unable to explain why clover and peas grown in soils to which no fertilizer was applied actually increased the soils' nitrogen content over the growth cycle, whereas this did not happen with wheat and oats. (The bacterial fixation of nitrogen in the root nodules of leguminous plants was demonstrated by Winogradsky only in 1890.[9]) By way of interpretation, Boussingault tentatively suggested that legumes might fix nitrogen directly. Liebig's alternative explanation was to suppose that nitrogen was fixed in the form of ammonia contained in the air. On the other hand, through analyses of manures applied and crops removed, Boussingault showed statistically that the source of carbon removed in crops could come only from carbon dioxide in the air, as the amounts extracted were more than were available in the soil and manures combined.

Thus, when Liebig published *Agricultural Chemistry* in 1840, most of

6. T. de Saussure, "Ueber die Ernährung der Pflanzen," *Annalen der Chemie*, 42(1842), 275–91, with Liebig's comments on pp. 291–7.
7. J. F. W. McCosh, *Boussingault* (Dordrecht,1984). This provides an excellent antidote to historians' overemphasis on Liebig and the controversies with Lawes and Gilbert. See also Richard Aulie, "Boussingault and the nitrogen cycle," *Proc. American Philosophical Society* 114(1970), 435–79; and his "The mineral theory," *Agricultural History* 48(1974), 369–82.
8. J. B. Boussingault, *Economie Rurale*, 2 vols. (Paris, 1843–4). Translated by G. Low as *Rural Economy* (London, 1845). Parts of the chemistry were omitted as unsuited to "the rural economist."
9. P. S. Nutman, "Centenary lecture," *Phil. Trans.* 37B (1987), 69–106.

the elements for a theoretical treatment of agricultural chemistry were all in place: photosynthesis, the atmospheric origin of plant carbon, criticism of the humus theory, a partial understanding of the origins and roles of minerals in plant metabolism. But a puzzle over nitrogen's source and the form of its fixation remained. The first part of *Agricultural Chemistry* essentially hinged on proposing a nitrogen cycle and a mineral theory as better explanations of plant nutrition than was organic humus.

The Corn Laws passed in 1815 severely restricted the importation of grain into Britain. This legislation kept corn prices high to placate landowners and farmers who, in the wake of the Napoleonic Wars, when prices escalated, now found prices tumbling. Agricultural depression encouraged them to expand the cultivation of wheat, barley, and clover and to make farm improvements. At the same time, the Anti–Corn Law League, a pressure group founded to lobby for the repeal of farming protectionist measures, was interested in lowering the cost of living of the country's working population. To that end, both the Anti–Corn Law League and farmers' lobbies were open to persuasion that chemical science applied to agriculture could increase yields and lower costs. The founding in 1838 of the Royal Agricultural Society of England (RASE) and other agricultural associations was a reflection of this interest in economic and efficient farm management and restructuring amongst wealthy landowners and of what economic historians call the High Farming movement. F. M. L. Thompson defines the movement as involving

> the technical changes of crop rotations and livestock improvement; it generally involved the physical changes of enclosure, and it embodied the economic changes of increased intensity of cultivation; that is, production functions were altered by an increase in the amount of both capital and labour that could be profitably employed in relation to land. Farmers clearly became more market-oriented than before, in deciding what crops to grow, when, and in what proportions. But the supply of the major part of the extra capital required came from landlords.[10]

In the 1840s, despite rapidly rising town populations, by far the majority of people still lived in the countryside and either worked on the land or were dependent on a huge variety of trades and businesses associated with agriculture.[11] Although agriculture was dominated by a number of wealthy aristocratic estates, the major part of the farming community was

10. F. M. L. Thompson, "The second agricultural revolution, 1815–1880," *Econ. History Review* 21 (1968), 62–77.
11. Vance Hall, *A History of the Yorkshire Agricultural Society 1837–1987* (London, 1987), Chap. 1.

made up of "small farmers working less than 50 acres."[12] These smallholders, with their limited capital and inexperience of farming methods other than those they had grown up with, were resistant to innovation or education and were content with the four-course, labour-intensive rotation systems. By the 1840s, however, market pressures were encouraging landowners and farmers to greater profitability, as more distant town and city markets demanded their produce.

The repeal of the Corn Laws by Robert Peel in 1846,[13] which again permitted the import of cheap grain from the Continent, caused prices to plummet and so led to an even greater concern amongst landowners and farmers for methods of increasing soil fertility and crop yields, as well as improving animal husbandry. Liebig's message that organic chemistry held the secret of successful agriculture therefore fell upon receptive British ears and was one of the factors in the setting up of the Royal College of Chemistry in 1845. A Parliamentary Select Committee established in 1833 "to enquire into the present State of Agriculture and of Persons employed in Agriculture in the United Kingdom" had recalled Davy's services to agriculture and recommended that the application of the sciences of geology and chemistry was desirable. It was no accident, therefore, that analytical chemists were beginning to offer their services to landowners and farmers or that books such as David Low's *Elements of Practical Agriculture* (Edinburgh, 1838) already contained some chemical advice well before Liebig's writings appeared in English. For example, the York Quaker chemist Joseph Spence was advising members of the Yorkshire Agricultural Society (YAS, founded 1837) from the beginning, that is, some years before he was made their official analytical chemist in 1844, the same year that Liebig's pupil Lyon Playfair was made consultant chemist to the RASE. One clause of the latter's charter (awarded in 1840) was that the society intended "to encourage men of science in their attention to the applications of chemistry to the general purposes of agriculture."[14] In Britain, Liebig found himself in a very receptive climate. As Rossiter noted, by hitting the crest of the wave of farming interest, Liebig became synonymous with agricultural chemistry.[15]

British ears had been opened during Liebig's first tour of England in 1837 when, at the Liverpool meeting of the British Association for the Advancement of Science (BAAS), he declared:

12. *Ibid.*, p. 27.
13. See letter from Liebig to Peel in Imperial College archives, quoted in Chapter 5 of this book.
14. *J. Roy. Agri. Soc. of England* 1(1840), clxx.
15. M. W. Rossiter, *The Emergence of Agricultural Science. Justus Liebig and the Americans, 1840–1880* (New Haven and London, 1975), pp. 25–6.

> It is . . . remarkable that in a country in which I now am, whose hospitality I shall never cease to remember, organic chemistry is only commencing to take root. We live in a time when the slightest exertion leads to valuable results, and, if we consider the immense influence which organic chemistry exercises over medicine, manufactures, and over common life, we must be sensible that there is at present no problem more important to mankind than the prosecution of the objects which organic chemistry contemplates. I trust that English men of science participate in the general movement, and unite their efforts to those of the chemists of the Continent, to further the advance of science which, when taken in this connection with the researches in physiology, both animal and vegetable, which have been so successfully prosecuted in this country, may be expected to afford us the most important and novel conclusions respecting the functions of organization.[16]

As Volhard reports, Liebig's paper was received with acclaim: Liebig was "the lion of the evening."[17] Not unnaturally, in view of Liebig's implied criticism of Britain's prowess at organic chemistry and prompted (one suspects, stage-managed) by Robert Kane, the general committee of the BAAS seized the opportunity of asking Liebig to write two reports: on isomeric bodies and on organic chemistry. The former was never accomplished, but neither, strictly speaking, was the latter. For, as Paoloni's bibliography makes clear, *Agricultural Chemistry* was actually written not as a report for the BAAS but as the "Introduction" to Liebig's *Traité de chimie organique,* which was published in Paris on 2 April 1840. This "Applications des principes de la chimie organique à la physiologie végétale et à l'agriculture" was then republished in German at Braunschweig under its more familiar title *Agricultural Chemistry* on 1 August 1840, the galley sheets being translated into English by Lyon Playfair, who was studying with Liebig at Giessen during the summer semester.

As we know from Liebig's correspondence with Wöhler in 1839 and 1840, his many literary activities at this time were taxing him unmercifully:

> The slightest tasks remain undone, the most pressing letters unanswered; so much has piled up that, like a ship in a storm, the waves engulf me, and there is no refuge in the cabin or on a warm rock. In such states man is not responsible. The main torture is the French edition of my organic chemistry, which cannot be got ready before Easter and whose contract dictates that for every month it is delayed I

16. *BAAS Reports* (1837), 38–41.
17. Volhard, vol. 1, pp. 14–15.

> lose 500 francs. It looks as if I can expect to work for a year without compensation.[18]

Although the German manuscript was delivered to Gerhardt for translation in March 1839, if we suppose that Masson & Cie, Liebig's French publisher, extracted a penalty, it may well have occurred to Liebig to add an "Introduction," which could be published separately to recoup the lost income:

> What couldn't I write to you as a Jeremiah, as a complainer about a life eaten up, of a life devoured by paper. The first volume of my organic chemistry is ready, and I was pressed to write a "Foreword" and "Introduction," which have cost me much time. (Liebig to Wöhler 3 March 1839)

> I've been unwell for some days and little inclined to write or think and in a very bad mood. Because I have a new assistant who doesn't yet know the ropes, my lectures are not as good as usual. I am losing a frightful amount of time. On top of this is the cursed book writing which brings on the greatest despair; never again will I write any more books [sic], even if mountains turn out to be made of diamonds.[19]

Even so, it was not until 17 March 1840, a month before the appearance of the French edition, that he wrote to Vieweg to suggest that his "revolutionary" "Introduction" deserved a separate edition to attract the doctors, economists, and landowners who were unlikely to read a chemistry textbook in French. Assuring Vieweg that it would "make a great impression," he asked for 1,500 copies to be printed.[20] The twentieth-century reader is staggered at the speed with which the German and English printings were done – four and five months respectively.

Playfair's translation was published in London by Taylor & Walton on 17 September, the very day the BAAS meeting opened in Glasgow. The book's German dedication is dated 1 August 1840 and the English translation 1 September 1840. Only the English edition was dedicated to the BAAS; the *Traité* was dedicated to Gay-Lussac, Liebig's former teacher in Paris, and the German edition was dedicated to his patron Alexander von Humboldt. To complicate matters, when Gerhardt issued the official translation of the German edition for farmers and agronomists in 1841 (*Chimie organique appliqué à la physiologie végétale et à l'agriculture*), Liebig dedicated it to his other former teacher Thenard. A valuable study

18. A. W. Hofmann, ed., *Aus Justus Liebig's und Friedrich Wöhler's Briefwechsel*, 2 vols. (Braunschweig, 1888), vol. 1, p. 135.
19. *Ibid.*, vol. 1, p. 146.
20. M. and W. Schneider, eds. *Justus von Liebig Briefe an Vieweg* (Braunschweig, 1986), pp. 92–94.

has yet to be made of the dedications (and motives for them) to be found in the many different editions and translations of Liebig's works.

On his return journey from England to Giessen in October 1837, Liebig spent seventeen profitable days in Paris, during which he and Dumas patched up their quarrels and agreed to form a common alliance on organic chemistry.[21] Historians have, however, previously overlooked the fact that, as Liebig informed Berzelius at the time, the collaboration was to extend to a book on the present state of organic chemistry. Although Liebig and Dumas soon fell out, as late as August 1840 Liebig saw the French version of *Agricultural Chemistry* as his contribution to the Dumas alliance.[22] Moreover, as the German dedication to Humboldt made explicit, Liebig gave his German readers to understand that the BAAS had asked both him and Dumas to report on the state of organic chemistry.

> At a meeting in Liverpool in 1837 the British Association gave me the noble task of preparing a report on the state of knowledge of organic chemistry. At my request the Association decided to ask Herr Dumas in Paris, who is a member of the Academy, to do the report together with me. This was the reason for the publication of the present work.[23]

Needless to say, this statement is neither made in the English edition nor is it referred to in any relevant report of the BAAS. In fact, there was room for doubt whether *Agricultural Chemistry* was quite what the BAAS had had in mind in 1837.

As correspondence with William Gregory shows, realizing that he had not honoured his commitment to the BAAS and weary with two years of literary creativity, Liebig had no inclination to submit reports on isomerism and organic chemistry. He therefore got Gregory to persuade the BAAS committee that, although published commercially as a book, *Agricultural Chemistry* really did represent the promised report that it had commissioned. Gregory then took it upon himself eloquently, if loquaciously, to persuade the BAAS to endorse Liebig and the contribution to science that he had made. There is even an implication in Gregory's letters that Liebig had hoped for a monetary reward.[24]

Gregory's gushing and profuse eagerness to please must have struck J. F. W. Johnson, a pupil of Berzelius, a founding member of BAAS, a leading

21. J. B. Dumas and J. Liebig, "Note sur l'état actuel de la chimie organique," *Comptes rendus* 5(1837), 569. See Chapter 3 in this book.
22. Liebig to Berzelius 26 November 1837, J. Carrière, ed., *Berzelius und Liebig ihre Briefe* (München, 1898), p. 134.
23. J. Liebig, *Die organische Chemie* (Braunschweig, 1840), unpaginated dedication.
24. See W. H. Brock and S. Stark, "Liebig, Gregory and the British Association, 1837–1842" *Ambix* 37(1990),134–47.

light of the Chemical Section B, and a significant figure in British agricultural circles. It was, in fact, Johnston who supplied the BAAS with its solicited report on isomerism.[25] Johnston was to become a powerful opponent of Liebig in both the squabble over protein in the mid-1840s and the great controversy between Liebig and Lawes and Gilbert over nitrogen fixation in the 1850s. Curiously, despite Gregory's statement to Liebig that a vote of thanks was to be given to Liebig, no record of this was published by the BAAS. Nor was an abstract of *Agricultural Chemistry*, which Graham reported to the association in 1840, ever printed. Even more curiously, the *Reports* for 1840 and 1841 continued to state that Liebig's reports were still awaited. Only in 1842, when Liebig published his sequel on animal chemistry, was *Agricultural Chemistry* read into the association's official record. On that occasion Playfair (and not Gregory, its translator) delivered an "abstract" of the work's contents at the Manchester meeting. Playfair, in his closing remarks, carefully referred to Liebig as "benefactor to his species, for the interesting discoveries in agriculture published by him in the first part of his report."[26] No doubt, despite Gregory's best efforts in 1840, it still rankled with some members of the association that Liebig had never formally honoured his commitment. If so, Playfair's solution was to abstract *Animal Chemistry* and claim that it was the second part of Liebig's extended report to the BAAS. Liebig's *ad hominem* dedication in the English *Animal Chemistry* was to support this pretence. As a letter from Gregory to Liebig in May 1842 makes clear, by then Liebig no longer saw the BAAS as useful to his aims and he directed Gregory not to dedicate the new volume to the organization.

Astutely, Gregory suggested instead that the book should be dedicated to the 1842 president of the association, Lord Francis Egerton. Gregory must have been particularly keen on this suggestion because Egerton was rector of King's College, Aberdeen, where Gregory was professor of chemistry. In view of Egerton's enormous wealth – it is said he was worth £90,000 per annum – Gregory hoped that such a dedication, together with his own plea for the endowment of Giessen type teaching laboratories throughout the United Kingdom, might touch this magnanimous aristocrat.[27] It is unfortunate that Liebig's side of this correspondence has not survived, because we do not know why he changed his mind between writing to Gregory on 15 May 1842 and penning the actual dedication on

25. J. F. W. Johnston, "On Dimorphous Bodies," *BAAS Reports* (1837), 173.
26. *BAAS Reports* (1842), 42–54.
27. W. Gregory, *Letters to the Rt. Hon. George, Earl of Aberdeen on the State of Chemistry in the United Kingdom* (London, 1842).

3 June 1842. *Animal Chemistry* was duly published on 18 June 1842, five days before the Manchester meeting of the BAAS began, with a three-page dedication to the association that included a handsome acknowledgement to Gregory's abilities as a translator. The dedication of the German edition to Berzelius also proved troublesome, as Liebig's correspondence with Berzelius and Wöhler shows (see Chapter 7).

Why had Liebig vacillated? We can only speculate. But it seems quite probable that he had hoped the 1842 meeting would be held, as was usual, in September when he would have been able to attend in person. As a letter from Liebig to Playfair suggests, he was annoyed that the meeting was to be held in June when teaching commitments at Giessen prevented his attendance and *Animal Chemistry* was still in press.[28] He had learned the reason for this from Playfair, namely that it was so that Egerton would be free to go on a spa tour in September. Surely this would have confirmed his feeling that the British Association was an organization of dilettantes and an unworthy a body to receive a second dedication from him. But then there was Playfair's promise to consider – that when Liebig came to England in September, Playfair would take him on a triumphal tour to meet the "gentlemen of science" who ran the BAAS. Was it simply this that prompted him to dedicate the book to the British Association after all?

In his dedication of *Agricultural Chemistry* to the BAAS, Liebig played chemical gatekeeper and described his book as a state of the art account of how chemistry, particularly organic chemistry, when applied to agriculture and physiology, led to an understanding of the causes of fermentation, decay, and putrefaction, the conversion of wood into coal; and "the nature of poisons, contagions and miasms and the causes of their action on the living organism" (p. iv).

> Perfect Agriculture is the true foundation of all trade and industry – it is the foundation of the riches of states. But a rational system of Agriculture cannot be formed without the application of scientific principles; for such a system must be based on an exact acquaintance with the means of nutrition of vegetables, and with the influence of soils and action of manures upon them. This knowledge we must seek from chemistry, which teaches the mode of investigating the composition and of studying the characters of the different substances from which plants derive their nourishment. (p. v)

Chemistry revealed that plants took nourishment only from inorganic substances, whereas vegetables formed the primary nutrients of even carnivorous animals.

28. T. W. Reid, *Memoirs and Correspondence of Sir Lyon Playfair* (London, 1899), p. 49.

The book was in two parts. In the first, Liebig examined plant nutrients and their metamorphoses within plants; in the second, which redeveloped an *Annalen der Chemie* essay of 1839 on fermentation, he analyzed plant decay and developed a chemical theory of putrefaction that he applied to the understanding of human disease. Insofar as this aspect of the treatise was ignored by the majority of English and American agriculturists, though not by doctors, it is best considered in the context of animal chemistry in the following chapter. It is worth stressing, however, that although historians commonly refer to *Agricultural Chemistry* and *Animal Chemistry* as if the two books were concerned with two distinct sciences, in fact, as the reference to physiology in its full title suggests, *Agricultural Chemistry* was an integrated treatment of chemical transformations within living systems, plants, and animals. We might go so far as to guess that initially Liebig was solely concerned to establish the sources and assimilation processes of food in plants and animals and the implications for healthy plants, beasts, and men. Faced by controversy over one aspect of his theoretical treatment, the mineral theory, however, Liebig found he had to take on the farming lobby in a seriously committed way.

The first chapter of Part I contained basic accounts of carbon, hydrogen, oxygen, and nitrogen, the latter being hallmarked by its inertness and its being "subject to the control of the vital powers" until death (p. 5), when it was assisted in the process of decay by resuming its inert character. It followed that a developing plant needed these elements, together with a soil that furnished other "inorganic matters which are likewise essential to vegetable life" (p. 7). It might be said that Liebig's theory adapted the ancient Greek model of the four elements: earth, air, fire (the sun), and water. Liebig established that atmosphere contained nitrogen, oxygen, carbon dioxide, water, and, in his opinion, ammonia that had been released into the air from decaying animal and vegetable matter. Admitting that ammonia could not be detected by ordinary eudiometric analysis, he pointed to its presence in rainwater (p. 9).

In the second chapter, he turned to soils and to a refutation of the humus theory, which as Playfair, his translator, wrote in a footnote, applied more to German than to English botanists and physiologists. Liebig's refutation consisted in a demonstration that humus was not absorbed by plant roots. Any appeal to solubility via a lime salt of humic acid was ruled out because there was insufficient lime in plants to saturate the humic acid present in humus, and by calculation there was insufficient rainfall to absorb the humate of lime to account for the total quantity of carbon found in plants within a certain acreage. Because soils did not yield the carbon of plants, or enough of it, "it can only be extracted from the atmosphere" (p. 20), otherwise, where had the first plants obtained their humus? Moreover, if millions of people and animals had been consuming oxygen for thousands of years, how could the quantity of oxygen in the air have remained constant at 21 percent unless it was replenished? These rhetorical questions, which were hammered home with further inventive calculations, had the function of driving the reader towards the concept of chemical cycles. For what happened to the billions of cubic feet of carbon dioxide produced in respiration?

For an answer, Liebig turned to Saussure's demonstration that the quantities of oxygen and carbon dioxide in air remained in a fixed relationship, namely, that for every volume of carbon dioxide absorbed by plants, an equivalent volume of oxygen was released. The fact that animals easily assimilate aliments such as sugars, starch, and gum (later he was to add protein), which were themselves products of animals, had misled physiologists into thinking vegetables must also feed on complex (i.e., organic) materials found in humus. Analogy was always a fertile source of error, he commented (p. 30). The reason the outstanding work of Priestley, Senebier, Ingenhausz, and Saussure had been ignored, Liebig thought, was that botanists and physiologists (the principal scientific writers on agricultural matters) had been far too concerned with morphology (form and function) to the neglect of plant chemistry. Admittedly, chemistry "in its pursuit of knowledge . . . destroys the subjects of its investigations" (p. 38). But dismemberment was an essential preliminary to the understanding of form, structure, and function. Their second error was not to experiment, "an art which can be learned accurately only in the chemical laboratory" (p. 39). These were much exaggerated criticisms, and Germany's leading academic botanist Mathias Schleiden, the promulgator of the cell theory, took umbrage at these accusations.[29]

Promulgation of photosynthesis and more evidence for the inadequacy of the humus theory followed in the third chapter. Anticipating a longer discussion in Part II, Liebig first declared that plant decay was a slow process of combustion, which he proposed to call *eremaucausis* (Greek, burning by degrees). This was easily demonstrated with rotting wood, which evolved carbon dioxide in the presence of oxygen, the decay ceasing if the oxygen was removed. From this it could be concluded that the role of wood humus was to bathe seeds and young plants in carbon dioxide.

29. Schleiden replied with *Herr Dr J. Liebig und die Pflanzenphysiologie* (Leipzig, 1842), as did Hugo von Mohl in *Dr. J. Liebig's Verhältnis zur Pflanzenphysiologie* (Tübingen, 1843). For a fairer picture of contemporary botany, see Reinhard Low, *Pflanzenchemie zwischen Lavoisier und Liebig* (München, 1977).

(The fact that this seemed to contradict his previous dismissal of humus – from where did the first young plants gain their nourishment? – went unremarked.)

Although Liebig frequently used vitalistic language when writing about the metamorphosis of plant foods into the plant's fabric (as when he referred to nitrogen's properties under the control of vital powers), he was in no doubt in 1840 that metamorphosis was an ordinary chemical transformation.

> Whatever we regard as the cause of these transformations, whether the Vital Principle, Increase of Temperature, Light, Galvanism, or any other influence, the act of transformation is a purely chemical process. (p. 52)
>
> We should not permit ourselves to be withheld by the idea of a *vital principle* from considering in a chemical point of view the process of the transformation of the food, and its assimilation by the various organs. This is the more necessary, as the views, hitherto held, have produced no results, and are quite incapable of useful application (p. 56)

Liebig's preferred explanation was that chemical affinities, or attractive forces (or bonding, as we would say today), were easily disturbed. Because chemists had been able to prepare *in vitro* many organic compounds found in living systems, why suppose that anything different happened within living subjects? A plant's waste products of chemical transformations (degradations) were returned to the soil where, by undergoing further decay, they formed renewing sources of nutrition, or humus. Left to themselves, plants did not exhaust a soil but formed a perpetual cycle of decay and renewal.

Hydrogen was assimilated by plants through water and its internal decomposition, which Liebig represented by nonsymbolic equations:

36 eq. carbonic acid	+ 36 eq. hydrogen derived from 36 equivalents water	= sugar, with the separation of 72 equivalents oxygen

Liebig was careful to warn that such balance-sheet equations should not be taken in any absolute sense, for "we do not know in what form the production of these constituents takes place" (p. 67). Much of twentieth-century biochemistry has been a successful attempt to explore the remarkable pathways and intermediates involved in such syntheses and degradations. Decay represented the opposite of such a sugar synthesis.

> In the process of putrefaction, a quantity of water, exactly corresponding to that of hydrogen, is again formed by the extraction of

> oxygen from the air; while all the oxygen of the organic matter is returned to the atmosphere in the form of carbonic acid. (p. 68)

Although expressed in a very individual manner, there was little new in Liebig's first four chapters. Originality came in the fifth chapter on the assimilation of nitrogen. Because there was no evidence in Liebig's view that plants absorbed nitrogen directly, as they did carbon dioxide, and because rainwater had been found to contain dissolved ammonia, rainwater was the most likely source. Moreover, ammonia was a recognized product of putrefaction and capable of chemical transformations into definite organic compounds such as urea. Argument by elimination also suggested this. Because farmers removed crops for money, there was in principle a net removal of nitrogen from the land each year:

> Yet, after a given number of years, the quantity of nitrogen will be found to have increased. Whence, we may ask, comes this increase of nitrogen? The nitrogen in the excrements cannot reproduce itself, and the earth cannot yield it. Plants, and consequently animals, must, therefore, derive their nitrogen from the atmosphere. (p. 72)

As we have seen, Boussingault, who was not mentioned, had already shown this conclusion to be simplistic, and it proved to be one of Liebig's most contentious, yet stimulating, conclusions.

Using some rather dodgy calculations again, Liebig had no difficulty in demonstrating that, although the amount of ammonia in air was undetectable by ordinary analysis, even if a pound of rainwater contained only a quarter of a grain of ammonia, a field of average size would receive upwards of 88 pounds of ammonia a year. Reporting his own experiments on distilling rainwater, Liebig confirmed ammonia's presence, and it had been confirmed by others by the time the second edition appeared. On the other hand, in the third edition (1843), he was to withdraw support for some experiments made by Edward Lukas in Munich in 1837. Lukas had claimed that plants could be grown on charcoal. More careful research showed that it had to be animal charcoal (rich in phosphates), and in 1851, Liebig's Scots pupil Thomas Anderson showed that charcoal did not absorb ammonia at all.

Liebig noted also how commonly ammonia was produced when flowers, herbs, and roots were distilled to extract medicinal ingredients, and in the manufacture of sugar from beets. The function of animal manure in promoting crop growth was, Liebig therefore concluded, solely to effect nitrogen assimilation through the absorption of ammonia following putrefaction. This accounted for the successful use of putrid urine as a fertilizer in many countries, for it contained more nitrogen than the solid excrements of animals and men.

> In a scientific point of view, it should be the care of the agriculturist so to employ all the substances containing a large proportion of nitrogen which his farm affords in the form of animal excrements, that they shall serve as nutriments to his own plants. (p. 83)

This was one of Liebig's few references to actual farming practice.

The successful use of gypsum (calcium sulphate) or calcium chloride of burned clay (marl) as fertilizers could now be explained as due to their ability to fix atmospheric ammonia into the soil where it might otherwise have evaporated; ammonia salts were key fertilizing agents. This is a point worth emphasizing in view of Liebig's later controversy with Lawes, for many people interpreted Liebig as saying that additional nitrogen manuring was superfluous. Indeed, he was to maintain this view between 1843 and 1856, though he tried to wriggle out of it. But what Liebig seems to have meant in the first edition was that nitrogen dressings only accelerated a natural process and were by implication a waste of a landowner's money unless he used farmyard manure. When farmers later pointed out that gypsum was effective only on clover and rye, Liebig explained it as due to an excessively dry atmosphere that prevented gypsum's reaction with ammonia.

Liebig then turned in Chapter 6 to the origin of the inorganic constituents of plants, as revealed in ash analysis. Plant roots absorbed minerals indiscriminately like a sponge and excreted those that were not required for assimilation. In 1832, Macaire-Princep had claimed that unwanted nutrients were exuded by the roots.[30] Rotation of crops, he deduced, was beneficial precisely because such "poisonous" excretions were essential nutrients for other plants. When other botanists were unable to confirm Macaire-Princep's experiments, Liebig was forced to allow plant roots a power of selection. In the 1850s, English agricultural chemists demonstrated that the power was in fact a function of the soil, which preferred ammonia or potash to other bases present and ignored free acids, allowing these to combine with limestone to form soluble salts.[31]

Minerals neutralized the many acids synthesized by plants, such as the tartaric acid found in grapes. But because Dalton's, and more recently in 1837 Graham's chemical investigations revealed that bases combined with acids in fixed proportions, the availability of minerals for the proper growth of plants "ought to be considered with the strictest attention both by the agriculturist and physiologist" (p. 94). Liebig then reinterpreted Saussure's percentage ash analyses on the different proportions of lime

30. J. Macaire-Princep, "Mémoire pour servir à la histoire des assolements," *Ann. Chem.* 8 (1833), 78–92.

31. Thomas Way, "On the power of soils to absorb manure," *J. Roy. Agri. Soc. of England* 11(1850), 313–79.

and potash found in plants growing on very different soils as showing, nonetheless, that they contained constant proportions of oxygen in their potash of lime. In other words, the quantities of alkaline bases or other equivalents absorbed by plants were invariable because they were required to saturate acids containing definite proportions of oxygen. "The progress of a plant must be wholly arrested when none are present" (p. 98). Liebig failed, however, to note that Sprengel had already reached this preliminary version of the law of the minima.[32]

The final three chapters of Part I discussed the implications of the foregoing for "the art of culture," – the rotation of crops and manuring. If natural philosophers could discover the conditions that promoted the development of plants, they would then be able to influence the plants' size and internal constituents.

> There is no profession which can be compared in importance with that of agriculture, for to it belongs the production of good for man and animals; on it depends the weight and development of the whole human species, the riches of states, and all commerce. There is no other profession in which the application of correct principles is productive of more beneficial efforts, or is of greater and more detailed influence. (p. 131)

Liebig clearly joined the hopes and aspirations of High Farming.

Why did methods of cultivation vary from country to country and district to district? The usual answer was that conditions depended on local circumstances like soil conditions. If so, retorted Liebig, why not remove surmise and ignorance by the investigation of the nature of these circumstances? Why not find out, for example, why manures work by investigating the nature of "the incomprehensible something which assists in the nutrition of plants, and increases their size?" (p. 132). Why do some plants grow on some soils and not on others? It must surely be because certain substances are present or absent in the soil or are supplied or not supplied by manuring. No rational system of agriculture was possible, he urged, unless such basic questions were investigated.

> The power and knowledge of the physiologist, or the agriculturist and chemist, must be united for the complete solution of these questions; and in order to attain this end, a commencement must be made. (p. 132)

In a rational agriculture we would know the appropriate nutriments for a specific plant or crop and be able to adjust the conditions of growth in

32. U. Schling-Brodersen, *Entwicklung und Institutionalisierung der Agrikulturchemie im 19. Jahrhundert* (Stuttgart, 1989).

order to obtain an "abnormal" development of those parts of the plant required either for food or for industry, just like the breeding of animals for haulage, food, or hunting. If all the alkaline salts were removed from a soil without replenishment, as had happened with the first colonists in Virginia, barrenness resulted. The rotations practised by Europeans, particularly in allowing cultivated fields to lie fallow every three years, permitted alkali within the soil to be set free in a soluble condition and to restore fertility. Careful tilling of such fallow land accelerated subsoil disintegrations. Even so, Liebig declared, "the fertility of a soil [for agricultural purposes] cannot remain unimpaired, unless we replace in it all those substances of which it has been . . . deprived" (p. 163).

This was the function of manures, which he divided into animal, urine, night soil, guano, and artificial. Because every constituent of man and animals was derived from plants, "all the inorganic constituents of the animal organism must be regarded . . . as manure" (p. 164). Analyses of dog, cow, and horse dung revealed only a minute nitrogen content, confirming that such excrements could not be a principal source of nitrogen for plants; but they did contain large proportions of calcium, magnesium, and sodium phosphates excreted from the animals' bones, as well as potassium silicate. Fields could be kept in a constant state of fertility only by replacing the minerals removed in cropping. To increase fertility, farmers should simply add more than they took away. Since it was a matter of indifference to plants what manures were used, assuming the appropriate elements were provided, this explained why the ashes of wood or the bones of animals were used successfully in Flanders and in Liebig's own Hessen-Darmstadt. Bone manure was a rich source of calcium and magnesium phosphates. Liebig advised, however, that for maximum efficiency bones needed to be very finely ground.

> The most easy and practical mode of effecting their division is to pour over the bones, in a state of fine powder, half of their weight of sulphuric acid diluted with three or four parts of water, and after they have been digested for some time, to add one hundred parts of water, and sprinkle this mixture over the field before the plough. In a few seconds, the free acids unite with the bases contained in the earth [i.e., ammonia], and a neutral salt is formed in a very fine state of division. (p. 174)

Liebig noted that such valuable materials (superphosphates dissolved in hydrochloric acid) were being thrown away by the ton at glue works. His translator Playfair observed that the same was true of the bran used in calico print works that contained calcium and sodium phosphates. Here lay the foundations of the superphosphate fertilizer industry, which the young aristocrat Sir John Bennet Lawes of Rothamsted was to patent in May 1842 and to begin to manufacture at Deptford in 1843.[33] James Murray (1788–1871), an Irish doctor and agriculturist who filed a virtually identical patent the same day, had used "vitriolised bones" on his Irish farms for thirty years. Both patents were granted, though Lawes subsequently (1846) bought the rights to Murray's process. Lawes's initial feedstocks were imported bones, complemented, from 1843, by phosphates imported from Spain. In the 1820s, the geologist William Buckland had identified coprolites in Gloucestershire as fossilized saurian dung and was advised by Liebig, during the 1842 tour of England, that if large deposits of it were found, they would make a source of valuable fertilizer.[34] Huge deposits were identified by the botanist John Henslow on the Cambridgeshire–Suffolk border during the 1840s. In 1847, therefore, Lawes switched to them as a cheaper source of phosphates. Coprolites were mined in East Anglia until the end of the century, when Lawes's company and others switched again to high-grade phosphate rock (appatite) from Norway.[35] By the beginning of the twentieth century, the medical officer of health for Rotherhithe, where the Lawes Manure Company had its works, could complain:

> In the mile length of Rotherhithe Street there are no less than nine factories for the fabrication of patent manure, that is to say, nine sources of foetid gases. The process gives out a stench which has occasioned headache, nausea, vomiting, cough, &c. Many complaints have been made by the inhabitants.[36]

In the 1850s Liebig might say dismissively that "the suggestion to dissolve bones in sulphuric acid . . . is from a scientific point of view of no greater importance than a useful recipe for boot polish."[37] But a decade later he was asserting that Lawes must have first gleaned the idea of superphosphates from *Agricultural Chemistry* in 1840. Yet there is documentary evidence that Lawes's experiments date back to 1839 when a

33. P. J. T. Morris and C. A. Russell, eds., *Archives of the British Chemical Industry 1750–1914* (British Society History of Science, 1988), pp. 124–5. See also a pamphlet by Lawes, *On Super-Phosphate of Lime* (circa 1856) in Rothamsted Archives, Letters, vol. 2 (1850–67).
34. W. Buckland, "On the causes of the general presence of phosphates in the strata of the earth," *J. Roy. Agri. Soc. of England* 10(1839), 520–25; Richard Grove, *The Cambridgeshire Coprolites Mining Rush* (Cambridge, 1976).
35. D. F. W. Hardie and J. D. Pratt, *A History of Modern Chemical Industry* (Oxford, 1966), pp. 48–53.
36. Henry Jephson, *The Sanitary Evolution of London* (London, 1907), p. 114.
37. Liebig, *Chemische Briefe,* 3rd ed. (Heidelberg, 1851), p. 643.

London friend asked him whether he could devise a use for an animal charcoal waste product. Lawes then discovered that bone charcoal proved an even better fertilizer when dissolved in sulphuric acid. Aware of Liebig's claims, Gilbert and Hofmann had a long, involved correspondence over the latter's report on artificial fertilizers for the 1862 Exhibition in London. Hofmann persisted loyally in maintaining that Liebig had been the first to invent superphosphate.[38]

In the same chapter on manures, Liebig also pointed out that although the excrements of animals contained relatively little nitrogen, human faeces and urine contained appreciable amounts. (He cited Berzelius's and Playfair's analyses of up to 5 percent.)

> In the faeces of the inhabitants of towns, for example, who feed on animal matter, there is much more of this constituent than in those of peasants, or of such people who reside in the country. (p. 178)

Playfair added notes on Chinese husbandry to the second English edition, and Liebig was to return to this subject in later editions. At the time of writing, guano (ammonium urate in bird droppings), which was used by South American coastal farmers as a fertilizer, had just begun to be imported into Liverpool. Liebig published an analysis by Voelckel in the *Annalen der Chemie* in 1841 but offered only cautious recommendation for it as a manure in the first two editions of *Agricultural Chemistry.* Later he was to view its importation as a form of cultural imperialism, while simultaneously laying claim to being the catalyst for its wide use. The latter claim was quashed by Gilbert.

Liebig finally turned to artificial (synthetic) manures, such as the ammoniacal waste from town gas works. Unfortunately, this waste contained ammonium sulphide, a plant poison, which had to be removed by precipitating it as ammonium sulphide with green vitriol (ferrous sulphate). He noted that his English friend Charles Daubeny had also recommended that it be mixed with gypsum to prevent loss of ammonia. Another English agriculturist Philip Pusey had recommended the use of sodium nitrate, though Liebig here doubted whether it acted solely "by virtue of the nitrogen which enters into its composition" (p. 197), since it seemed not to be uniform in its effects. In other words, it might be working because of its sodium content where fields were alkali deficient.

In the second edition of 1842, Liebig added a long chapter on the chemical constituents of soils that contained little not previously stated but that did include forty-seven analyses made or reported by Karl

38. Rothamsted archives, *Papers and Correspondence 1860–67*, vol. 5. A. D. Hall, *The Book of Rothamsted Experiments* (London, 1905), p. xxiii.

Sprengel. Liebig noted that it was not necessary to describe how such analyses were made (how different his attitude was from that of Davy), "for this kind of research will never be made by farmers, who must apply to the professional chemist, if they wish for information regarding the composition of their soils" (p. 208). It was only here that Liebig acknowledged Sprengel's priority in advocating what was to become known as "the mineral theory." Dr Sprengel was, said Liebig, "a chemist who has unceasingly occupied himself for the last twenty years in endeavouring to point out the importance of the inorganic ingredients of a soil for the development of plants cultivated upon it" (p. 243).

Despite this acknowledgement, in Britain and America at least, the mineral theory was perceived as Liebig's insight. (It is intriguing to recall that Liebig's spurious doctoral thesis of 1823 had been titled "On the Relation Between Mineral and Plant Chemistry.") The mineral theory may be summarized as claiming that the principal nutrients of plants are, besides carbon, hydrogen, and nitrogen, a clutch of essential minerals, including sulphur, phosphorus, iron, calcium, magnesium, and silicon, and in some plants, sodium and potassium. By ensuring that such ingredients were present in a soil or added by manuring, rotations could be abolished and land use and yields increased and maximised. Liebig's claim coincided completely with the aspirations of High Farming; yet it was found wanting in ignoring or underplaying the role of large doses of nitrogen.

Agricultural Chemistry, despite extravagant praise from reviewers who were often Liebig's pupils, cannot have been an easy book to read. Designed initially to promote chemistry's status, it succeeded brilliantly, if dauntingly, in underlining the subject's potential political and economic significance. Excessively repetitious and digressive, demanding some chemical knowledge (in particular, an ability to think in terms of atoms, molecules, and their rearrangements), demanding acceptance of the premise that laboratory experimentation explained and rationalised practice, and mixing ideas and knowledge of botany, agriculture, physiology, chemistry, and medicine together indiscriminately, the book was as it stood far too theoretical to attract the practical farmer, whereas doctors and sanitarians probably needed the reinforcement of *Animal Chemistry* in 1842 to grasp the full implications and possibilities of Liebig's message about disease. As Liebig told his publisher Vieweg in June 1840, "The book is not a guide to practical agriculture."[39] This did not prevent

39. Margarete and Wolfgang Schneider, eds., *Justus von Liebig, Briefe an Vieweg* (Braunschweig/Wiesbaden, 1986), p. 98.

farmers and landowners from "trying both to implement and verify Liebig's theories at the same time," with often disastrous results for Liebig's reputation.[40] It was as if the ideal, abstract world of mathematical points and circles were identified with the real world of friction.

It was also a much less finished book than Boussingault's *Economie rurale,* and this was reflected in Liebig's need to revise it eight times in German, quite apart from the specific local alterations he made in English and French translations. Although multiple editions and translations reflected Liebig's changing views, they helped breed hostility, because critics were frequently confused by Liebig's ability to outsmart them with texts. Much of the prolonged dispute of accusations and counteraccusations, quotations and counterquotations between Liebig and Lawes and Liebig and Gilbert hinged upon confusion between editions and subtle changes of wording and meaning, or failures to specify the edition used – "mutations of originals," as Gilbert called them.[41] A magnificent analysis of this verbal warfare has been made by Vance Hall but unfortunately remains unpublished.

The single most important change occurred in the third edition of 1843 when Liebig appeared to plump completely for the mineral theory and to dismiss the need for additional ammonia or nitrate fertilizers on the grounds that a plant's mechanism for obtaining atmospheric nitrogen was completely adequate. The key passages of the English text, as noted by Lawes in 1847, were:

1840 (1st ed.), p. 85	*1843 (3rd ed.), p. 54*
Cultivated plants receive the same quantity of nitrogen from the atmosphere as trees, shrubs, and other wild plants; and this is *not sufficient* for the purposes of agriculture.	Cultivated plants receive the same quantity of nitrogen from the atmosphere as trees, and other wild plants; and this is *quite sufficient* for the purposes of agriculture.
Agriculture differs essentially from the cultivation of forests, inasmuch as its principal object consists in the production *of nitrogen under any form capable of assimilation;* whilst the object of forest culture is confined principally to the production of carbon.	Agriculture differs essentially from the cultivation of forests inasmuch as its principal object consists in the production *of the constituents of the blood;* whilst the object of forest culture is confined principally to the production of carbon.

40. Rossiter, *Emergence of Agricultural Science,* p. 30.
41. Gilbert to H. J. Thompson, 31 October 1863, Rothamsted archive.

> *But the presence of ammonia does not suffice for the production of nitrogeneous ingredients. Other conditions* [viz, minerals] *likewise are essential.*

Two things had caused Liebig to change his mind. In the first place, the agricultural tour of England and Scotland in 1842 led him to conclude erroneously that what British soils most lacked were minerals, not ammonia. This conclusion was then applied to the whole of less intensively farmed European agriculture. But as Lawes expostulated in 1847:

> Many of the errors into which Liebig has fallen, have, I think, arisen from his not sufficiently considering what agriculture really is. Practical agriculture consists in the artificial accumulation of certain constituents to be employed either as food for man or other animals, upon a space of ground incapable of supporting them in its natural state. This definition of agriculture is, I think, important, as distinguishing English agriculture at least from the system pursued in various parts of the world, where the population is small and the land of little value. . . . If Liebig had sufficiently considered the distinction he would not have *assumed* [my italics] that certain [nitrogeneous] substances employed as manures are of little value, because plants and trees, in their *natural* state, are incapable of obtaining them in sufficient quantity for their use.[42]

The state of British agriculture was not the sole reason for the change, however. More important was that since 1840 Liebig's prime research interest had become animal chemistry. As we shall see in the following chapter, research at Giessen implied that plants contained the protein molecule that formed the basis of animal blood and muscle. Animal protein also contained phosphorus and sulphur and perhaps other inorganic minerals which, it was deduced, must also have originated from the assimilation of plants by animals. Since such plant minerals must have come from the soil, their significance in husbandry had to be underlined.

Although this mineralogical view, which had such powerful significance for rational agriculture, was perfectly correct, by contextualising it in 1843 with a dismissal of nitrogen fertilizers, Liebig laid himself open to attack and ridicule by practising farmers. Stubbornly refusing to accept Boussingault's as well as Lawes's and Gilbert's empirical evidence to the contrary, it was another twenty years before Liebig was to concede the value of additional nitrogen fertilizer. Perhaps he thought that a conces-

42. J. B. Lawes, "On some points connected with agricultural chemistry," *J. Roy. Agri. Soc. of England* 8(1847), 227–28.

sion on this point would rekindle the notion of humus? What appalled Lawes and Gilbert was that Liebig then made the concession by sleight of words – that the mineral theory had *always* incorporated ammonia fertilizers because, as any chemist knew, ammonia was a mineral, not an organic compound.[43]

Berzelius, who was still in 1840 regarded by Liebig as a father figure, had been sent a copy of *Agricultural Chemistry* on publication, and Liebig waited anxiously for his reaction. He was genuinely shocked by the Swede's accusation that Liebig's account was a multicoloured soap bubble of speculation. Even when Liebig was able to provide Berzelius with empirical evidence that plants could be grown on a diet of ammonium sulphate, Berzelius insisted that humus must nevertheless supply other organic nutrients.[44]

In the United States Liebig's *Agricultural Chemistry* was described as a "thunderbolt." Margaret Rossiter, in a detailed portrayal of the reception of Liebig's work and of the activities of his American pupils, concluded that Americans were enthusiastic because they "did know so much about agricultural chemistry. [Hence] they recognized and appreciated the breakthrough he had made. To the farmer and layman, the book seemed to explain for the first time the success of certain empirical farming practices."[45] Americans also had the advantage over Europeans in that the third, altered, edition of Liebig's book was never printed in the States. Even so the usual doubts remained.

> Mr Justus Liebig is no doubt a very clever gentleman and a most profound chemist, but in our opinion he knows as much of agriculture as the horse that ploughs the ground, and there is not an old man that stands between the stilts of a plough in Virginia, that cannot tell him of facts totally at variance with his finest spun theories.[46]

43. For a full discussion, see Liebig's papers "On some points in agricultural chemistry," *J. Roy. Agri. Soc. of England* 17(1856), 284–326, and "Mr J. B. Lawes and the mineral theory," *ibid.*, 25(1864), 502–8, which answer Lawes and Gilbert "On some points connected with agricultural chemistry," *ibid.*, 16(1855), 411–98, and "Further report of experiments with different manures on permanent meadow land," *ibid.*, 24(1863), 504–9.
44. "Even living bodies themselves cannot produce a great portion of materials out of purely inorganic substances but require in addition the products of other living bodies as material for their processes. So, for example, the plants of one season live from the remains of the preceding one." J. J. Berzelius, *Lehrbuch der Chemie*, vol. 6 (Leipzig, 1837), pp. 3–4.
45. Rossiter, *Emergence of Agricultural Science*, p. 11.
46. *Southern Planter* 5(1845), 23, as quoted by Albert L. Demaree, *The American Agricultural Press 1819–1860* (New York, 1941).

In France, Liebig was immediately embroiled in a bitter priority dispute with Dumas and Boussingault. This quarrel is best discussed, however, in the context of animal chemistry in the next chapter. We should note, though, that in 1853 Boussingault showed that Liebig's estimation of the amount of ammonia in rain water was fallacious, a result that was confirmed by George Ville's daily readings for a whole year.[47] This French work stimulated Lawes and Gilbert to demonstrate that the amount of atmospheric ammonia that found its way into the soil was well below the amounts of nitrogen to be found in crops.[48] Thus, by the mid-1850s, although Liebig's nitrogen cycle was not doubted, his mechanism for nitrogen absorption was queried if not completely scotched.

In view of his contribution to agricultural chemistry, it is surprising to find that in Germany Liebig was scarcely involved at all in the institutionalization of the subject in private and state research institutes of Saxony (e.g. Möckern, Leipzig,1851) and elsewhere, and its eventual appearance as a university discipline at Halle under Julius Kühn in 1861. At the time of Liebig's death in 1873, the united Germany had some twenty-five agricultural research stations, and the movement was to be particularly influential in the United States.[49] Although several members of this growing academic agricultural community were former Liebig students (e.g., Wilhelm Henneberg), Liebig complained that the teachers and students of agriculture were ignorant of scientific chemistry. What he really meant was that, like Lawes and Gilbert in England, they had come to oppose his teachings from a more practical perspective.

As Mark Finlay has shown, farmers, not chemists, tended to have control of these research stations.[50] For his pains in promoting Liebig to farmers in the most practical form, Julius Adolph Stöckhardt (1809–86), who recognized the need for nitrogen fertilizers as well as minerals, was

47. J. B. Boussingault, "Mémoire sur le dosage de l'ammoniaque contenus dans les eaux," *Ann. de chim.* 39(1853), 257–91; G. Ville, *Recherches expérimentales sur la végétation* (Paris, 1853), pp. 1–5.
48. J. B. Lawes and J. H. Gilbert, "Reply to Baron Liebig," *J. Roy. Agri. Soc. of England* 16(1855), 411–98.
49. U. Schling-Brodersen, "Liebig's role in the establishment of agricultural chemistry," *Ambix* 39(1992), 21–31; Rossiter, *Emergence of Agricultural Science.*
50. Mark R. Finlay, "The German Agricultural Experimental Stations and the beginnings of American agricultural research," *Agricultural History* 62(1988), 41–50. Finlay also reveals a family rift insofar as Liebig's youngest son Hermann enthusiastically supported the German agricultural station movement. See Finlay, "Science and practice in German agriculture: Justus von Liebig, Hermann von Liebig, and the agricultural experimental stations" in W. R. Woodward and R. S. Cohen, eds., *World Views and Scientific Discipline Formation* (Dordrecht, 1991), pp. 309–20.

viciously reprimanded by Liebig in his *Über Theorie und Praxis.*[51] Although Theodor Reuning (1807–76), Saxony's agricultural minister and a personal friend of Liebig's, was convinced of the value of basic agricultural research, when Emil von Wolff (1818–96) came to direct the Möckern facility's scientific section in 1851, he complemented Rothamsted's attacks on Liebig. The latter issued a forty-five-page "Supplement" to his *Die Grundsätze der Agricultur-Chemie*[52] attacking Wolff, who continued relentlessly to expose Liebig to further German ridicule in *Die Mineralstoffe und die Stickstoffler in der Landwirtschaft* (Stuttgart 1858) and *Die Wirkung des Dungers und Liebig's neue Behauptungen* (Berlin 1858). To Gilbert, who maintained close contacts with Wolff and other German agriculturists, this was all grist to the Rothamsted mill.

It transpired, therefore, that the institutionalization of agricultural chemistry in Germany owed more to the need to refine Liebig's precepts, models, and methodology than to any direct action by him. In any case, in Munich by the 1850s, Liebig was concerned to promote agricultural chemistry less for its intrinsic value than as a propaganda weapon in his commitment to the social and political relevance of applied science.

In Britain, besides the important propaganda provided by his *Chemical Letters* (Chapter 10, this book) and his visits of 1842 and 1844 (when the banquet was given at Glasgow in gratitude for his contribution to agricultural chemistry), Liebig's ideas were propagated by his pupils of chemistry, especially Playfair and Gregory, and amongst landowners through the country's four leading agricultural societies: the Royal Society of Agriculture, the Yorkshire, the Bath and West, and the Highland and Agricultural Society of Scotland. These societies proved two-edged swords, for which reason Liebig soon found himself embroiled in polemic. Believing passionately that the use of extra nitrogenous dressings was a waste of effort and money and that the mineral theory really would revolutionize agricultural practice, he determined to offer a practical proof with an artificial fertilizer of his own design. If this made money as well, so much the better. As we have seen (Chapter 5), this venture proved disastrous. The mineral theory now literally implied, whatever Liebig was to say later, the use of nonnitrogenous mineral fertilizers and the abandonment of rotation.

The Yorkshire Agricultural Society's (YAS) Committee on Agricultural Geology commented on Liebig's work in 1841. Although sympathetic and excited by Liebig's message that farmers should understand the scientific

51. J. Liebig, *Über Theorie und Praxis in der Landwirtschaft* (Braunschweig, 1856).
52. J. Liebig, *Die Grundsätze der Agricultur-Chemie* (Braunschweig, 1855).

reasons for their actions, like Berzelius, the committee disagreed with Liebig's claim (shared by Daubeny and Johnston) that the carbonaceous matter of manures (as humus) made no contribution to plant growth. This echo of the old humus theory and unfamiliarity with Saussure's work on photosynthesis was to be reinforced on the Continent by the publications of the Dutch chemist Jan Gerrit Mulder (1802–80). Initially much admired by Liebig for his postulation of the protein molecule (Chapter 7), they fell out over both agricultural and animal chemistry during the 1840s. Mulder was to argue passionately for humic acid as a nutrient.[53] It was particularly galling for Liebig when pupils of Mulder's, such as the Scotsman James Weir Johnston and the American Samuel Johnson (who had studied with Liebig at Munich), sided with the Dutch chemist and continued to advocate an organic theory of plant nutrition.[54] Nor was Liebig best pleased to find Pusey praising the work of Dumas and Boussingault.[55] Liebig had been made an honorary member of the Royal Agricultural Society at its Liverpool meeting in 1842. However, the general index to the first twenty-five years of the society's *Journal* is notable for the length of its subheading "Objections" under the entry "Liebig."

Vance Hall, who has paid close attention to the British reception of Liebig's agricultural ideas in provincial newspapers and other publications, concluded that "from the very start there was a body of opinion critical of Liebig's ideas on how chemistry could aid (indeed, revolutionize) farming in England.[56] Some, like Henry Stephen Thompson (1809–74), a founder member of both the YAS and the RASE, began field trials to test Liebig's ideas. By 1845 Thompson had discovered that soils had the property of absorbing alkaline bases. Drainage water from land manured with ammonium and potassium salts contained no (or little) ammonia or potash. This immediately put in doubt the efficacy of the manures patented by Liebig and Muspratt that year. Following trials, Thompson rejected them as worthless in 1847 in the *Transactions* of the YAS. Liebig's mineral theory seemed wanting in both theory and practice. Meanwhile, as Hofmann warned Liebig by letter, the powerful editor of the *Journal* of the RASE, Philip Pusey (1799–1855), was turning the national society

53. H. A. M. Snelders, ed., *The Letters from Gerrit Jan Mulder to Justus Liebig (1838–46)* (Amsterdam, 1986). See also E. Glas, "The Liebig–Mulder controversy," *Janus*, 63(1976), 27–46.
54. Johnston was responsible for the English translation of Mulder's deluded polemic, *Liebig's Question to Mulder Tested by Morality and Science* (Edinburgh, 1846).
55. See Pusey's editorial postscript to Lawes and Gilbert, "On agricultural chemistry . . . ," *J. Roy. Agri. Soc. of England* 12(1851), 1–40 at p. 40.
56. Hall, *Yorkshire Agricultural Society*, p. 68.

against Liebig. As Hall has shown, Thompson was also deeply involved in this anti-Liebig campaign.[57]

Until his death in 1855, Pusey ensured that none of Liebig's papers appeared in the *Journal* of the RASE on the principle that it was a mistake "to suppose that men can be made farmers by teaching them doubtful chemistry."[58] Although under Thompson's editorial control Liebig was allowed to publish three papers in the *Journal,* the tone of the three leading British agricultural societies, the RASE, the YAS, and the Bath and West, remained decidedly pragmatic, anti-Liebig, and pro-Rothamsted. As Liebig came to realize, British agriculture, unlike that on the Continent, was dominated by a clerisy of country gentlemen.

Their combined attitude of contempt for "book learning" was explicitly and forcibly expressed in Pusey's most famous and important article, "On the Progress of Agricultural Knowledge During the Last Eight Years."[59] By then, inspired by Boussingault's approach, a British concordat between land ownership, science, and practice seems to have been achieved, with Pusey, Thompson, Acland, and others coordinating nationwide field research programmes on the use of fertilizers and other aspects of husbandry. For example:

> If the farmer will assist the chemist with carefully conducted trials, the chemist will be enabled to afford such deductions as will render the subject capable of definite explanation. It is the combination of practice with theory, that alone can improve the science of agriculture, and it is impossible that either the chemistry of agriculture, or the practice of the art, can be carried on successfully without combination.[60]

This message had also reached Liebig, who, in 1844, as we have seen, purchased a portion of land in Giessen (later known as Liebig Heights). This is not to be confused with the strip of land, opposite the Giessen laboratory, in which, as he told Berzelius in May 1841, he planted potatoes that flourished on ammonium and magnesium phosphate dressings, asparagus on ammonium carbonate, and wheat on calcium phosphate and potassium silicate.[61] However, Liebig was unable to afford the

57. *Ibid.*, p. 216.
58. Pusey, "On the progress of agricultural knowledge," *J. Roy. Agri. Soc. of England* 11(1850), 381–438 at p. 392.
59. *Ibid.*
60. T. H. Barker, "The chemistry of manures," *Trans. Yorkshire Agri. Soc.* 5(1842), 31–52. Cf. Hall, *Yorkshire Agricultural Society,* p. 71.
61. Liebig to Berzelius, 17 May 1841, in J. Carrière, ed., *Berzelius und Liebig ihre Briefe* (München, 1898), p. 231. The vegetable garden was tended by the laboratory *Familus,* Aubel.

expense of developing the land for experimental purposes, and it was sold within a couple of years.

The epitome of the "English Chemistry of Farming" (as Thomas Dyke Acland dubbed it in 1857[62]) was the experimental farm set up by the baronet Sir John Bennet Lawes (1814–1900) at Rothamsted in Hertfordshire. Lawes, who had attended Daubeny's lectures on chemistry at Oxford, transformed an old barn on his estate into a laboratory. To aid his research, from 1843 he employed one of Liebig's former pupils Joseph Henry Gilbert (1817–1901). The criticisms emanating from Rothamsted were to cause Liebig a good deal of bother. Lawes's attack began in the *Journal* of the RASE in 1847 when he contrasted his own evidence that nitrogeneous manures were essential for a good wheat crop with the fiasco of Liebig's patent manure. Aware, too, that Liebig's views had changed between 1840 and 1843, Lawes "regretted that Liebig should have altered, in the third edition of his work, so many of the views and opinions laid down in the first."[63] Lawes was supported by Pusey in the same issue with the evidence of the value of superphosphate manures. In the next few years the English agricultural community, although recognizing the theoretical importance of Liebig's work, turned to Rothamsted for practical guidance.

By 1856 Liebig's theoretical conviction that the addition of minerals to the soil would expand crop yields had been seriously challenged by Lawes and Gilbert's practical demonstrations at Rothamsted. Their initial scepticism of 1847 was expressed even more forcibly in the *Journal* of the RASE in 1851 and hinged a good deal on disagreement with Liebig's supposition that atmospheric ammonia was sufficient to supply a plant's nitrogen.

The Rothamsted experimental plots had shown the following:

1. Nitrogen fertilizers were by far the most significant form of manure for increasing yields;
2. After nitrogen, phosphates were the most important nutrients;
3. Alkalis and alkaline earths were of much less significance than Liebig had argued.
4. The addition of soluble silicates had no significant effects at all.

In Table 6.1, the keys are as follows: (a) no fertilizer; (b) minerals (P, K, Na, Mg) only; (c) minerals plus 43 lbs N per acre as ammonium sulphate; (d) minerals plus 86 lbs N per acre; (e) minerals plus 129 lbs N per acre; (f) farmyard manure only.[64]

62. Hall, *History of Yorkshire Agricultural Society*, p. 71.
63. J. B. Lawes, "On agricultural chemistry," *J. Roy. Agri. Soc. of England* 8(1847), 226–60 at p. 227.
64. Nutman, "Centenary lecture," p. 53.

Table 6.1. *Grain yields of wheat on Broadmackfield, Rothamsted (in bushels per acre; 1 bushel = 36.37 litres)*

	a	b	c	d	e	f
1844	15					22
1845	23			32		32
1846	18			28		27
1847	17			26		30
1848	15			19		26
1849	19			33		31
1850	16			27		28
1851	16			29		30
1852	14	17	21	27	28	28
1853	6	10	18	24	23	19
1854	21	24	34	45	49	14
1855	17	18	28	37	31	35
1856	15	19	28	37	39	36

As well as demonstrating the value of farmyard manure (f) and the large reserves of nitrogen in the soil (a), the explanation of which had to await soil microbiology, the long-term sequence of experiments, which have continued into the twentieth century, clearly showed the advantages of using nitrogeneous fertilizers. Rothamsted was leading the protesting vanguard of critics. In his presidential address to the BAAS in 1856, even the Liebig stalwart Charles Daubeny was forced to concede:

> The practical question remains whether, allowing for theoretical proof of Baron Liebig's position, a larger expenditure of capital will not be required for bringing a given farm into a condition to dispense with ammoniacal manures than by procuring those materials which contain that ingredient ready for use. And here experimental researches such as those conducted on so extended and liberal a scale by Mr Lawes and Dr Gilbert come in aid of theory. They stand, as it were, midway between the abstract principles which science points out to the farmer, and the traditional usages with respect to his art, which have been handed down to him from one generation to another.[65]

His research on the chemistry of food out of the way, Liebig had time to return to agricultural chemistry in 1851, as the third edition of his *Chemical Letters* shows. Refused by Pusey the right to reply to his English critics

65. C. Daubeny, *BAAS Reports*, 1856, p. lvi.

and to refute Lawes and Gilbert's practical results, Liebig retorted in 1855 with a 100-page polemical monograph, *Die Grundsätze der Agricultur-Chemie*. This was translated as *Principles of Agricultural Chemistry, with special reference to the late Researches in England* (London, 1855) by his faithful henchman William Gregory. (Even Gregory omitted some of the wilder accusations, however, such that Lawes had co-opted Pusey against Liebig in order to boost the sales of his own superphosphates against Liebig's artificial manures.[66])

Not to be worsted and freely able to use the RASE's *Journal* as their mouthpiece, Lawes and Gilbert replied with a long refutation the same year, but it was not published until January 1856.[67] They also arranged for a German translation to be made and widely disseminated, as well as for an English translation of Emil von Wolff's contemporaneous replies to Liebig. Liebig was now beside himself with rage, as he revealed in a private letter to one of his few reverential English farmers, John Mechi:

> How ignorant and stupid and devoid of all good sense must be the great mass of agricultural people to allow such a set of swindlers to lead them in all these questions. If you ask any scientific man about the theoretical and practical value of their [Lawes and Gilbert] papers . . . it is all humbug, most impudent humbug. . . . Lawes and Gilbert hitch onto me like a vile vermin and I must get rid of them by all means. There is a cowardice among the scientific men in England which I am unable to understand and it is an offence against the public welfare that they have not the courage to take a public stand in that most degrading controversy between science and ignorance.[68]

In his white heat Liebig overlooked the fact that Gilbert had been a pupil and that only a year before, those same scientific men whom he now berated had presented him with testimonial silverplate for his service to British agriculture (see Chapter 5). Pusey's death in 1855 led to a slight relaxation of the RASE's editorial policy, allowing Liebig to reply again to Lawes and Gilbert in English.[69] Liebig now viewed the controversy as a way of showing that theory must always lead practice.

66. Voelcker to Gilbert, 7 September 1856, is a vitriolic attack on Liebig's morals. Rothamsted archives.
67. Lawes and Gilbert, "Reply. . ." *J. Roy. Agri. Soc. of England* 16(1855), 411–98. A note by (the then deceased) Pusey makes it clear that the paper was not published until January 1856.
68. Liebig to Mechi, 7 November 1863, Rothamsted archives.
69. J. Liebig, "Some points . . . ," *J. Roy. Agri. Soc. of England* 17 (1856), 284–326. This article appeared in German as the pamphlet *Über Theorie und Praxis in der Landwirtschaft* (Braunschweig, 1856).

> It seems to me that this [Lawes's] pretended criticism of science by practice can furnish an excellent opportunity of showing agriculturists how important it is for them not to deviate from the true method when their experiences have the object the appreciation, the verification of a scientific principle.[70]

Clearly, Liebig saw no contradiction in maintaining as a principle that a plant must be supplied with all its theoretical alimentary needs against Lawes's demonstration that in practice this was not the case. In the long term, Liebig was proved right, and what turned out to be really significant about the Rothamsted observations was that – although again theoretically correct about nitrogen feedstocks – Liebig had misidentified the principal source of nitrogen fixation as air and had underestimated the significance of nitrates and phosphates. In Gilbert's view "agricultural chemistry has suffered much from having a Pope – believed to be not only infallible, but even beyond the reach of question by any mere mortal!"[71]

In 1855, aware that Liebig was to be present, Playfair attempted a reconciliation by persuading Gilbert to attend the Glasgow meeting of the British Association. Gilbert, suspecting that Glasgow would be packed with Liebig's supporters such as Gregory, wavered until Playfair pointed out:

> As to your remarks about Personalities, Liebig only uses them when he has the pen in his hand. I suppose the gall in the ink produces them, for in conversation and *viva voce* discussion, he is the most amicable and good natured of men.[72]

Gilbert did attend and delivered another paper that pulled no punches, but all references to any public debate were expunged from the BAAS record and neither side was converted.

As historians of agriculture have pointed out, the practical approach to agriculture practised by Boussingault and Lawes held little attraction for Liebig.[73] His message was much more "scientific" – you must learn the language of chemistry if you want to understand agriculture. And it was philosophical, too, insofar as by the mid-1850s he had come to believe that his message had broader social, economic, and political implications than the initial one of 1840 for the improvement of the social status of chemists.

70. Liebig, "Some points . . . ," p. 284.
71. Gilbert to the editor of *Farmer's Magazine,* 17 May 1856, Rothamsted archives.
72. Playfair to Gilbert 6 August 1855, Rothamsted archives. See Lawes and Gilbert, *Chemical Gazette* 13(1855), 415–20, which was their British Association paper.
73. J. B. Dumas, "Liebig et son empreinte sur l'agronomie moderne," *Revue d'histoire des sciences* 18(1965), 73–108.

The new social thesis formed the basis for his passionate *Naturgesetze des Feldbaues,* which in 1862 formed the second volume of the seventh German edition of *Agricultural Chemistry.* It was translated separately into English (with much toning down of *ad hominem* polemic) by Liebig's Irish pupil John Blyth as *The Natural Laws of Husbandry* (London, 1863).[74] (Gregory had died in 1858.) Lawes and Gilbert soon became aware that Blyth had *not* translated part of the first volume, that is the *Einleitung,* or "Introduction," to the new edition. In fact, the whole of the edition's first volume had been rejected as libellous by Liebig's English publisher Walton, who even destroyed the copy in his possession. So, with Gilbert's connivance, Liebig's former pharmaceutical pupil Benjamin Paul published extracts from the *Einleitung* in *Chemical News.*[75] Gilbert also helped Paul compile the definitive history and natural history of manures for Watts's *Chemical Dictionary* in 1864.[76] It is interesting that even Vieweg, Liebig's German publisher, was worried by the "strong" tone of the first volume, and he asked Liebig's friend Mohr for an opinion. The latter was far more disturbed by Liebig's reference to a Creator (which he thought made Liebig hostage to ridicule by Moleschott and the materialists; see Chapter 11) than by his bitter denunciation of German and English agriculturists.[77]

Finlay has argued convincingly that Liebig completed the new seventh edition to help restore his flagging reputation in Germany, as well as to provide himself with money.[78] In the revised edition, armed with a knowledge of how important the work of Malthus had been to Darwin, Liebig employed the Malthusian spectre of starving millions and blamed the collapse of past civilizations on the *Raubsystem* (robbery system) of agriculture, in which soil fertility was slowly lost through a failure to replace vital ingredients of the soil. Using the speculative evidence of the German historian Friedrich Schlösser (1776–1862), he claimed the demise of the Roman and Spanish empires had been caused by their spoliation practices. In fact, as we now know from the American dustbowl experiences of the 1930s and more recent disasters, changes in the soil structure by overintensive cultivation do play crucial roles in causing infertility. Hence, although Liebig was correct in theory about the removal of minerals from the soil in explaining the rise and fall of human populations, we must also

74. Blyth to Gilbert, 3 January 1863, Rothamsted archives.
75. *Chemical News* 7(1863), 256–8, 268–70, 292–4, 302–5. Paul drew attention to Liebig's slipshod method of quotation.
76. H. Watts, *Dictionary of Chemistry,* 7 vols. (London, 186–75).
77. Mohr to Liebig, 8 August 1862, *Briefe,* pp. 174–9.
78. Mark R. Finlay, "The rehabilitation of an agricultural chemist: Justus von Liebig and the seventh edition," *Ambix* 38(1991), 155–69.

consider the soil conditions as factors, as well as the effects of war and disease.

By blaming the "robbery" system, Liebig was also accusing European nations of something far worse than sloppiness and indifference to recycling, namely, the way in which rich and powerful nations such as Britain, with its blinkered aristocracy, robbed the manurial resources of other countries to feed their own. In a famous angry and violent passage that was much reproduced in British newspapers, he accused Britain of plunder:

> Great Britain deprives all countries of the conditions of their fertility. It has raked up the battle-fields of Leipsic, Waterloo, and the Crimea; it has consumed the bones of many generations accumulated in the catacombs of Sicily; and now annually destroys the food for a future generation of three millions and a half of people. Like a vampire it hangs on the breast of Europe, and even the world, sucking its life-blood without any real necessity or permanent gain for itself.[79]

Because the earth was not inexhaustible, the population of European countries was greater than their soils could feed without tonnages of imported fertilizer increasing food yields. Agriculture had reached a stage at which it could be maintained only if one of two alternatives happened. Either by a miracle soils would recoup their fertility that the lack of intelligence had led men to ravage, or further deposits of fertilizer would be discovered that would be as rich as British coal deposits. As neither alternative looked likely, Liebig predicted Malthusian disaster:

> For their self-preservation, nations will be compelled to slaughter and destroy each other in cruel wars. These are not vague and dark prophecies nor the dreams of a sick mind, for science does not prophesy [sic], it calculates. It is not if, but when, that is uncertain.[80]

We shall see Liebig's solution to this bleak scenario in Chapter 9 and their entrepreneurial consequences in Chapter 8.

Liebig's jeremiads resonate well with our contemporary ecological concerns, though many agriculturists would deny Liebig's assumption that the earth is not inexhaustible. Subsoils contain an astonishing reserve of phosphates and potash, and forty years after Liebig's death, nitrogen was to be won by Fritz Haber from the abundant air. No doubt were he alive today, though, Liebig would be very concerned with energy, rather than material, reserves.

79. Translated from *Einleitung*, vol. 1, p. 125. See also *Letters on Modern Agriculture* (London, 1859), Letter 11.
80. From *Daily Telegraph* press cutting, October 1862, in Rothamsted archives.

In the seventh German edition of *Agricultural Chemistry,* Liebig made several quiet concessions to his German and English critics, and he was decidedly deferential towards farmers whose "triumphs of art" now eclipsed those of science. He freely confessed that his own ignorance had led to the failure of his patent manure, and he now accepted the Rothamsted evidence that additional nitrogen dressings were useful and even necessary. He also accepted the evidence of Thompson and others, culminating in the work of Thomas Way in 1850, that soils possessed a remarkable ability to absorb soluble salts differentially.[81] This explained why he had been so wrong in making sparingly soluble manures, as he now admitted. But it also showed that a soil's fertility could never be improved artificially if it lacked the particular absorptive powers necessary to extract particular salts. Effectively, he conceded that chemistry alone could not increase yields. Agricultural science would depend as much on knowledge of plant physiology and soil structure as chemistry. Intriguingly, Liebig seems to have been personally persuaded by Gilbert of the validity of Way's findings at the BAAS meeting in Glasgow in 1855. Unfortunately he capped all these concessions by cheekily implying, to Gilbert's disgust, that he had been aware of soils' absorptive powers before Way's convincing demonstration.

The cynical historian will note that the seventh conciliatory edition not merely restored Liebig's position as the elder statesman of agricultural science and scientific policy but that it brought his name back to public attention in time for his final burst of successful entrepreneurial activities. The latter (Chapter 9) not only illustrated the usefulness of chemistry but projected him as the world's saviour. Meanwhile, within Germany, Liebig's work received adulation again: The King of Saxony gave him a medal, he was asked to chair a meeting of agriculturists, and above all, German farmers collectively subscribed to a Liebig Medal to be awarded annually to an agricultural chemist. The seventh edition of 1862 was in every way as important an event as the publication of the first in 1840.

Liebig's message was popularised not only by himself in translation but by others such as Ville in France, Stöckhardt and Petzholdt in Germany, Norton in America, and Mechi in Britain.[82] Probably the most significant aspect of this message was the impetus it gave to the foundation of an artificial fertilizer industry. His own commercial attempt at creating this

81. Liebig described these soil experiments in *Letters on Modern Agriculture* (London, 1859), Letter 5, p. 7.

82. J. A. Stöckhardt, *A Familiar Exposition of the Chemistry of Agriculture Addressed to Farmers* (London,1855). Translated by Arthur Henfrey, whose editorial notes were critical of Liebig.

industry was a failure (Chapter 5), but by the time of his death it was a multimillion-pound world industry, vertically integrated through sulphuric acid manufacture with the rest of chemical manufacturing.

Liebig's agricultural chemistry was not entirely original, but it was a forceful example of the power of the Lavoisier–Gay-Lussac analytical method of weighing, measuring, calculating, and concluding when coupled with the idea of chemical cycles and metamorphoses. Disparate elements of theory and empirical practice were synthesized into a plausible and logical explanation of plant nutrition and husbandry. Even if Liebig was not the founder of the subject, the historian has no choice but to regard his 1840 book as its milestone.[83] Although primarily theoretical in approach, it did have the practical effect of helping to inhibit hunger, want, and suffering in the world by eliminating worries concerning the necessities of life. In this sense, from the mid-1850s, when Liebig restructured his theory in terms of what he perceived as urgent social problems, *Agricultural Chemistry* is a possible example of what German sociologists have called finalization theory, that is, when a discipline's development is no longer primarily dictated by conceptual and experimental priorities but by social, political, and economic factors. His dogmatic assurance complemented the mood of English High Farming.[84] One day, Liebig promised:

> The farmer will be able to keep an exact record of the produce of his fields in harvest, like the account book of a well-regulated manufactory; and then by a simple calculation he can determine precisely the substances he must supply to each field, and the quantity of these, in order restore their fertility.[85]

His work galvanised the fertilizer industry and in the United States, if not in Europe, it stimulated the creation of agricultural research stations. Although the law of the minimum was not original with Liebig, the publicity he gave to the idea that the growth of plants is determined by the nutrient available in the least amount could also be applied to animal nutrition and to the understanding of bacterial growth later in the century. Today, agricultural science still has a humus theory, because deeper research into soil mechanics, which was stimulated by Liebig, has shown that fertile soils with the appropriate microbiological fauna do require the

83. Schling-Bodersen, "Liebig's role in agricultural chemistry," p. 21.

84. W. Krohn and W. Schäfer, "Agricultural chemistry. The origin and structure of a finalized science," in Wolf Schäfer, ed., *Finalization in Science. The Social Orientation of Scientific Progress* (Dordrecht, 1983). However, their thesis is spoiled by their dating this to 1840 while using only the 7th edition of 1862.

85. J. Liebig, *Familiar Letters on Chemistry,* 1st ed. (London, 1843), p. 171.

presence of organic matter from decaying vegetation. Today we have a much more sophisticated *dynamical* view of the soil than Liebig's simple static model. Although Sprengel and Liebig were correct in rejecting the original theory that plant nutrients were organic, the organic compounds contained in humus do actually provide an environment for microorganisms that play a vital role in generating the "crumb structure" of healthy soils.[86] Liebig's contention that plants absorb nitrogen directly from the atmosphere – the point so elaborately contested by forty years of elaborate experiments at Rothamsted – turned out to be right after all for the special case of leguminous plants. Both Lawes and Gilbert appreciated the irony.

The agricultural revolution in which Liebig participated is epitomised by a statement made by J. F. W. Johnston in 1848. Ten years previously, he recalled,

> the strongest of the agricultural periodicals that ever touched upon the subject at all for the most part undervalued the worth of natural science to the farmer, and ridiculed the pretended value of chemistry. [Now in 1848] the weekly journal was considered badly conducted in which every number did not embody some scientific and especially some chemical information. Scarcely a provincial paper which boasted of an agricultural corner, but indulged freely in chemical nomenclature as being more agreeable to the taste, and within the easy comprehension of almost every farmer.[87]

Despite immense provocation, Lawes and Gilbert retained the utmost respect for Liebig. In 1856, in the middle of their dispute, Lawes was able to say in the presence of Prince Albert:

> It is to an illustrious countryman of His Royal Highness, to Baron Liebig alone, that is due the merit of having effectively roused the attention of British farmers to the importance of applying chemical science to advance the practice of their art.[88]

That was said at least a decade before Liebig was able to agree with Lawes's qualification:

> It has of late years been too much the custom to explain everything connected with the practice of agriculture by means of chemistry alone. I would, however, venture to claim for agriculture the rank of an independent science, in close alliance, it is true, with chemistry, botany, physiology, and some other sciences, but still distinct from them all. Its truth can alone be investigated and explained by experi-

86. K. Scharrer, "Liebig and today's agricultural chemistry," *J. Chem. Educ.* 26(1949), 515–18.
87. Quoted in Hall, *Yorkshire Agricultural Society*, p. 82.
88. *J. Roy. Soc. Arts* 4(1855–56), 282 (5 March 1856).

> ments conducted in the field with all the aids and refinements which modern science affords.[89]

Gilbert, who appreciated the irony of Hellriegel and Wilfarth's demonstration that leguminous plants do feed directly off atmospheric nitrogen,[90] was also splendidly appreciative in his presidential address to Section B of the BAAS, given after Liebig's death in 1880:

> If the part which has fallen to my own lot in these discussions qualified me at all to speak for others as well as myself, I would say that those who, having themselves carefully investigated the points in question, have the most prominently dissented from any special views put forward in [Liebig's works], will, – whether they be agricultural chemists, vegetable physiologists, or animal physiologists – be the first to admit how important has been the direction, given to research in their own department, by the masterly review of their existing knowledge, and the bold and frequently sagacious generalizations of one of the most remarkable men of his time.[91]

As Vance Hall has noted in quoting this passage, Gilbert's tribute rings more true than the formal eulogies of Hofmann and Thudichum because he had been so deeply involved in controverting Liebig's theories and methods. It is therefore agreeable to report that the Rothamsted archives contain a chatty letter from Liebig to Gilbert written on Christmas Day, 1870, which opens, "My dear friend." As Playfair had said, Liebig's bark was worse than his bite.

89. *Ibid.*, p. 283.
90. H. Hellriegel and H. Wilfarth, "Untersuchungen über die Stickstoffernährung der Gramineen und Leguminosen," *Zeitschrift der Vereins für die Rübenzuckerindustrie* (1888), 1–234.
91. *BAAS Reports*, 1880, pp. 507–8.

7

Liebig and the Doctors: Animal Chemistry

> Organic chemistry is the child of medicine, and however far it may go on its way, with its most important achievements, it always returns to its parent.[1]

Although *Animal Chemistry* was a logical extension to Liebig's book on agricultural chemistry, the initial impetus for writing it had come from the work of the Dutch chemist Gerrit Mulder. In 1838 Mulder concluded from organic analyses that certain nitrogenous ("albuminous") materials in plants and animals (plant albumin, and animal fibrin, casein, and albumin) were identical in composition. It appeared to follow that animals obtained their nitrogen ready formed from plant albumin. Between 1838 and 1842, Liebig had many of his students repeat and extend Mulder's work, which had been praised by Berzelius. It was the latter who suggested that the nitrogenous molecule that plants and animals shared and the composition of which Mulder had estimated as $C_{40}H_{62}N_{10}O_{12}$ should be called the *proteine* radical.

The exploration and extension of Mulder's work in Liebig's laboratory meant that for the first time advanced students were working with him on a common problem – which is why Liebig's work of that time is often seen as a benchmark for the emergence of the modern graduate school.[2] In July 1840 Liebig told Wöhler excitedly:

> The grand work on fats which six chemists in my [advanced] laboratory are busy with forges ahead and has brought to light a mass of earlier incorrect analyses. It is certainly a bit of luck to have such forces at my disposal.[3]

These forces included two British pupils, John Stenhouse and Lyon Playfair, as well as Franz Varrentrap, who with Heinrich Will perfected a

1. J. L. W. Thudichum, "On the discoveries and philosophy of Liebig," *J. Roy. Soc. Arts* 24(1876), 141.
2. See J. S. Fruton, *Contrasts in Scientific Style* (Philadelphia, 1990).
3. A. W. Hofmann, ed., *Aus Justus Liebig's und Friedrich Wöhler's Briefwechsel*, 2 vols. (Braunschweig, 1888), vol. 1, p. 162.

better method for estimating nitrogen. As a result of this common work, between 1841 and 1842 Liebig published two major papers in *Annalen der Chemie*.[4] These in turn formed the basis for *Animal Chemistry*, which, dedicated to Berzelius, was published by Vieweg in June 1842. The English translation by William Gregory appeared a month later.[5]

Whereas Mulder had seen protein only as a radical that could be subtly modified in plants and animals by the addition of sulphur and phosphorus to give their particular nitrogenous constituents, Liebig concluded in 1841 that the nitrogenous animal matter was identical to that in plants. "One could truly say," he told Wöhler, "that animal organisms only need to fashion the form of blood."[6] It followed therefore that, if vegetables provided the vital constituents of blood and flesh in animals, the fats and sugars that were also assimilated from plants played no role in the formation of animal matter. The function of sugars and fats, Liebig concluded in the manner of Lavoisier, was solely to be oxidized to generate animal heat. Accordingly, in *Animal Chemistry* he divided foodstuffs into the two broad categories of nitrogenous or plastic (plastische Nahrungsmittel) and nonnitrogenous or respiratory (Respirationsmittel), with only the former being flesh-formers.[7]

The Eskimo, he explained, fed chiefly on fat and tallow because of their dire need to keep out the cold, whereas the pampas gaucho could live more or less on dried meat, and like the rower or boxer trained on beefsteaks and porter, the gaucho required little nonnitrogeneous food to maintain his temperature but plenty of plastic food to meet his extensive muscular demands. In the later *Chemical Letters*, Liebig transformed this suspect anthropology into a Carlylean aphorism:

> Our clothing is merely an equivalent for a certain amount of food. The more warmly we are clothed the less urgent becomes the appetite for food, because the loss of heat by cooling, and consequently the amount of heat to be supplied by the food, is diminished.[8]

4. J. Liebig, "Über die stickstoffhaltigen Nährungsmittel des Pflanzenreiches," *Ann. Chem.* 39(1841), 129–60; "Die Ernährung Blut- und Fettbildung im Thierkörper," 41(1842), 241–85.
5. J. Liebig, *Die organische Chemie in ihrer Anwendung auf Physiologie und Pathologie* (Braunschweig, 1842); *Animal Chemistry in its Applications to Physiology and Pathology* (London, 1842; reprint, Johnson Corporation, New York, 1964). Later editions, also simultaneously in German and English, were 2nd (1843) and 3rd (retitled *Tierchemie* to follow the English model), 1846. The Philadelphia complete works printing of 1856 reprinted the 1842 edition without comment.
6. Hofmann, *Liebig's und Wöhler's Briefwechsel*, vol. 1, p. 185.
7. J. Liebig, *Animal Chemistry*, 1842, p. 92.
8. J. Liebig, *Familiar Letters on Chemistry* (London, 1843), p. 69.

Reflecting its origins as papers on fat metabolism, protein degradation, and fermentation, *Animal Chemistry* was divided into three sections; nevertheless, it forms an integrated whole rich in details and controversial consequences.

Metabolism and the Origin of Fats

In the first untitled part of *Animal Chemistry,* Liebig developed a *chemical,* as opposed to vitalistic, theory of metabolism. Much of it was an elaboration of a paper he had published in 1841 in typically controversial circumstances. If Liebig had hoped to escape fruitless theoretical debates with Dumas by moving into physiological chemistry, he was to be mistaken. Neither Dumas nor Boussingault seems to have been best pleased when Liebig published *Agricultural Chemistry* in 1840. In response, Dumas, who was lecturing on the relationships between plant and animal chemistry at the Paris Medical School in 1841, collaborated with Boussingault on an *Essai de statique chimique des êtres organisé,* which included a summary of the contrasts between the vegetable and animal kingdoms as perceived by contemporary French chemists and physiologists.

Animal	*Vegetable*
Is a combustion apparatus.	Is a reduction apparatus.
Can locomote.	Is immobile.
Burns carbon, hydrogen, ammonia.	Reduces carbon, hydrogen, ammonia.
Exhales carbonic acid, water, ammonium oxide, nitrogen.	Fixes carbonic acid, water, ammonium oxide, nitrogen.
Uses oxygen, neutral nitrogenous substances, fats, starches, sugars, gums.	Produces oxygen, neutral nitrogous substances, fats, starches, sugars, gums.
Produces heat and electricity.	Absorbs heat and electricity.
Restitutes its elements to air and earth.	Borrows its elements from air and earth.
Transforms organic matters into mineral matters.	Transforms mineral matters into organic matters.

This dichotomy might have been accepted passively by Liebig but for the two French authors' claims, supported by documentary evidence in appendices, that all these ideas had been publicly stated or published by them before 1839 and certainly *before* the appearance of Liebig's *Agricultural Chemistry.* Provoked by this explicit priority claim concerning the fixation of nitrogen, Liebig pressed ahead with experimental work on the analysis of the products of animal metabolism, publishing a preliminary report in January 1842 on the grounds that this was a necessary precaution to avoid further French plagiarism.[9] Dumas immediately retorted that Liebig's latest paper contained nothing new and nothing that he, Dumas, had not already taught.[10] Not unexpectedly, Liebig's angry six-page reply was scornful.[11] Boussingault was resentful, writing to Dumas on the latter's election as dean of the Parisian faculty of sciences:

> Now it is essential to organize great resources for research at the faculty; people have produced enough hot air over it. It is truly shameful for France, and I am never so good a Frenchman as when I am on the banks of the Rhine; it is truly shameful that an evil hole like Giessen is a focal point of science, a place where all of the chemical world of Europe meet, and that one would only come to Paris to see our scientists as one comes there to see the zoo in the botanical gardens, purely for curiosity.[12]

Although historians agree impartially that both the French and the German camps had arrived at their positions independently by standing on the shoulders of previous workers like William Prout (whom Liebig little acknowledged), the ferocity of the conflict between them was to affect French willingness to accept anything of Liebig's as correct. In fact, the French and the German positions differed fundamentally, insofar as Dumas and Boussingault were emphasizing the *differences* between the vegetable and animal kingdoms, whereas Liebig was working his way towards seeing a fundamental unity rendered by the common ability of plants and animals to synthesize complex molecules from simpler ones.[13]

French chemists, beginning with Lavoisier and continuing with Gay-Lussac, Thenard, and Chevreul, had made many significant contributions to vegetable chemistry in the first three decades of the century. By the 1830s, it seemed that both plants and animals contained sugars, fats (which Chevreul had shown to be compounds of glycerol and fatty acids), and albuminous (nitrogenous) substances. From this position, it was but a

9. J.Liebig. "Der Lebensprocess im Thiere, und die Atmosphäre," *Ann. Chem.* 41(1842), 189–219.
10. J. B. A. Dumas, "Essai de statique des êtres organisés," *Ann. Chim.* 4(1842), 115–26.
11. "Antwort auf Hrn.Dumas's Rechtfehrung wegen eines Plagiats," *Ann. Chem.* 41(1842), 251–7. Unfairly, Liebig blamed the young Swiss chemist Marignac for revealing Giessen secrets when he moved from Giessen to Dumas's laboratory in Paris in December 1841. Marignac, who became a distinguished analyst, suffered no damage from this accusation.
12. F. L. Holmes, *Claude Bernard and Animal Chemistry* (Cambridge, Mass., 1974), p. 42. Holmes provides a blow-by-blow discussion of the controversies between Paris and Giessen.
13. D. C. Goodman, "Chemistry and the two kingdoms of nature in the 19th century," *Medical History* 16(1972), 113–30.

small step for Liebig (and others) to conclude that plant materials simply served as the source of the same principles in animals. As Liebig wrote:

> How beautifully and admirably simple, with the aid of these discoveries, appears the process of nutrition in animals, the formation of their organs, in which vitality chiefly resides! Those vegetable principles, which in animals are used to form blood, contain the chief constituents of blood, fibrin and albumen, ready formed, as far as regards their composition. All plants, besides, contain a certain quantity of iron, which reappears in the colouring matter of the blood. . . . Vegetables produce in their organism the blood of all animals, for the carnivora, in consuming the blood and flesh of the graminivora, consume, strictly speaking, only the vegetable principles which have served for the nutrition of the latter.[14]

But whereas Liebig interpreted animals as "a higher kind of vegetable" that used vegetable materials to synthesize the complex tissues, membranes, nerves, and brain, Dumas and Boussingault insisted that the power of synthesis in animals was extremely limited. For them, vegetables were "the great elaboratory of organic life," and *all* the sugars, proteins, and fats found in animals had been assimilated from plants, because only the latter were able to synthesize them.[15] In effect, vegetables are synthesized, whereas animals are degraded.

In the case of fats, Liebig was convinced immediately from comparative analyses of animals' foods and excreta that they must synthesize fats from sugars and starch. This seemed to follow naturally from the way the "stall-fed animal eats, and reposes merely for digestion" (p. 82). As such captive animals took in more food than was required for reproduction or was expelled in their excretions or oxidized for warmth, excess carbon formed fat. Suggestive, too, was the fact that the proportion of carbon to hydrogen was the same in both sugars and fats – a point already made by the English physiological chemist William Prout but unacknowledged by Liebig.[16] It followed – or so Liebig asserted –

> that sugar, starch, and gum, by the mere separation of a part of their oxygen, may pass into fat, or at least into a substance having exactly the composition of fat. If from the formula of starch $C_{12}H_{10}O_{10}$, we take 9 equivalents of oxygen, there will remain in 100 parts –
>
> C_{12} 80.4
> H_{10} 10.8

14. J. Liebig, *Animal Chemistry,* 1842, p. 8.
15. J. B. A. Dumas and J. B. Boussingault, *The Chemical and Physiological Balance of Organic Nature,* 3rd ed.(London, 1844), p. 6.
16. W. H. Brock, *From Protyle to Proton* (Bristol, 1985), Chap. 6.

O 10.8

The empirical formula of fat, which comes nearest to this, is $C_{11}H_{10}O$, which gives in 100 parts –

C_{11}	78.9
H_{10}	11.6
O	9.5

According to this formula, an equivalent of starch, in order to be changed into fat, would lose 1 equivalent of carbonic acid, CO_2, and 7 equivalents of oyxgen.[17]

And he went on triumphantly to demonstrate how the loss of 4 equivalents of water and 31 of oxygen from 3 equivalents of milk sugar [3 $C_{12}H_{12}O_{12}$ = $C_{36}H_{36}O_{36}$] left the molecule $C_{36}H_{32}O$, which corresponded with cholesterine, the fat of bile previously identified by Chevreul.

As to the fate of the excess oxygen liberated, Liebig supposed that it was used to produce animal heat by the oxidation of carbon compounds (including stored fat) to form the carbon dioxide exhaled in the breath. Whatever the exact mechanisms involved, it was undeniable that

> the herbs and roots consumed by the cow contain no butter; that in hay or the other fodder of oxen no beef suet exists; that no hog's lard can be found in the potato refuse given to swine, and that the food of geese or fowls contains no goose fat or capon fat.[18]

The one possible exception, beeswax, which could conceivably have been gathered by bees directly from plants, had been shown by the naturalist F. W. Gundlach to be formed from honey. Gundlach's *Natural History of Bees* was conveniently published at the beginning of 1842 in time to be cited by Liebig.[19] Unfortunately, later research showed that waxes are not fats, being esters of nonglycerol alcohols. Much of the conflict between the French and German chemists lay in differing contemporary definitions of fats.[20]

None of this convinced Dumas, so that Liebig returned to the formation of fats in 1843, using solubility in ether as his test for their presence.[21] Only minute portions of potatoes and hay responded to the test, and yet these foods clearly fattened animals. Moreover, he asked rhetorically, because marine vegetation seemed deficient in fats, how else but by synthesis could one explain the presence of fish oil in fish? In a discussion of a letter

17. J. Liebig, *Animal Chemistry*, p. 82.
18. *Ibid.*, p. 82.
19. F. W. Gundlach, *Naturgeschichte der Bienen* (Cassel, 1842).
20. Holmes, *Bernard*, Chap. 3.
21. J. Liebig, "Ueber die Fettbildung," *Ann. Chem.* 45(1843), 112–28.

from Liebig on the subject read at the French Academy of Sciences, Dumas retorted that Liebig's analyses were faulty and that "the hay eaten by Liebig's cow was richer in fat than he thinks."[22] Dumas and Boussingault were now joined by other collaborators, such as Anselme Payen and Henri Milne Edwards, all of whom remained intractably opposed to Liebig's position. Nevertheless, by 1845 it had to be conceded that Liebig had won his case.[23]

Liebig's argument pointed future research in two directions. On the one hand, as Liebig's disciple in physiological chemistry Carl G. Lehmann put it in 1849, a dynamical rather than a statical approach was needed:

> We cannot, it is true, arrive at any conclusion regarding the working of the process [of synthesis, or metabolism] by a mere juxtaposition and quantitative comparison of the ingesta and excreta of the animal organism. . . . It need scarcely be observed that science should not rest satisfied with a knowledge of the final results of chemical processes in the animal body . . . but should be made to enter more deeply into the course of the separate processes, and into the causal relations of phenomena. Here the statistical [sic, statical] method cannot of course afford any satisfactory solution to our enquiries; for when we have ascertained by this experimental method that fat is formed on the animal body, we must learn from other methods the manner in which this substance is formed.[24]

This was to be the approach of Claude Bernard, who in 1857, using vivisections of animals fed on sugar-free diets, showed that animals did synthesize sugar, which was then stored in the liver as glycogen. As he observed at the time, this finding demonstrated that animals, like plants, could perform synthesis. By implication, too, fats were no longer a problem.[25]

The alternative research strategy was to follow up the limited animal-feeding experiments that Liebig had made at Giessen with controlled and long-term experiments in animal husbandry. Not surprisingly, this was the path chosen by Liebig's British agricultural adversaries Lawes and Gilbert. Between 1848 and 1853, they conducted experiments on some 500 animals, using foods of known composition and analyses of slaughtered animals for the presence of fat. Their results more than confirmed Liebig's

22. J. B. Dumas in discussion, *Comptes Rendus* 16(1843), 559–60.
23. J. Liebig, "On the formation of fat in the animal organism," *Lancet* 1845, ii, 568–9. See Selik Soskin, "The formation of fat in the animal body," *J. Roy. Agri. Soc. of England*, 38(1897), 355–67.
24. C. G. Lehmann, *Physiological Chemistry*, 3 vols. (London, 1851–54), vol. i, p. 14. German original, 1849.
25. For a comprehensive account, see Holmes, *Bernard*.

position on the matter. Nevertheless, in 1865, at a meeting of agricultural chemists in Munich, Carl Voit reported on respiratory experiments with dogs that had been fed on flesh (i.e., nitrogenous diets) and suggested that fat was being derived from this source. If so, proteins could also be a source of fats in herbivora. Moreover, if fat did come solely from carbohydrates, as Liebig had maintained, then, Voit argued, experiments ought to show that the amount of fat deposited in an animal was greater than "that supplied by the food *plus* that which could be derived from the transformation of albumin."[26] Aware of this, it was a fairly simple matter for the Rothamsted experimentalists to reassess their herbivore data and to report their conclusions at the meeting of the British Association in 1866.[27] More fat was laid down, they showed, than could have been derived from the degradation of nitrogenous matter alone. They allowed, however, that nitrogenous foods might be a source of fat if animals were fed either an excessively nitrogenous diet or one in which there was a lack of non-nitrogenous foods.

This twist to the fats debate in the 1860s was, in fact, made within the context of the second part of Liebig's *Animal Chemistry,* which, unlike the other parts, consisted of numbered paragraphs.

The Metamorphosis of Tissues

As F. L. Holmes has shown, Liebig's *Animal Chemistry* was written in the tradition of Lavoisier's argument at the end of the eighteenth century that respiration was a slow form of combustion in which food carbon was oxidized to form carbon dioxide expelled in the breath and that the heat generated by the reaction was the source of animal heat. In Liebig's analogy of the age of steam, he wrote:

> The animal body acts . . . as a furnace, which we supply with fuel. It signifies nothing what intermediate forms food may assume, what changes it may undergo in the body, the last change is uniformly the conversion of its carbon into carbonic acid, and of its hydrogen into water; the unassimilated nitrogen of the food, along with the unburned or unoxidized carbon is expelled in the urine or in the solid excrements. In order to keep up in the furnace a constant tempera-

26. J. B. Lawes and J. H. Gilbert, "The feeding of animals for the production of meat, milk, and manure, and for the exercise of force," *J. Roy. Agri. Soc. of England* 56(1895), 47–146.

27. *BAAS Reports* 1866, p. 41; *Phil. Mag.* 32 (1866), 55–64. The work was confirmed independently by Hermann von Liebig in 1873. See his letter to Gilbert, 3 May 1878, Rothamsted archives.

ture, we must vary the supply of fuel according to the external temperature, that is, according to the supply of oxygen.[28]

This respiratory model, which had been refined in detail by Lavoisier's French and English successors, suggested a bank balance approach to living systems. We have already noticed Boussingault making effective use of this method in his agricultural experimentation. By quantitatively determining imputs (aliments and gases imbibed) and outputs (excretions and gases exhaled) and heat output, the model was demonstrated. In November 1840, for example, in one of his only explicit physiological experiments, Liebig obtained permission to measure the food intake and excrement of 855 soldiers forming the bodyguard of the grand duke of Hessen-Darmstadt over a period of a month. The excess of carbon imbibed over carbon excreted was, he argued, the quantity of carbon that had been burned daily in respiration – though as Liebig had no way of measuring the carbon dioxide output of a regiment, the experiment was clearly flawed. But perhaps it had the purpose of convincing Liebig that Lavoisier's balance method was right and that the only source of animal heat was the oxidation of carbon. Liebig himself confessed that the results "have surprised me no less than they will others."[29] He quietly ignored the fact that in 1824 both C. Despretz and P. Dulong had found that, during respiration chamber experiments, the heat combustion of carbon was less than that calculated from the carbon dioxide output. The implication that there were other possible chemical sources of animal heat was, however, ignored by Liebig. Instead, to this basic respiratory model of carbon balance, Liebig added a nitrogen balance.

Mulder's demonstration between 1837 and 1838 that vitally important constituents of the blood shared a common molecule – protein – with plant albumin, casein, and gluten provided Liebig with the idea that plants synthesized protein, which was then with only minor modification taken directly into the animal system to form flesh and blood. To Liebig's delight, as he seems to have first demonstrated in his lectures in the winter semester of 1841, if one assumed that fats were absorbed by the liver to form bile secretions that were later oxidized in the blood to carbon dioxide and heat, then it was possible to balance the *internal* degradation products with the constituents of blood. The result was impressive equations such as the following:

5 atoms protein	$5(C_{48}N_6H_{36}O_{14}) = C_{240}N_{30}H_{180}O_{70}$
15 atoms starch	$15(C_{12}\ \ H_{10}O) = C_{180}\ \ H_{150}O_{150}$

28. J. Liebig, *Animal Chemistry*, p. 20. For an even more explicit steam-engine analogy, see p. 250.

29. *Ibid.*, p. 125.

12 atoms water	$12(HO)$	=	$H_{12}O_{12}$
5 atoms oxygen		=	O_5
The sum is			$C_{420}N_{30}H_{342}O_{237}$
which is equivalent to the degradation products			
9 atoms choleic acid	$9(C_{38}NH_{33}O_{11})$	=	$C_{342}N_9H_{297}O_{99}$
9 atoms urea	$9(N_2H_4O)$	=	$C_{18}N_{18}H_{36}O_{18}$
3 atoms ammonia	$3(NH_3)$	=	N_3H_9
60 atoms carbonic acid	$60(CO_2)$	=	$C_{60}O_{120}$
The sum is			$C_{420}N_{30}H_{342}O_{237}$

(Choleic acid was a bile secretion product.)

These assumptions were supported, Liebig argued, by the analytical evidence that soda was present only in blood and bile. Because no soda was found in tissue, he supposed that when blood formed tissues, its soda combined with bile. (Conveniently, some soda was found in excreta.) It was this unstable soda–bile compound that was oxidised to produce animal heat. This was all complete invention. As Claude Bernard was able to show later, the function of bile is to break down food fats (triglycerides) into simpler compounds that are then resynthesised into human fats.[30]

Nevertheless, providing that one accepted Liebig's assumption that,

> if we subtract from the composition of blood the elements of the urine, then the remainder, deducting the oxygen and water which has been added, must give the composition of the bile [i.e., choleic acid],[31]

such equations or balances seemed enormously impressive. Again, merely by adding ammonia, water, or oxygen to protein, Liebig generated the materials of bones, tendons, cartilage, hair, and arteries: Pr = protein $C_{48}N_6H_{36}O_{14}$, as shown by Table 7.1.

Even more impressive was Liebig's demonstration of how *pathological* degradation products could form where there was an insufficiency of oxygen, or how uric acid, rather than urea, came to be the principal excretory product of nitrogen metabolism in birds. It did not matter that this was done by an extreme Pythagoreanism of the kind "add 9 and multiply by 5."

According to Liebig's speculative model, fats and sugars played no role in the formation of blood and tissues, so it followed that muscular work was to be accounted for as the degradation of muscles when protein separated into a large carbon molecule and a large nitrogenous one. Both

30. T. O. Lipman, "Vitalism and reductionism in Liebig's physiological chemistry," *Isis* 58(1967), 173–5.

31. J. Liebig, *Animal Chemistry*, p. 126.

Table 7.1

	Protein	Ammonia	Water	Oxygen
Fibrin, albumen	Pr			
Arterial membrane	Pr		2 HO	
Chondrine (cartilage)	Pr		4 HO	2 O
Hair, horn	Pr	NH_3		
Gelatinous tissue	2 Pr	3 NH_3	HO	7 O

J. Liebig, *Animal Chemistry,* 1842, p. 121.

of these were released into the blood stream, where they were further oxidized and degraded. The carbonaceous molecule produced carbonic acid that was exhaled in the lungs, any excess of the carbon being stored as choleic acid in the liver and bile, to be used if the animal's temperature required boosting. Meanwhile, the nitrogenous portion was excreted through the kidneys as urea, or uric acid. It followed, as Liebig himself pointed out, that the urea content of urine was a measure of muscular work.[32] Here was scientific support for the common belief that strenuous occupations called for plenty of beef. Liebig named the whole process of tissue degradation *Stoffwechsel.* The harder a man worked, the higher the nitrogen content of his urine would be. If one follows Sir Karl Popper's account of scientific method, here was a classic case of Liebig's providing a benchmark of good science: Here he was offering a direct challenge to his speculative system. If the nitrogen content of urine turned out not to be balanced by muscular activity, his scheme would be falsified.

However, before following this point further, something should be said about the reception of *Animal Chemistry* in 1842.[33]

Because the only major previous synthesis of knowledge concerning animal chemistry had been made by Berzelius, Liebig was naturally anxious to have the master's reaction.[34] Having guessed the speculative tendency of Liebig's enterprise from the advance articles in *Annalen,* Berzelius had already shown reluctance to having the German edition of

32. *Ibid.,* p. 233.
33. The definitive discussion is Holmes's edition of *Animal Chemistry* (New York, 1964).
34. Berzelius's original Swedish text on animal chemistry appeared in 1806–08 but was never translated. Much of it was later absorbed into his chemical textbook, which Wöhler translated as *Lehrbuch der Chemie,* 4 vols. in 8 (Dresden, 1825–31). See A. J. Rocke, "Berzelius's Animal Chemistry," in E. M. Melhado and T. Frängsmyr, eds., *Enlightenment Science in the Romantic Era. The Chemistry of Berzelius and Its Cultural Setting* (Cambridge, 1992), Chap. 5.

Animal Chemistry dedicated to him.[35] Berzelius's "devastating critique" in the *Jahresbericht* for 1843 is reminiscent of his dismissal of Thomas Thomson's work on atomic weights nearly two decades before:

> For animal chemistry the time has come . . . in which chemists, without understanding the need to have a deep, special, and comprehensive knowledge of the anatomical aspects of physiology, sketch for us in rapid strokes the chemical phenomena which occur in vital processes. This facile kind of physiological chemistry is created at the writing table, and is the more dangerous, the greater the genius with which it is carried out, because the majority of readers will not be in a position to distinguish what is right from the mere possibilities and probabilities, and will be misled into taking probabilities for truths, which, when they have once become imbedded in physiological chemistry, will require great effort to be eradicated.[36]

Berzelius was certainly not being unfair in describing Liebig's work as "probability physiology," though this description did overlook the fact that scientific progress can often be accelerated by the appearance of faulty world views that have the power to stimulate research. He might well have been ruder still and labelled it a return to the crude iatrochemistry and iatrophysics of the seventeenth century. Berzelius was particularly irritated by the equations as, whatever the Giessen school claimed, there was not unanimity concerning the formulas of proteins and other degradation products; nor was Berzelius convinced that choleic acid, Liebig's bile secretion, was a homogeneous compound. Privately he told Wöhler that he thought Liebig's book was drivel (*Radoterie*).

Wöhler toned down Berzelius's remarks when translating from the Swedish; even so, Liebig was never to forgive Berzelius or his pupil Eilhard Mitscherlich, whom he believed must have turned Berzelius against him during a visit to Stockholm. Liebig's public reaction was extreme, "lashing out at all the world," as Wöhler described it. In May 1844 Liebig tore into Berzelius in *Annalen,* accusing the grand old man of being out of touch and out-of-date and writing that none of Berzelius's analyses of animal products had ever assisted physiologists in the way that Liebig's were doing.[37] Berzelius's reply was surprisingly temperate, but he did not

35. J. Carrière, ed., *Berzelius und Liebig ihre Briefe* (München, 1898), p. 134.
36. J. J. Berzelius, *Jahresbericht* 22(1843), 535; trans. Holmes, *Animal Chemistry* (1964), p. lx. Compare Berzelius's "writing desk" attack on Thomson, *Jahresbericht* 8(1827), 244. Berzelius renewed the attack on Liebig in *Jahresbericht* 23(1844), 575–82.
37. J. Liebig, "Berzelius und Probabilitätstheorien," *Ann. Chem.* 50(1844), 295–335; see excerpts in M. Teich, *A Documentary History of Biochemistry* (Leicester, 1992), 471–73. Liebig's anger also led to *Bermerkungen über das Verhältniss der Thier-Chemie zur Thier-Physiologie* (Heidelberg, 1844).

withdraw his doubts. In consequence the hitherto regular correspondence between the two men ceased.

Like *Agricultural Chemistry, Animal Chemistry* polarized chemists. On the one hand, of which Berzelius's reaction was the most extreme, were those who felt Liebig had given chemistry too high a profile in physiological investigations. Others, particularly Liebig's British pupils such as Lyon Playfair, William Gregory, and Henry Bence Jones, and his German friend Friedrich Mohr, accepted the book enthusiastically and uncritically. Gregory and Playfair, for example, publicized the book extensively in long abstracts that were in effect trailers, advertisements. and popularizations.[38] At one stage the *Dublin Medical Review* complained of writers who simply hitched themselves to Liebig's name to publicize their own dubious work.[39] Even the uncritical, however, found the work useful, particularly when it came to medicine. More interestingly, a majority held to a middle ground. Although recognizing that the book had serious flaws, they recognized that it opened up new questions and new possibilities for the use of chemistry in physiology. Effectively, in this route lay a new discipline, biochemistry.

Liebig himself, although he never admitted the justice of Berzelius's criticisms, very quickly realized that he had been overquick to accept the validity of Mulder's assumption of the protein radical as common to plants and animals and as not having sulphur or phosphorus among its constituents. Rather than blaming himself, he took his revenge on Mulder (who in turn took Berzelius's side) in a series of attacks in *Annalen der Chemie* culminating in *The Chemistry of Food* in 1847.[40] What had gone wrong?

Perhaps as a hangover from his pharmacy days with Geiger, Liebig often had his advanced students redo the analyses asserted by other chemists before allowing their appearance in *Annalen.* Partly due to this and partly because a full programme of research into the constitution of animal materials was under way in the 1840s, not surprisingly, Liebig's most skilled assistants began to have trouble confirming the existence of a

38. L. Playfair, *London and Edinburgh Monthly Journal of Science* 2(1842), 768–72; W. Gregory, *Quarterly Review* 1842, 98–128; anon. (possibly Gregory), *Medico-Chirurgical Review* 37(1842), 337–71. See W. H. Brock and S. Stark, "Liebig, Gregory and the British Association," *Ambix* 37(1990), 134–47. For a rave review of the third edition, see *Lancet,* 1846, ii, 455–6.
39. *Dublin Quarterly Journal Medical Science* [3]2(1846), 479–90, with reference to John Leeson, *Liebig's Physiology applied in the Treatment of Functional Derrangement and Organic Disease* (London, 1846), which was also condemned in *Lancet* 1846, ii, 560–2 and in *Medico-Chirurgical Review* [3]4(1846), 501–13.
40. H. A. M. Snelders, "The Mulder-Liebig Controversy elucidated by their correspondence," *Janus* 69(1982), 199–221.

common protein molecule $C_{48}N_6H_{36}O_{14}$ in albumin, fibrin, and casein. In particular, two students Nicholaus Laskowski (1816–71) and Theodor Fleitmann (1828–1904) were unable to free protein from sulphur and phosphorus in many products.[41] Mulder had always claimed that when sulphur and phosphorus were detected, they were compounds of the protein. When the students' results were announced in the *Annalen der Chemie,* Mulder went crazy – there is no other word to describe his reaction. He first demanded that Liebig publish a public retraction of the following unreasonable kind:

> I have prayed to God with all my heart to pardon me for the great number of faults that I have committed in bruising men of science continually under the immoral pretext of promoting *the truth.* I pray all those who have been the victims of my obsession during the last twenty years, to excuse me, and I promise that in future that nothing will regulate my way but that of being a man of honour.[42]

Naturally, Liebig ignored this "blackmail," which reads like the General Confession of the Church of England, whereupon Mulder retorted publicly in Dutch with a strange pamphlet, *Liebig's Question to Mulder Tested by Morality and Science* (Rotterdam, 1846). This was translated into German by Mulder's assistant Augustus Voelcker, who was to make his career in England as an agricultural chemist, and into English by his Dutch student Pieter Fromberg. Accusing Liebig of establishing a tribunal at Giessen "before which Liebig is at the same time accuser, witness, public prosecutor, advocate and judge," Mulder complained bitterly of how, having once been told by Liebig that everything he did was wonderful, he was now told it was all wrong. "Put a stop to your injustice," he pleaded, "above all, be just, for justice towards men, is the foundation of truth."[43]

At first Liebig sensibly did not respond to this onslaught, which must have set the chemical world buzzing in 1847; privately, though, he had harsh things to say, referring to Mulder as either a swindler or a chemical dilettante. By the summer of 1847, Liebig could stand it no more and in an unpaginated appendix, which is not always to be found bound in the July issue of *Annalen,* he published all of Mulder's correspondence with a debunking commentary. To put it mildly, he made Mulder seem insane.[44]

41. J. Liebig, "Some experimental researches tending to disprove the existence of protein," *Lancet* 1846, i, 206–7.
42. Snelders, "The Mulder–Liebig controversy," p. 212.
43. *Ibid.*, p. 213.
44. J. Liebig, "Zur Charakteristik des Hrn Prof Mulder in Utrecht," *Ann. Chem.* 62(1847), follows p. 128. Liebig actually directed that the article was *not* to be included in the bound volume.

In retrospect it is easy to see that both men were right and wrong. There is not a homogeneous protein but a whole series of proteins, some containing sulphur and phosphorus, others not. This was not well understood before the early twentieth century. Mulder, who was undoubtedly a skilled analyst, cannot be dismissed as a dilettante for interpreting his results in terms of the prevailing radical theory. Liebig at first had done the same. The impartial historian cannot but think that Liebig was unfair to Mulder. To disguise his own change of mind and in a murderous mood from his reception by Berzelius, he made poor Mulder a convenient scapegoat who turned out not to take the scourging lying down. English reviewers of Mulder's pamphlet, although recognizing Liebig's genius, were in no doubt that Mulder had been cruelly mistreated.[45]

The major effects of Berzelius's criticisms and the controversy with Mulder seem to have been twofold. In the first place, although Liebig published a third edition of *Animal Chemistry* in 1846, which was shorn of its speculative equations, unlike *Agricultural Chemistry,* no further editions were ever produced.[46] With Mulder disproved, the whole of the former Part 2 was undermined and had to be replaced by the incorporation of some further conclusions that were also separately published in 1846 as *Chemistry and Physics in Relation to Physiology and Pathology.* The new edition was announced as Part 1, but Part 2 was not published, unless *Chemistry of Food* was it. One reason there was no fourth edition may well have been that Liebig found it easier to incorporate his teachings into revisions of the popular *Chemical Letters.* The third edition of the latter in 1851 included a great deal of animal chemistry. A second, important consequence, which was very different again from *Agricultural Chemistry,* was that Liebig admitted that until chemists were trained in physiology, no further progress could be made on the finer details of organic metamorphosis in animal systems. In the meantime, the only thing that could be done was to invite the collaboration of a physiologist.

45. E.g., Liebig's conduct was "overbearing and capricious," *Medico-Chirurgical Review* [3]5(1847), 247–8. See M. Rossiter, *The Emergence of Agricultural Science* (1975), pp. 105–7, for visitors' perceptions of Mulder's state of mind at this time.

46. The second edition of 1843 was more or less a reprint of the first edition. In the third English edition (1846), the original Part 2 was replaced by a didactic essay, "Chemistry and Physics in Relation to Physiology and Pathology," which Liebig had published in *Deutsche Vieteljahrschaft* (1846, Heft 3), 169–243. This was serialized in rival translations by *Monthly Journal Medical Science* 7(1846–7), 262–70, 337–46, and the more rapid *Lancet,* 1846, ii, 352–4 et seq. This was then published separately in English by Ballière in 1846 and was reproduced in America in the inappropriately titled *Complete Works of Liebig* (Philadelphia, 1852). The third German edition was titled *Die Thier-Chemie* (Braunschweig, 1847) and appeared after the English version.

Germany's principal medical teacher Johannes Müller, who had probably been influenced by Liebig's views on vitalism, had been surprisingly warm towards Liebig's *Animal Chemistry,* hailing it in the fourth edition of his *Handbuch der Physiologie des Menschens* (Berlin, 1842) as providing new insights into the relations between respiration and nutrition. As Holmes has noted:

> Müller understood . . . that Liebig's work heralded such a thoroughgoing change that it could not simply be incorporated into the older synthesis which he had established. Physiology based on comparative anatomy would give way to physiology based on chemistry and physics.[47]

Müller himself was unwilling to make such an alteration to his research path; it was one of his students, Theodor Bischoff (1807–82), who took up the challenge. When Bischoff reviewed *Animal Chemistry* in Müller's *Archiv für Anatomie, Physiologie, Wissenschaftlich Medizin,* Liebig was impressed by the way he had succeeded in balancing perceptive criticism with enthusiasm for the overall approach.[48] Therefore in 1843 he wrote to Bischoff, who was teaching at Heidelberg, inviting him to list weak spots in his argument so that they might be remedied in a future edition. When Giessen's conservative professor of anatomy J. B. Wilbrand, who still denied the circulation of the blood, published a tract against the use of chemistry in physiology, Liebig seized his chance. He wrote to Vice-Chancellor Linde in May 1843 stating that medical education in Giessen had become the laughingstock of Europe and was in great danger of losing its attraction for students. He said further that because his own *Animal Chemistry* had obviously demonstrated the direction that future medical research should take, what Giessen most needed was a physiologist who could work with Liebig in furthering its programme. He strongly recommended that Bischoff should be called from Heidelberg.[49] It says much for the weight that Liebig's views then carried with the rector and with the Ministry of Education that he had his way. Bischoff joined the medical school in November 1843.

Unfortunately, because Wilbrand had so neglected the subject at Giessen, Bischoff had to spend several years organizing and building an anatomical-physiological institute. He and Liebig were therefore unable

47. Holmes, *Animal Chemistry* (1964) p. lxix.

48. *Ibid.,* p. lxx.

49. Liebig to Linde, 4 May 1843, in E. M. Felschow and E. Heuser, eds., *Universität und Ministerium im Vormärz* (Giessen, 1992), p. 171.

to begin any serious collaboration until 1850.[50] Interestingly, this was the same year that Liebig's son Georg, who was working in Berlin, showed that isolated frog muscle could contract for a longer time in the presence of oxygen than in other gases and concluded that respiratory exchanges took place in muscle rather than in the capillaries of the arterial-venous system.[51] This work implied that rather more was happening when muscles worked than the degradation of nitrogeneous flesh, but this implication was noticed neither by his father nor by Bischoff. The latter suggested that they should test and refine Liebig's claim that urea formation was directly related to the rate of tissue consumption (*Stoffwechsel*). The immediate problem for Liebig was to devise an easier and faster way of analyzing the urea content of urine. He solved this problem ingeniously in 1851 with an extremely influential quantitative analytical technique based upon the titration of urine with mercuric nitrate.[52]

During the interim, a number of new investigations seemed to suggest that Liebig's assumption that *all* alimentary nitrogen was used to make blood and tissues was incorrect. In an elaborate series of food-excretion exchange experiments made at the faraway German University of Dorpat in Latvia, Friedrich Bidder, a professor of physiology, and Carl Schmidt, a former student of Liebig's, concluded that Liebig was wrong to assume that all the urea came from the degradation of tissue nitrogen. The urea output of fasting animals was well below that of animals fed on a nitrogenous diet, forcing them to conclude that nitrogenous materials in excess of that used by fasting animals was directly oxidized in the blood, just like fats and carbohydrates. This *Luxusconsumption,* as they named it,[53] was disturbing to Liebig and Bischoff but was not confirmed by Bischoff's own experiments with a large dog. In practice, however, Bischoff had to make a number of assumptions and to admit that just possibly there were other pathways by which nitrogen was metabolized.

By the time his work was published, Bischoff had lost his collaborator, for Liebig had gone to Munich in October 1852. Liebig, however, was determined to continue the relationship. After cajoling both King Max-

50. F. L. Holmes, "The formation of the Munich school of metabolism," in W. Coleman and F. L. Holmes, eds., *The Investigative Enterprise. Experimental Physiology in Nineteenth-Century Medicine* (Berkeley, Cal., 1988), pp. 179–210; *idem.*, "The intake-output method of quantification in physiology," *Hist. Studies Phys. Bio. Science* 17(1987), 235–70.

51. G. Liebig, "Ueber die Respiration der Muskeln," *Arch. Anat. Phys.* (1850), 393–416.

52. J. Liebig, "Ueber einige Harnstoff Verbindungen und eine neue Methode zur Bestimung von Kochsalz und Harnstoff im Harn," *Ann. Chem.* 85(1853), 289–329; W. H. Brock, ed., *Liebig und Hofmann in ihren Briefen* (Weinheim, 1984), p. 134. See T. L. W. Bischoff, *Der Harnstoff als Maass des Stoffwechsels* (Giessen, 1853).

53. F. Bidder and C. Schmidt, *Die Verdauungssaefte und die Stoffwechsel* (Mitau, 1852).

imilian and Bischoff (who was naturally reluctant to leave Giessen after building up a decent department), Bischoff was appointed professor of physiology at the new prestigious Physiological Institute in April 1855. Liebig's role in the further development of this Munich School of Metabolism, as Holmes has called it, was muted; he had returned to agricultural chemistry and was busy with mirrors, socializing, and enjoying being chief scientific advisor to the Bavarian court. Bischoff therefore took on a new collaborator, Carl Voit (1831–1908), who was both medically and chemically trained (he had worked with Wöhler and had listened to Liebig's lectures in Munich). Voit rapidly outstripped Bischoff in energy and commitment, and within a decade he had become the research school's moving force.

Following a lengthy series of ingeniously varied animal experiments, Bischoff and Voit were satisfied by 1860 that Liebig had been correct – urea really was a satisfactory measure of *Stoffwechsel*.[54] Still not satisfied that the conclusion was watertight until the balance sheet of imput and output was made to include the gaseous exchanges, with King Maximilian and Liebig's support and with the aid of the hygienist Max Pettenkofer, a huge respiratory chamber was built in the Physiological Institute. The results in 1863 were once more a triumphant vindication of Liebig's distinction between plastic and respiratory foods and the concept of the *Stoffwechsel*'s correlation with urea output.

But no sooner had this "high point of efforts to build an experimental science of nutrition upon the speculations of Liebig's *Animal Chemistry*" appeared, than the edifice began to collapse.[55] In part this was due to the work of Claude Bernard in France and a growing realization that "the method of balances" did not enable one to say with confidence anything about the internal fate of nutrients. In part, too, it was a result of a growing consciousness of the principle of conservation of energy, which had been first announced by Helmholtz in 1847.[56] Was it likely that one set of chemical reactions (*Stoffwechsel*) would produce muscular work and another (respiration exchanges) would produce animal heat? In other chemical reactions where work was done (as in the explosion of gunpowder), heat always accompanied the reaction. Why should organic systems be any different?

More immediately, however, the edifice's collapse was caused by Voit's decision to probe the deduction that increased muscular work ought to

54. T. L. W. Bischoff and C. Voit, *Die Gesetze der Ernährung des Fleischfressers durch neue Untersuchungen festgestellt* (Leipzig, 1860).
55. Holmes, *Animal Chemistry* (1964), p. xcvi.
56. H. Helmholz, *Über die Erhaltung der Kraft* (Berlin, 1847).

increase urea output. A dog put to work on a treadmill did not increase its urea output. Although Voit was able to explain this outcome by assuming that muscles degrade continuously, their energy being stored electrically until it was needed, others, like Moritz Traube (1826–94), felt there was a serious flaw in the model. The English physician Golding Bird (1814–54) had pointed out as early as 1843 that if uric acid was the first product of nitrogeneous tissue degradation, followed by urea, the determination of uric acid in human urine would be a measure of a patient's degree of oxidation (or metabolic rate). It followed that in patients suffering from anemia, where *less* oxygen was circulated, there ought to be more uric acid in the urine. Correspondingly, in fever cases there should be less. Bird claimed to find that the opposite of the predictions occurred in both clinical situations. In any case, he observed, why did birds, with their high metabolic rate, produce uric acid to the exclusion of urea? He therefore concluded that Liebig's account of metamorphosis was flawed in detail, though not in principle.[57]

In 1861 Moritz Traube, a former student of Liebig's whose personal circumstances had forced him into the wine trade, where he could devote only his leisure time to science, focused on an overlooked but obvious zoological difficulty.[58] Animals that were commonly used for haulage, such as horses and oxen, and that expended huge amounts of energy, were all herbivores. Yet herbivores ate little protein. A devastatingly simple calculation showed that a working horse could not possibly derive its energy from nitrogenous aliments alone. Stalemate ensued, but five years later Traube's point was put to human test by a chemist, Johannes Wislicenus, and a physiologist, Adolf Fick, when they climbed the Faulhorn in the Bernese Oberland in 1866. Their experiment had actually been designed by Fick's brother-in-law, the English chemist Edward Frankland, but at the last minute he was unable to join the climb. Knowing the height of the mountain, they could calculate the work done during the ascent. From the nitrogen content of the urine they excreted and collected during the climb, they could calculate the protein equivalent, which, according to Liebig, Bischoff and Voit, must have been degraded. And from this protein equivalent they could calculate an energy equivalent. Inevitably they had to make many assumptions, but their result was clear; they had done *more* work than could have been derived from plastic foods alone. This demonstration was reinforced by Frankland's determination of the heat content

57. G. Bird, "Remarks on Prof. Liebig's views," *London Medical Gazette,* 31(1843), 761; see N. G. Coley, "The collateral sciences in the work of Golding Bird," *Medical History* 13(1969), 363–76.

58. M. Traube, "Ueber die Beziehung der Respiration zur Muskelthätigkeit," *Arch. Pathol. Anat.* 21(1861), 386–98.

of staple foods in a bomb calorimeter – an instrument that rapidly provided a new methodology for nutritional research.[59]

Although Voit remained unconvinced for a time, using his electrical hypothesis to save the phenomenon, by the late 1860s he had come to accept that metabolism was far more complex than Liebig's original programme had allowed. Liebig, who respected Wislicenus and especially Frankland, agreed in 1870 in his final word on animal chemistry that "we have lost urea as a measure of work."[60] Liebig also agreed that some alimentary protein might be oxidized during respiration, just as fats and carbohydrates were. But he would not waver from his contention that the degradation of tissue protein was the sole source of muscular work. To explain this point, he fell back on Voit's suggestion that animals stored energy electrically but specifically in the compound creatine that he had first discovered in muscle in 1847. To the cynically minded, this creative explanation had the convenience of implying that it enhanced the value of his Extract of Meat (see Chapter 8).

Unfortunately, as Liebig had so often done, he went on to spoil a beautiful friendship. In criticizing Voit's growing doubts about the overall validity of Liebig's system and what appeared to Liebig to be the growing anarchy of physiological research without a model, he mortally offended Voit, just as he had done Mulder three decades before. Like Mulder, Voit replied with a tirade. Released from his inhibitions and honest doubts, he destroyed Liebig's complete system from the plastic-respiratory divide to the claim that the metamorphosis of nitrogenous tissue compounds was the only source of muscular work. And just as Liebig had whiplashed Berzelius for being an old man out of touch, Liebig now found himself scorned by a younger generation:

> [Liebig] stands still on the ground he created 25 years ago; from chemical experiments he attempts to draw conclusions and analogies about the processes within the animal body; he has brought forth a great effect through his ideas, and from his creative effort has developed the entire movement to study the decompositions within the animal body. But he has forgotten, to the sorrow of those who know and value his . . . high services to science better than do his flatterers, that these are all mere ideas and possibilities, whose validity

59. A. Fick and J. Wislicenus, "On the origins of muscular power," *Phil. Mag.* 31(1866), 485–503; E. Frankland, "On the origin of muscular power," *Phil. Mag.* 32(1866), 182–99.

60. J. Liebig, "Ueber die Gährung und die Quelle der Muskelkraft," *Ann. Chem.* 153(1870), 1–46, 137–57, 157–228. The first two parts on alcohol and acetic acid fermentation had begun as lectures to the Bavarian Academy on 9 May 1868 and 5 November 1869. The complete paper was also offprinted (Leipzig, 1870).

> must first be tested through investigations on animals; and this is the basis on which I have placed myself.[61]

To rub this in, Voit had the cheek to quote Liebig's similar criticism of Berzelius twenty-six years before. Liebig did not reply, though he leapt to the defence of his Extract of Meat, the supposed nutritious qualities of which had been undermined by the force of Voit's criticisms.

The Phenomena of Motion in the Animal Organism

In the second part of *Agricultural Chemistry* (1840 edition), Liebig had proposed a purely chemical explanation of fermentation: decay (eremacausis), and putrefaction. Liebig appealed to Berzelius's adoption of catalysis, though not of catalytic force, as an explanation of why, in certain reactions, one substance in contact with another caused a decomposition, whereas either one by itself did not. For example, platinum did not decompose nitric acid, but once it was made into the form of a platinum-silver alloy, it readily did so. After summarizing several examples, Liebig concluded:

> Now no other explanation of these phenomena can be given, than that a body in the act of combination or decomposition enables another body, with which it is in contact, to enter into the same state. It is evident that an active state of the atoms of one body has an influence upon the atoms of a body in contact with it; and if these atoms are capable of the same change as the former, they likewise undergo that change; and combinations and decompositions are the consequence. (p. 262)

By way of explanation, Liebig looked to the Newtonian tradition of Laplace and Berthollet and to disturbances of equilibria by motion. (Liebig's model has resonances both with our contemporary theories of bonding and chemical change and with the theory of fermentation given by Stahl in the early eighteenth century.) Organic compounds, because of the large numbers of atoms they contained, were, he believed, inherently unstable and, as isomerism showed, subject to rearrangements. The fermentation of sugar, which stood for the type of all fermentations, consisted in the rearrangement "of its atoms, and by their union with the elements of water" (p. 282). This much had been stated by Lavoisier in 1789. Yeast was merely a contact trigger and took "no appreciable part in the transposition of the elements of the sugar; for in the products resulting

61. C. Voit, "Ueber den Einfluss der Kohlehydrate auf den Eisweissverbrauch," *Zeitschrift für Biologie* 6(1870), 303–401. By being published in a biology journal, Voit's critique may have escaped the attention of chemists.

from the action, we find no component part of this substance" (p. 283). Instead, Liebig saw yeast, or ferment, as a substance whose atoms were in "the act of transposition," in other words, instability. It had "all the characters of a compound of nitrogen in the state of putrefaction and eremacausis" (p. 285). He said further:

> Yeast is a product of the decomposition of gluten; but it passes into a second stage of decomposition when in contact with water. On account of its being in this state of further change, yeast excites fermentation in a fresh solution of sugar, and if this saccharine fluid should contain gluten (should it be *wort,* for example) yeast is again generated in consequence of the transposition of the elements of the sugar exciting a similar change in this gluten. After this explanation, the idea that yeast reproduces itself as weeds [do], cannot for a moment be entertained. (p. 366)

This last cryptic comment was a thrust at the microscopists Charles Cagniard-Latour and Theodor Schwann, though Liebig made no mention of them by name or of their efforts to give yeast a biological role in fermentation. He did, however, give a reference (p. 380) to the notorious satire by Wöhler that had been published in *Annalen der Chemie* in 1839:

> They have a stomach and an intestinal canal, and their urinary organs can be readily distinguished. The moment these animals are hatched they begin to devour the sugar in the solution, which can be readily seen entering their stomachs. It is then immediately digested, and the digested product can be recognized with certainty in the excreta from the intestinal canal. In a word, these infusoria eat sugar, excrete alcohol from their anuses, and carbonic acid from their urinary organs. The bladder, when full, is the shape of champagne bottle, when empty it resembles a little ball.[62]

Liebig's adoption and advocacy of a chemical model of fermentation was standard and had the full support of Berzelius. Like Mitscherlich, however, Liebig did not like the term "catalytic force," which sounded mysterious, and both men opted for the necessity of mechanical contact rather than an "influence," which implied vitalism. It is impossible to tell how familiar, if at all, Liebig was with brewing practice. But if he was, it

62. S. C. H. Windler [F. Wöhler], "Das Geheimnis der Gärung," *Ann. Chem.* 29 (1839), 100–10. This was in response to a paper by P. J. F. Turpin, *Annalen der Chemie* 29(1839), 87–99. There is a good modern English translation of this satire, "On the riddle of vinous fermentation," in the *J. Chem. Educ.* 67(1990), 552–4. For the controversies over fermentation, see J. S. Fruton, *Molecules and Life* (New York, 1972), pp. 22–86; and R. G. Anderson, "Yeast and the Victorian brewers," *J. Institute Brewers* 95(1989), 337–45.

would probably have been with Czech-Bavarian bottom fermentation in which the quantitative growth of yeast during fermentation is less obvious than in English top brewing. British brewers, including Liebig's pupil Heinrich Böttinger (1820–74), who worked for Allsopp's brewery in Burton, appear to have been convinced of the living nature of yeast in the 1850s.

Eremacausis was a slow decaying process caused by gradual oxidation of organic material in air. In some cases, oxidation occurred with the evolution of carbon dioxide; in others, no carbonic acid was involved. In both cases, Liebig pictured decomposition and rearrangement of atoms as caused by a conflict of affinities promoted by an external cause, namely, the presence of air. For example, the juices of berries were perfectly stable until exposed to the atmosphere. This explained the success of methods of preserving vegetables and animal food by heating them to boiling point in "vessels from which the air is completely excluded." Oxygen caused the decomposition by converting the atoms from a state of rest into motion. Once this "intestinal motion" began, the presence of oxygen was no longer necessary. It acted as a trigger only. Another trigger was contact with another substance that was already in an eremacausic state. Decaying wood caused fresh wood to decay. This was an example of "true contagion." A final chapter of *Agricultural Chemistry* therefore looked forward to *Animal Chemistry* insofar as it discussed the connections between eremacausis and disease.

If yeast reproduced itself as a result of "exciting" the decomposition of sugar, other "exciters" might act to reproduce contagia, their multiplication being a consequence of the decomposition of the liquids they excited. If putrefying blood, pus, and other animal products came into contact with fresh wounds, "vomiting, debility, and at length death" occurred. (Here Liebig noted the dangers of small cuts in anatomical dissections, and in later editions, he referred approvingly to the obstetric work of Semelweiss in Vienna.)

Hitherto, said Liebig, a principle of life had been ascribed to contagious matter, or the "morbid virus" – "a life similar to that possessed by the germ of a seed, which enables it under favourable conditions to develop and multiply itself" (p. 371). The analogy was fine, he thought, except that the same interpretation applied to the majority of reactions in organic chemistry. "Life would, according to that view, be admitted to exist in every body in which chemical forces act" (p. 372). Accordingly, it was preferable to view diseases as "chemical processes dependent upon the common chemical forces." In any case, "life" implied form and organization, which were not characteristic of contagia, the influence of which was solely chemical in action. He wrote in *Agricultural Chemistry:*

> All the supposed proofs of the vitality of contagions are merely ideas and figurative representations, fitted to render the phenomena more easy of apprehension by our senses, without explaining them. These figurative expressions, with which we are so willingly and easily satisfied in all sciences, are the foes of all inquiries into the mysteries of nature; they are like the *fata morgana* [a mirage seen on the Calabrian coast of Sicily], which shows us deceitful views of seas, fertile fields, and luscious fruits, but leave us languishing when we have most need of what they promise. (p. 381)

Liebig's views had obvious implications for doctors and sanitarians, because in *Animal Chemistry,* he interpreted blood as the body's chemical factory for synthesis and degradation. It was, therefore, a vulnerable site for morbid decomposition. Normal, healthy blood contained materials easily affected by externally induced contagia. Children and youths were particularly susceptible to attack because their blood contained substances not present in nongrowing adults. Fortunately, the diseases were benign if excreted, because the substances affected were usually not essential for life. But if the function of secreting organs was impeded, the skin, lungs, and other organs might seek to eliminate those secreting organs and hence spread contagion. Such possibilities could be affected by

> the manner of living, or by the nutriment taken by an individual. A superabundance of strong and otherwise wholesome food may produce them, as well as a deficiency of nutriment, uncleanliness, or even the use of decayed substances as food. (p. 387)

Because such conditions for contagion were accidental, they were also preventable.

In an abstract chemical sense, reproduction of a contagion depended upon the presence of two substances, one of which became completely decomposed, communicating its own state of transformation to the second. The latter thus thrown into a state of decomposition was the newly formed contagion. If the patient was lucky, the body was able to assimilate the transformed constituent, albeit abnormal secretions (e.g., in the urine) might accompany the process of convalescence. This explained why cowpox vaccination worked against smallpox. The cowpox vaccine destroyed an accidental constituent of blood by changing it into a harmless molecule, whereas the same material was decomposed by smallpox virus into a malignant contagion. Revaccination was necessary, however, in case the same accidental substance might emerge in the blood at a later date.

Venous blood in the lungs was particularly vulnerable to eremacausic attacks from miasma (the products of decay and putrefaction of animal and vegetable substances, including dung) or the chemicals carried with it. Because the smell of miasmas was largely ammoniacal, and ammonia was produced in hospital rooms containing patients affected by contagious diseases, the evaporation of hydrochloric acid, acetic acid, or nitric acid into the air to neutralize miasmic smell was bound to be beneficial.

As can be seen from this brief summary, once again Liebig seemed to have succeeded in bringing together and explaining comprehensively and simply a wealth of experiences associated with disease and to have proffered a programme of action from vaccination to disinfection, the removal of rubbish and sewage heaps to the nutrition of individuals.

Liebig's treatment of disease as a dynamic chemical process promoted by morbid poisons and not as moral judgement had a very considerable influence on British doctors. Its attraction was to be all things to all men, insofar as it could appeal to strict contagionists as well as to miasmatist sanitarians. He could be criticized for not being medically experienced, just as he was for not being a farmer, but the case he made for interpreting diseases as *chemical* processes seemed worth following up, now that organic analysis had reached maturity. Liebig's close connection with English medical sanitarians through his pupils and *The Lancet,* as well as with Edwin Chadwick himself, and his influence on the adoption of the argument by analogy during the course of the long controversies over the nature of cholera, ensured that his contribution to the public health debate in Britain was very significant.

Much of Liebig's work was conveyed to British medical men through *The Lancet,* the great weekly reforming engine of medicine edited by Thomas Wakely. Not only did it continually review Liebig's work; it also carried translations of his chemical lectures. Between 1842 and 1843, it ran a series of articles by the dispensary surgeon Henry Ancell (1802–63) that made Liebig's views on animal chemistry more assimilable. Ancell's account was even translated into German.[63] For Wakely, Liebig was simply grist to his mill of establishing "a new era of medicine," but it did mean that the journal was genuinely supportive of the venture to establish in 1845 a Giessen type of laboratory in London, the Royal College of Chemistry.

Perhaps the most direct application of Liebig's theory of disease was that made by his pupil Henry Bence Jones (1813–73). While reading medicine at University College, London, where he learned chemistry from another former Liebig pupil George Fownes, Bence Jones had become interested in the analysis of bladder stones. Upon qualification in 1841, he

63. H. Ancell, "Liebig, his chemistry and reviewers," *Lancet* (1842–3), i, 230–35; M. Pelling, *Cholera, Fever and English Medicine 1825–1865* (Oxford, 1978).

went to Giessen, where Liebig's developing views on chemical physiology "appeared like a new light where all had been confusion and incomprehensible before."[64] At one point in *Animal Chemistry*, Liebig had noted that the mulberry calculus (calcium oxalate), which formed in patients who suffered from bladder stone, were found in the type of patient who took little exercise and who therefore suffered from oxygen deficiency. These patients' low rate of metabolism – as we would now say – brought about a pathological degradation of nitrogenous tissue during their limited exercise and work. This diagnosis suggested a treatment to Bence Jones, namely, the application of oxygen therapy, which he tested accordingly in the wards of St. George's Hospital in London. In his treatise, *On Gravel, Calculus and Gout: Chiefly an Application of Professor Liebig's Physiology to the Prevention and Cure of These Diseases* (London, 1842), he asserted the value of his and Liebig's insights without, however, submitting them to experimental test. As we have seen, his fellow physician Golding Bird was more circumspect.

In 1837, William Farr (1807–83), who had studied medicine in London and Paris, became the compiler of abstracts at the newly created Registrar-General's Office. Within a couple of years, he had drawn up a classification of diseases for death certificates that would, he hoped, enable meaningful statistical analyses to be made. This scheme included the category of "epidemic, endemic and contagious" diseases that "were of a specific nature, propagated in a peculiar manner and known by experience to become epidemic in unhealthy places and among the sickly classes, at greater or less intervals of time."[65] Farr, who was a friend of Robert Dundas Thomson, a nephew of Thomas Thomson's who had studied with Liebig in 1840, was obviously struck by the putrefaction explanation of this group of diseases that Liebig provided in 1840 and 1842. Hence, when he revised his nosological scheme in 1842, he renamed the epidemic group "zymotic diseases" after *zymosis*, a Greek word for fermentation. Farr had clearly accepted that this group of diseases, which included cholera, typhoid, typhus, and scarlet fever, were all blood diseases that arose from specific chemical poisons derived either from without or generated within the victim's body.

The Liebig–Farr category of zymotic diseases was to remain a feature of the Registrar-General's statistical returns for the remainder of the century. It is no accident that the first free ferment, or enzyme, which was isolated

64. N. G. Coley, "Henry Bence Jones, MD, FRS," *Notes and Records Royal Society* 28(1973–4), 31–56, quoted at p. 35.

65. *Parliamentary Papers* 1839, XVI. *Annual Report Registrar-General,* First Report, Appendix, p. 67, discussed by John M. Eyler, *Victorian Social Medicine. The Ideas and Methods of William Farr* (Baltimore and London, 1979), Chap. 5.

by Eduard Büchner in 1897, was named *zymase*. The advantage of the zymotic typology was that it could promote both the sanitarians' environmental campaigns and at the same time accommodate the contagionists' models of zymotic particles in the air or water that acted as exciters in Liebig's sense. As Margaret Pelling has demonstrated, theories of disease between the 1840s and 1880s, when the bacteriological germ theory became firmly established, cannot be understood unless it is realized that doctors and physiologists were continually using Liebig's chemical process model as a theme for their own variations.[66]

This is true, for example, of John Snow (1813–58) and William Budd (1811–80), who are both renowned for suggesting that cholera (and in Budd's case, typhoid fever as well) were waterborne diseases. Snow's connection with Liebig is seen immediately in his phrase "morbid matter," the material he supposed was swallowed by healthy people in water that had been contaminated by the evacuations of cholera victims. Because Snow's theory was right when viewed from the vantage point of bacteriology, it has been easy to see him as a pioneer germ theorist. But his initial, purely speculative view, which was first made known in 1849, was clearly derived from Liebig. His principal originality was in claiming water, rather than air, as the vector of transmission.

Following the 1854 cholera epidmic in London, during which Snow demonstrated the probability that particular local outbreaks could be attributed to a well in Soho or to the water supplied by a particular water company in Lambeth, he referred to the morbid matter as organized. However, this description has to be interpreted in terms of a little-known essay he published the year before with the Liebigian title, *Continuous Molecular Change* (London, 1853). Although, as Pelling has shown, this moved far from Liebig's position, its main point being the impossibility of separating vital from chemical phenomena since "all changes of composition . . . whether taking place within the living body or not, are alike the result of the attraction or affinity which exists among the ultimate atoms or molecules of matters."[67] So when Snow referred to morbid matter in his monograph on cholera in 1854, it may be assumed that he viewed it as alive in his own peculiar chemical sense.

But how could such an entity be detected analytically?

The way forward seemed to lie in the detection of nitrogen in water. Liebig's pupils Angus Smith and R. D. Thomson had searched for decom-

66. M. Pelling, *Cholera, Fever and English Medicine* (Oxford, 1978), p. 93; and C. Hamlin, *What Becomes of Pollution* (PhD thesis, University of Wisconsin, 1988; reprint Garland, New York, 1988), pp. 105–27.

67. Pelling, *Cholera,* p. 211.

posing matter in water in the form of unstable nitrogenous compounds in the 1850s; and a decade later their study became the basis of Edward Frankland's remarkable work in which he alerted sanitary authorities to water pollution with the concept of "previous sewage contamination."[68]

The tendency of the Liebigian process model to push doctors towards a search for specific morbid poisons reached its climax in 1854 when Liebig publicized the work of his son-in-law, the Munich surgeon Karl Thiersch. In his experiments Thiersch allowed the stomach contents of animals to putrefy. The material was then soaked in filter papers and fed to mice in weighed amounts over a period of four days. Whereas fresh stomach contents did the mice no harm, the putrefied materials caused severe discomfort in fifty-six mice, fourteen of which died. For the first time there was a clear demonstration of the transmission of a cholera type of morbidity from a digestive source. Assuming that such cholera putrefaction gained entry to water supplies, Snow's argument suddenly seemed more plausible.

Liebig reported the results to Hofmann in London and urged him to publicize them, which was duly done in the *Medical Times*.[69] Liebig saw this episode as a remarkable vindication of his *Animal Chemistry*.

> Other things have surprised me more than that my theory of contagion should triumph on this occasion, and I was wholly convinced that these diseases are generated and propagated by means of a kind of ferment. It is nonetheless particularly gratifying, when one has participated in the emergence of a doctrine.[70]

As a consequence, during the final English cholera epidemic of 1866, Liebig's Anglo-German pupil John Thudichum, who was employed by John Simon, the Privy Council's medical officer, was asked to examine the fermentative state of cholera excretions. Imaginatively, Thudichum used the latest method of spectroscopic analysis, which was to become the hallmark of his research on brain chemistry. Inevitably, however, his findings were negative; microscopical rather than spectroscopical technology was to be necessary.

One major difficulty with Liebig's conception of disease was self-consistency. As Hamlin has shown, when Liebig wore his agriculturist's hat, he seemed to advocate the beneficent effects of human sewage as a manure (Chapter 9), but as a chemical physiologist, he identified decom-

68. See C. Hamlin, *A Science of Impurity. Water Analysis in Nineteenth-Century Britain* (Berkeley, 1990).

69. J. Liebig, "Etiology of Cholera," *Medical Times* (1854), ii, 515; Sir James Clark to Hofmann, 2 November 1854 in Brock, *Liebig–Hofmann Briefwechsel*, p. 183.

70. Hofmann, *Liebig's und Wöhler's Briefwechsel*, vol. 2, pp. 20–1.

posing nitrogenous materials as potential morbid ferments. One way of reconciling these opposites was achieved in the early 1850s by Philip Gosse, the naturalist, and Robert Warington, the pharmacist, who independently developed the aquarium. Although no one has been able to show that Liebig's *Animal Chemistry* was the inspiration for the aquarium, it was a most elegant demonstration of Liebigian cycles and of nature's overall bounty. In this respect, it is ironic that Liebig and Wöhler's satire against vitalistic theories of fermentation in 1839 had hit upon the key to the aquarium, namely, the balance achieved through having living scavengers restore the chemical equilibrium by devouring the excretions of others, their own excretions (together with oxygen) then forming the foods of fish or anemones or other plant life. This is a pointed reminder of how difficult it was to believe Pasteur's contention that bacteria and fungi were the causes of fermentation and disease; it was just as easy to believe that they were opportunist scavengers that helped the processes of decomposition that Liebig had postulated.[71]

Liebig was wonderfully struck by Warington's work and popularized aquaria in Munich, using Hofmann as his British agent. Pragmatically, the aquarium supported the efforts of sanitarians to pass human refuse through piped sewerage systems and to recycle it onto the land where the soil would absorb its potentially dangerous morbid qualities. The aquarium also suggested the power of biological self-purification that rivers and seas possessed.

It was not until the 1860s that Louis Pasteur turned his attention to the idea that the microorganisms that he had found responsible for the spoiling of wines and beers might also be the causes of animal and human diseases. Almost incidentally as part of the controversy he unleashed as to whether spontaneous generation was possible, he had demolished Liebig's contention that fermentations and putrefactions (and hence "diseases") depended upon the excitation of a nitrogenous molecule into a pathologically dangerous decomposition product. He did this simply and elegantly by demonstrating that yeast and other microorganisms could be grown in media that were devoid of nitrogen; moreover, he showed that acetic acid would not ferment to vinegar except in the presence of the organism *Mycoderma aceti.*

Liebig began a counterattack in public lectures in Munich in 1868, publishing his formal reply in a long paper in 1870. He now accepted that yeast was a living organism, but he refused to budge from his original position on the essential chemical nature of fermentation.

71. Hamlin, *Science of Impurity,* Chap. 4.

> I assumed that the breakdown of the fermentable substance to simpler compounds must be explained by a process of cleavage residing in the ferment. . . . The rearrangement of the sugar atoms in the sugar molecule is therefore a consequence of the decomposition or rearrangement of one or several constituents of the ferment, and occurs only when they are in contact.[72]

And, he went on,

> it may be that the physiological process stands in no other relation to the fermentation process than the following: a substance is produced in the living cells which, through an operation similar to that of emulsin on salicin and amygdalin, leads to the decomposition of sugar and other organic molecules; the physiological process would in this case be necessary to produce this substance but would stand in no further relation to the fermentation.[73]

As for acetic acid, he lambasted Pasteur with the evidence of Döbereiner in 1831 that vinegar could be prepared from acetic acid in the presence of a platinum black catalyst. No life was involved.

Pasteur's cool reply, composed in the middle of the Franco-Prussian War, mocked Liebig's position as Voit had done, and he left it to posterity to decide whose viewpoint was correct.[74] Posterity decided, as so often in such cases, that both parties were right. Indeed, it is unfortunate that Liebig never explained what he meant by the term "ferment" as opposed to the yeast itself. Given its obvious vitality, the yeast could hardly have been dying or decomposing. As early as 1858, Moritz Traube had commented:

> Even if all putrefactions depended on the presence of infusoria or fungi, a healthy science would not block the road to further research by means of such an hypothesis; it would simply conclude from these facts that the microscopic organisms contain certain substances which elicit the phenomenon of decomposition. It would attempt to isolate these substances, and if they could not be isolated without changed properties, it would only conclude that all the separation methods had exerted a deleterious effect on these substances.[75]

Twenty years later, following Liebig's death, Traube, by now admired as a professional scientist, reiterated his point:

72. J. Liebig, "Ueber die Gährung" *Ann. Chem.* 153(1870), 1. See G. E. Hein, "The Liebig–Pasteur controversy," *J. Chem. Educ.* 38(1961), 614–19.
73. J. Liebig, "Über die Gährung," *Ann. Chem.* 153(1870), 6; see H. Finegold, "The Liebig–Pasteur controversy," *J. Chem. Educ.*, 31(1954), 403–6.
74. L. Pasteur, "Note sur un mémoire de M. Liebig rélatif aux fermentations," *Comptes rendus* 73(1871), 1419–24.
75. Cited by Fruton, *Molecules and Life*, p. 62.

> The ferments are not, as Liebig assumed, substances in a state of decomposition, and which can transmit to ordinary inert substances their chemical action, but are chemical substances, related to the albuminoid bodies [i.e., proteins] which, although not yet accessible in pure form, have like all other substances a definite chemical composition and even evoke changes in other substances through definite chemical affinities. Schwann's hypothesis (later adopted by Pasteur) according to which fermentations are regarded as the expressions of the vital forces of lower organisms is unsatisfactory. . . . The reverse of Schwann's hypothesis is correct: Ferments are the causes of the most important vital-chemical processes, and not only in lower organisms, but in higher organisms as well.[76]

But this support came too late for Liebig who, according to Volhard, was "shattered" by Pasteur's reply.[77]

From a twentieth-century perspective, nearly everything in *Animal Chemistry* was wrong. Although it is still useful to classify foods into carbohydrates, fats, and proteins, this classification was due to Prout, not Liebig. Emil Fischer's work at the beginning of the twentieth century revealed that proteins differed widely in structural composition, their only common feature being that they were composed from a small range of amino acids, some of which had been first identified by Liebig. Since the plant or meat aliments we imbibe in food are not simply absorbed into the flesh if nitrogenous or oxidized if nonnitrogenous but are all degraded during digestion into simpler entities such as amino acids before being reassembled into the proteins of flesh, blood, and bones, the dichotomy of plastic and respiratory foods is redundant. Liebig's bile theory proved to be complete invention. Muscle is now known to oxidize carbohydrate rather than protein during exercise and to oxidize its own tissue only *in extremis,* such as during starvation. Indeed, the only supposition of Liebig's that was correct was that animals possessed the power of making fats – though even here, Liebig was not original, for it had been an assumption made by William Prout in his equally speculative chemical physiology of 1834. Nor do biochemists solely correlate fat synthesis with respiratory rate.

What, then, did Liebig achieve with *Animal Chemistry*?

By providing a model of the basic chemical processes of digestion, respiration, assimilation, and degradation in animal organisms, he stimulated the interest of chemists, physiologists, doctors, and sanitarians who

76. M. Traube, "Der chemische Theorie der Fermentwirkungen," *Berichte* 11(1878), 1984–92; Fruton, *Molecules and Life,* p. 61.
77. Volhard, vol. ii, pp. 83–93.

were thereby encouraged, in McCollum's words, "to formulate better experimental techniques for the purpose of deciding the soundness of his views."[78] As such, it was a model of scientific method – the construction of a hypothesis that was then slowly refined by trial and error until little remained of the original stimulating assumptions. Yet, as Holmes has concluded in his outstanding commentary on *Animal Chemistry*,[79] this was achieved, first, because Liebig's outstanding genius and experience allowed him to imagine what the chemistry of organisms ought to be like; second, because his deep knowledge of organic compounds enabled him to erect a plausible mechanism of metabolism that seemingly revealed to physiologists for the first time phenomena that had always appeared mysterious, or "vital." In this respect, third, just as he had accomplished for agricultural chemistry, Liebig aimed to redirect physiology and to demonstrate that the anatomical, comparative, pathological, and microscopical approaches to animal systems that had been tried hitherto were all ancilliary to the fundamental insights that chemistry alone offered for understanding health and disease. In this respect, through his teaching, writing, students, and disciples, he paved the way for the discipline of biochemistry, which began to be institutionalized at the time of his death. Finally, although Liebig himself never practised experimental physiology and the vivisection techniques it involved, it may be claimed that the chemical approach to physiology that he advocated made physiologists more receptive to other techniques for revealing and understanding metabolic processes.

As Liebig had foreseen in the gatekeeping "Preface" to his *Animal Chemistry:*

> My object . . . has been to direct attention to the points of intersection of chemistry with physiology, and to point out those parts in which the sciences become, as it were, mixed up together. . . . In the hands of the physiologist, organic chemistry must become an intellectual instrument, by means of which he will be enabled to trace the causes invisible to bodily sight; and if, among the results which I have developed or indicated in this work, one alone shall admit of useful application, I shall consider the object for which it was written fully attained. The path which has led to it will open up other paths; and this I consider as the most important object to be gained. (pp. xxxv–xxxvi)

78. E. V. McCollum, *A History of Nutrition* (Boston, 1957), p. 37.
79. Holmes, *Animal Chemistry* (1964), pp. cxv–cxvi.

8

Liebig on Toast: The Chemistry of Food

> Wohl ziemst's den Frauen nicht, Toaste bringen,
> Doch, wenn bei diesem Wahl die Glaser klingen:
> Dann auf! dem Namen Liebig's groß und hehr,
> Ein Freudig Hoch! Und sieh', mein Glas ist leer![1]

Just as Liebig was not the first chemist to write about agricultural or animal chemistry, he was not the first to domesticate science by writing about the chemistry of cookery. As with agricultural and animal chemistry, where his confident theorising often overrode practice and yet simultaneously forwarded discipline and practice, his chemistry of cookery and doctrines of nutritional science overturned culinery beliefs, and simultaneously it assisted in the industrialisation and commercialisation of diet and the institutionalisation of cookery and nutrition in colleges of household science and schools of cookery.

These processes had begun long before Liebig's time. Drawing upon his interest in chemistry and pharmacy, Dr. Johnson once threatened to write a simple cookery book based upon philosophical principles. At about the same time, Count Rumford investigated how the comforts and enjoyments of the poor might be improved by adjustments to kitchen ranges and cookery pots. In 1821, following his sensational exposé of the horrifying extent of contemporary food adulteration and death in the pot, and his even more sensational dismissal from the Royal Institution's employment for defacing library books, Friedrich Accum staved off starvation by writing a potboiler, *Culinery chemistry exhibiting the scientific principles of cookery with concise instructions for preparing good and wholesome pickles, vinegar, conserves, fruit jellies, marmalades and various other alimentary substances employed in domestic economy, with observations on the chemical constitution and nutritive qualities of different kinds of food.*

1. Women aren't keen to propose toasts; yet, when the great and grand name of Liebig is proposed: such a cheerful toast! H. Davidis, *Kraftküche von Liebig's Fleischextrakt für hohere und unbemittelte Verhaltnisse erprobt und verfasst* (Braunschweig, 1870).

"The subject may appear frivolous," admitted Accum, "but let it be remembered that it is by the application of the principles of philosophy to the ordinary affairs of life, that science diffuses her benefits and perfects her claim to the gratitude of mankind."[2] And he went on to make the obvious point that the art of preparing food is a branch of chemistry: "The kitchen is a chemical laboratory . . . the boilers, stew-pans, and cradle spit of the cook correspond to the digestors, the evaporating basins, and the crucibles of the chemist."[3] And so on. More abstractly, during the 1820s, after analysing milk into its component parts of sugars, fats, and nitrogeneous compounds, William Prout drew the nutritional conclusion that a well-balanced diet should contain each of the principles, because milk was "the universal pabulum."[4]

All that goes to show is that some had brought science to the kitchen before Liebig, who in a reminiscence of the 1860s, tells us that the kitchen was really where his whole career had begun – learning French from the wife of one of the cooks of the Duke of Hesse-Darmstadt's and being fascinated by the preparation of food that he witnessed. "Hence," he reminisced, "I have retained a taste for cooking, and in leisure hours occupy myself with the mysteries of the kitchen."[5] Indeed, in December 1866, annoyed by some continental cookery writers who were using information on food that he had incorporated in his *Chemische Briefe*, he told his publisher Friedrich Vieweg that he might himself try to make some money by writing a commentary to a cookery book. Although this was not to be, as we shall see, cookery books were greatly influenced by Liebig's long paper of 1847, "Ueber die Bestandtheile der Flussig keiten des Fleisches" (On the Ingredients of Meat Juices).[6] Liebig's work in animal chemistry was to have a marked effect on what might be called "scientific cookery."

The Study of Meat

The burden of Liebig's paper, as Gregory explained in his English translation, was that the chemistry of life involved strictly chemical changes, and every chemical that was differentiated and identified in an organism had a

2. F. Accom, *Culinary Chemistry* (London, 1821), preface.
3. *Ibid.*
4. W. H. Brock, *From Protyle to Proton* (Bristol, 1985), Chap. 3.
5. J. Liebig, "A cup of coffee," *Pharmaceutical Journal* 7(1865–66), 412–16, from *Dingler polytechnisches Journal* 3(1866), 466.
6. J. Liebig "Über die Bestandteile der flüssigkeiten des fleisches," *Ann. Chem.* 62(1847), 257–369.

precise function and use. While working up a second edition of his *Animal Chemistry,* Liebig had been led to investigate the "nature of the organic acid diffused through the muscular system," and from this to an investigation of the creatine already extracted by Chevreul in 1835 and thought to be responsible for the smell of meat (osmazome) during cooking. Liebig purified creatine by crystallization and showed that strong acids transformed it into creatinine $C_8N_3H_7O_2$ [methylguanidyl-acetic acid], which he speculated might be an amide of caffeine $C_8N_2H_5O_2$:

caffeine	$C_8N_2H_5O_2$
amide	$N\ H_2$
creatinine	$C_8N_3H_7O_2$

Both substances were found in urine (as Pettenkofer had shown somewhat inconclusively in 1844),[7] which, Liebig suggested, was an easier source than meat for extraction. When creatine and creatinine were boiled together, Liebig identified a new base he called sarcosine $C_6NH_7O_4$ [methylglycine], as well as inosinic acid $C_{10}N_2H_7O_{11}$. Finally, he confirmed once and for all, after decades of chemical argument, that the acidity of muscle juice was due to lactic acid.

All this was in itself interesting chemistry, though to a considerable extent it merely refined and extended the earlier work of Chevreul. More important for the sequel, however, was the confirmation of Chevreul's demonstration that besides these nitrogenous compounds (including gelatine), meat juices contained inorganic alkaline salts such as sodium and potassium phosphates. From this finding, Liebig developed a nutritional theory.

The reason that gelatine was not nutritious, as Magendie had demonstrated by feeding dogs with it, was that gelatine contained more nitrogen but less carbon than other proteins such as fibrin and albumin. This finding enabled Liebig to argue that gelatine was not a flesh-former, or plastic, foodstuff, though he agreed that it was somehow laid down in bone and the gelatinous parts of tendons. As Drummond and Wilbraham commented in *The Englishman's Food,* this conclusion was rather disturbing for those who were attempting to follow Liebig's lead and to estimate the tissue-forming value of diets from the diets' nitrogen contents; they were forced to admit that one important nitrogen-containing constituent of meat had no tissue-forming value.[8] Indeed, this was the beginning of difficulties for the theory of plastic versus respiratory foods.

7. M. Pettenkofer, "Vorläufige Notiz über einen neuen stickstoffhaltigen Körper im Harne," *Ann. Chem.* 52(1844), 97–100.
8. J. C. Drummond and A. Wilbraham, *The Englishman's Food* (1939; repr. 1991), p. 350.

When meat was boiled, it separated effectively into soluble and insoluble constituents. The water in which it was boiled contained dissolved gelatine (hence, the formation of a gelatinous liquid on cooling), creatine, creatinine, alkaline phosphates, lactates, isosinates, and magnesium phosphate; the boiled flesh consisted mainly of fibrin and the insoluble phosphate of calcium. Therefore, for a nutritious meal, following which human flesh was to be formed from the food via the bloodstream, it was vitally important to eat the boiled meat *with* its gravy or soup. As Gregory, Liebig's translator, commented, the advantage of stewing or casseroling meat was that the whole concoction was eaten. Thus, despite the fact that they were nonnitrogenous substances, and indeed, nonrespiratory, Liebig concluded that inorganic chemicals were essential nutrients for the formation of flesh. Here was another mineral theory.

In this extraordinary paper, Liebig therefore argued that the essential nutrients in meat were not contained in the muscle fibres of the flesh but in the muscle fluids, which were lost during boiling or roasting. It was essential, therefore, to minimise the loss of these vital salts if nutritional quality was to be maintained. The way to do this, suggested Liebig, was to heat the surface of a joint very rapidly (searing) by plunging the meat into boiling water so that the albumin (protein) was coagulated sufficiently to form a crust that prevented the penetration of further water into the flesh. The meat juices would thereby be retained. By a similar procedure of roasting a piece of meat close to the fire, and then continuing the cooking with the meat farther away, the juices would again be retained. As Harold McGee has pointed out, this suggestion turned centuries of traditional cooking procedures on their heads.[9]

Given the publicity that this information received, after the reappearance of the paper as a book in 1847, its English translation by Gregory in the same year, publicity by Hofmann in his *Annual Reports on the Progress of Chemistry,* and by Liebig himself in the revised German and English editions of *Chemische Briefe,* it was not surprising that "modern" cooks went the way of Liebig. The *Lancet* reviewer hailed the book for revealing "the true principles of cookery."[10] In the same journal, William Benecke, a former pharmaceutical student of Liebig's who owed his appointment as physician to the German Hospital in north London to him, urged rational diets based upon Liebig's nutritional theory as the solution

9. H. McGee, *On Food and Cooking. The Science and Lore of the Kitchen* (New York, 1984), pp. 113–14.
10. J. Liebig, *Research on the Chemistry of Food,* trans. by W. Gregory (London, 1847) from *Chemische Untersuchung über das Fleisch* (Heidelberg, 1847); A. W. Hofmann, ed., *Annual Report of the Progress of Chemistry,* vol. 2 (London, 1850), pp. 338–40; J. Liebig, *Familiar Letters on Chemistry,* 3rd ed. (London, 1851), Letter 32.

to public health.[11] During the late 1840s and 1850s, and before Mrs. Beaton, one of the most popular recipe books amongst middle-class families was Eliza Acton's *Modern Cookery for Private Families*.[12] Acton was much plagiarised and was undoubtedly a sourcebook for many other compilers. Up to and including the third of the three editions, published in 1845, Mrs. Acton cooked meat the traditional way – boiling it after initial immersion and washing in cold water and then slow roasting it a comfortable, nonsearing distance from the fire. In the third revised edition of 1855, significantly subtitled "In which the Principles of Baron Liebig and other eminent writers have been as much as possible applied and explained," Mrs. Acton hardened the meat's surface by plunging it into boiling water or searing it as a roast.

There were some sceptics. Fifty years before, Rumford had demonstrated that a slowly roasted leg of lamb roasted at a lowish temperature weighed more and was juicier and tastier than one roasted directly over the fire. Fifty years later, at the University of Missouri in 1930, it was shown experimentally that seared roasts really lose more juice than those cooked slowly at a relatively low temperature. Again, as Harold McGee has pointed out, the "water-soluble components on whose retention [Liebig] places so much stress are minor products of muscle metabolism and nutritionally negligible." In any case, the seared crust on a roast or a piece of boiled mutton is not a barrier to water penetration and the resulting solution of Liebig's supposedly desirable nutrients. Nevertheless, even today many cooks sear meat, particularly for pot roasts when the meat glaze produces an additional flavour, as well as enhancing the dish's appearance. Cooks do not in general think like scientists but are happy to use the methods of discredited practice for their own aesthetic and serving ends. Liebig's overturning of centuries of empirically correct knowledge clearly satisfied and still satisfies "a desire for logic and reason"; it also, suggests McGee, provided cookery writers with competitive novelty in the 1850s and enabled them to proclaim the up-to-dateness, modernism, and rationality of their approach.[13] Overall, it is another striking illustration of a pattern in Liebig's career of undermining tacit knowledge by explaining it "scientifically" and demanding that praxis follow theory, rather than theory deriving from praxis.

Although there is no evidence that Liebig ever specified how vegetables should be cooked, it is interesting that Eliza Acton, no doubt by analogy

11. William Benecke, "On extractum carnis," *Lancet* (1851), i, 6–8.

12. E. Acton, *Modern Cookery for Private Families* (1865; repr. Elek Books, 1966, ed. by Penelope Farmer).

13. McGee, *On Food*. p. 114. I thank Valerie Mars for help with this and the following paragraph.

with the retention of meat juices, advocated the heavy boiling of vegetables. We now know that this quickly destroys their vitamin content.[14] Liebig's intervention in cookery demonstrates a middle-class shift from food being seen solely as the means for the assuagement of hunger to a more scientific concern with food as nutrition and as a key to health, efficiency, and racial strength. Acton commented in her new 1855 "Preface" to *Modern Cookery,* "It is from these [middle] classes that men emanate to whom we are chiefly indebted for our advancement." They had to be fed properly to ensure the "recruitment of their enfeebled powers [after] the toils of the day."

> It is the want of a scientific basis which has given rise to so many absurd and hurtful methods [of cookery].[15]

As Penelope Farmer has noted, it was not so much that before 1855 cookery writers had ignored science – most of them, for example, claimed that osmazome was the chemical that flavoured meat – but that nutritional wastage occurred when scientific principles were ignored. Liebig was undoubtedly a major factor in this shift of attitude, though we must not ignore the contemporary exposure of adulteration, which suggested that a well-informed cook was a safe cook; and also the sanitarians' exposure of poverty and its effects that underlay the poor diets of the majority.

The phenomenon of osmosis had been only briefly noted in *Animal Chemistry,* and no issue was made of it. When he was writing some of the historical "letters" for a revised edition of his *Familiar Letters on Chemistry,* Liebig had discovered the eighteenth-century work of the English chemist and physiologist Stephen Hales (1677–1761), whose *Vegetable Staticks* (1727) contained a series of imaginative experiments on the ascent and descent of sap in plants. Now, while investigating food in 1846 and inspired by Hales, Liebig realised that the motion of foodstuffs through the animal gut and its absorption were largely unexplored problems in physiology. In France since about 1826, René Dutrochet had been designing experiments which showed that if a tube containing a sugar solution and covered by an animal membrane (e.g., a bladder) was placed in water, the water slowly passed *into* the tube, causing the sugar solution to rise. Simultaneously, however, the water in the vessel became sweet from sugar passing from the tube through the membrane. This process continued until equilibrium was reached and the two solutions

14. Acton, *Modern Cookery,* pp. xiv, 308.
15. *Ibid.,* Preface.

had reached a common density (actually, a common concentration). Dutrochet named the inward motion *endosmose*, the external, *exosmose*, and the whole phenomenon, *osmosis*. On reading about these experiments, Liebig realised that osmosis was of physiological significance for the absorption of aliments into the bloodstream. There followed a clever series of experiments on the permeability of membranes that he summarised in a monograph, *On the Motions of the Juices in the Animal Body*, which Gregory dutifully translated in 1848.[16] Like Dutrochet, Liebig remained unclear whether osmosis was an electrical or a purely mechanical effect, and despite the stimulus that Liebig's publication gave to Thomas Graham's work on osmosis, the physiological significance of osmosis was to remain mysterious until the emergence of physical chemistry in the 1870s.[17]

Boswell once likened a page of his *Journal* to a cake of portable soup: "a little of which may be diffused into a considerable portion." Indeed, one of the proposed selling points of Papin's digester (an early form of pressure cooker) in the seventeenth century had been its potential usefulness for preparing nutritious portable bone extracts for armies on the march. Such gelatine "beef" tablets were well known in the eighteenth century, as Boswell's remark shows. During the French Revolution, the chemists D'Arcet, Pelletier, and Cadet improved the method of boiling bones to produce gelatine, and with a growing interest in physiological experimentation, a "gelatine Commission" of French scientists argued in 1814 that this was a valuable nutritious and cheap food for the poor as well as for marching armies. Only in 1841 did the physiologist Magendie show that dogs fed on gelatine would not survive and that only if vegetables were added to the gelatine soup (vegetable hotpot) would it be safe to use in workhouses and prisons. Even so, as late as 1845, the Society of Arts in London awarded a gold medal "for the preparation and importation from Australia of an essence of beef in tablet form"; tests on voyages between Australia and South America soon showed, however, that the product became rancid.[18] Meanwhile, in 1821, two other French chemists Proust and Parmentier had developed a more tasty and nutritious soup that involved "extracting" the essences from meat by standing it in cold water for several hours. The evaporated thickened brew was then used directly as a clear soup or as a stock. The basic process for making "extract of

16. J. Liebig, *Untersuchungen über einige der Säftebewegung im thierschen Organismus* (Braunschweig, 1848); translated by W. Gregory as *Researches on the Motion of the Juices in the Animal Body* (London, 1848).
17. See J. R. Partington, *A History of Chemistry*, vol. 4 (London, 1964), pp. 650–53.
18. Derek Hudson and Kenneth W. Luckhurst, *The Royal Society of Arts* (London, 1954), p. 299.

meat" was therefore well known before 1847, as Liebig conceded. However, despite its reportage in French cookery books in the 1820s, its use seems to have been largely confined to invalids.

Liebig's 1847 paper was extremely important for his entry into commerce in the 1860s. Because he believed that meat juice was highly nutritious and that an essential plastic component of muscles was diminished by use, he concluded – or, rather, confirmed folk wisdom – that a cold effusion of meat (traditionally known as beef tea) was a valuable restorative. If such a solution were boiled down to form a portable extract, it might easily be reconstituted into a nutritious soup. As we have noticed in Chapter 4, one of Liebig's closest British friends was the Irish alkali manufacturer James Muspratt (1793–1886). Liebig and Muspratt had met in 1837 during Liebig's first tour of the British Isles, and when Liebig had lectured to the British Association meeting in Liverpool. Muspratt sent all four of his sons to the University of Giessen for their chemical education and he himself frequently stayed with the Liebigs at Giessen or, after 1852, Munich. In the winter of 1853–54 the Liebigs were host to Muspratt's seventeen-year-old daughter Emma, who had been sent to Munich to learn German with the Liebigs' daughters Nanny and Marie. Munich in 1854, the year of its great Industrial Exhibition, was still undrained and full of cesspools. Unfortunately, after several weeks' fun of dances, operas, theatres, and horse-riding, in November Emma fell seriously ill with what was diagnosed as scarlet fever and she lay at death's door for three months.

Her brother Edmund, who was attending medical classes in Munich that semester, later recalled how the Liebig family doctor Pfeufer, despairing of saving her life because she was too weak to take nourishment, warned Liebig that the only chance of recovery lay in finding a food that Emma could assimilate. Liebig, feeling a dreadful responsibility to Emma's parents, who were unable to reach Munich because of the severity of the European winter that year, decided to experiment with beef tea. In this traditional invalid remedy, meat was suspended in boiling water. According to Liebig's theories, this merely led to the coagulation of the meat albumin and so prevented the flesh's nutrients from dissolving in the water. The answer, as he had said in 1847, was to prepare the beef tea as a cold effusion. The next day, therefore, Liebig presented Dr. Pfeufer with a potion in which minced chicken had been suspended for several hours and to which a few drops of hydrochloric acid had been added to soften the meat. After Emma was given this effusion every half hour, she had made a complete recovery by the time her parents reached Munich. They held a celebratory dance for Emma in which, in Edmund's words, she "danced the first waltz with dear fat old Dr Pfeufer, who, however, danced beautifully."[19] Patients suffering from typhoid fever need to drink water generously, and since the intestines are the focus of this bacterial attack, predigested food (or peptonised food) is appropriate. Liebig's peptonised beef tea was no doubt useful in Emma's case for being easily assimilated.

Subsequently, Pfeufer, together with Dr von Gieli, made use of the remedy in the Munich hospitals, endorsing Liebig's account of its preparation and use in the *Annalen*.[20] And when one of Liebig's own daughters fell sick a few years later, he assured Emma in a letter that she had survived on the cold soup for many weeks. In 1861 Emma married Dr. George Harley (1829–96), who had studied in both Germany and Paris. The Harleys remained good correspondents with Liebig, who was Emma's godfather. The friendship was continued by the Harleys' daughter Ethel Brilliana, who as Mrs. Alec Tweedie became a well-known author and her father's biographer.[21]

This liquid remedy is not to be confused with Liebig's *extractum carnis*, or solid beef tea, which he had also discussed in detail in his *Annalen* food article in 1847, returning to the matter in the third edition of his *Familiar Letters on Chemistry* in 1851. Although he encouraged several German pharmacists, including the uncle and nephew Franz and Max Pettenkofer at Munich and Karl Frederich Mohr at Coblenz, to prepare the remedy for sale, production must have been small.[22] Costs were in any case high for pharmacists had to buy their beef directly from the local meat markets. Mohr, however, was enthusiastic, writing to Liebig how good it was with scrambled eggs! In three weeks Mohr and his wife had made five pounds of extract and still had not been able to satisfy demand by consumptives and convalescents in Coblentz. To this Liebig replied in March 1853 that the extract was also frequently in demand in Munich and as *extractum carnis* was about to be included in the new Bavarian pharmacopeia. "Doctors know it works." He added, "I've put myself to the trouble of encouraging many factories in Buenos Ayres [sic] and Australia to make the extract for sale." This was nine years before the creation of the Liebig Company. Such altruism was to prevent the company from ever claiming an exclusive monopoly on the product.

Prior to the 1860s, the remedy was clearly something only the affluent

19. E. K. Muspratt, *My Life and Work* (1917), Chap. 5.
20. J. Liebig, "Eine neue Fleischbrühe für Kranke," *Ann. Chem.* 91(1854), 244–46.
21. Mrs A. Tweedie, *George Harley, FRS, The Life of a London Physician* (London, 1899), includes a number of letters from Liebig to Emma (Muspratt) Harley. For some chatty reminiscences of Mrs Tweedie (d. 1940), see her *My Tablecloths* (London, 1916).
22. J. M. Scott, "Karl Friedrich Mohr," *Chymia* 3(1950), 191–203; G. W. A. Kahlbaum, ed., *Justus von Liebig und Friedrich Mohr in ihren Briefen*, 20 and 24 March 1853, pp. 124–26.

could afford. By the 1860s, to judge from C. H. F. Routh's *Infant Feeding,* Liebig's beef tea, as well as rival products such as Hogarth's essence of beef and Brand's essence of meat, were available from a few London pharmaceutical companies for use as a weaning food or as a substitute for a mother's breast milk, or as strengthener or tonic for nursing mothers. Routh commented that babies seemed to dislike Liebig's beef tea, adding that Liebig's artificial milk or malt extract, which Liebig devised in 1865 (see later discussion), was far better for infants.

Extract of Meat

In the spring of 1862 the German railway engineer Georg Christian Giebert (d. 1874), who had been building railways in Brazil and Uruguay, drew Liebig's attention to the cattle industry of the River Plate area, where most of the animals were slaughtered for their hides only; only a little of their flesh was exploited commercially in the production of *tasajo* (or charque, salted beef) for export to the West Indies or for commercial distribution to the indigenous populations of Brazil and Uruguay. Overexpansion and competition between *saladeros* (beef-salting plants) led to a dramatic fall in cattle prices in the early 1860s. On furlough in Europe in 1862, Giebert read *Familiar Letters on Chemistry* and was intrigued by Liebig's reference to the waste of cattle meat in Australia and South America. Giebert examined the pharmaceutical laboratory production of the extract carried out in Pettenkofer's Royal Pharmacy at Munich and by Mohr at Coblenz and came to the conclusion that the process could be worked on a larger industrial scale.

Such a possibility had already occurred to Liebig's friend, the Scots-Australian wine grower James King (who had emigrated to New South Wales in 1826 and struck up a correspondence with Liebig in 1848).[23] In a letter to Liebig in October 1850, when Liebig was working on a new edition of *Chemical Letters* (the third German-English edition of 1851), King had alluded to the excellence of New South Wales pasture for rearing beef cattle and had forecast that there was a fortune to be made if the meat of cattle slaughtered for their hides and tallow could be extracted and shipped back to European markets. Liebig had quoted from this letter in a footnote to support his textual contention of the nutritional importance

23. Correspondence in Staatsbibliothek, München, and Mitchell Library, Sydney. Muspratt, *My Life,* p. 107, gives 1858 as the date of their first contact. See I. Jack and C. A. Liston, "A Scottish immigrant in New South Wales – James King of Irrawang," *Journal Royal Australian Historical Society* 68(1982), 92–106.

A typical Liebig image used in publicity by the Liebig Extract of Meat Company. From the front cover of a *Haushaltungs-Kalender* issued by the company in 1904. (Courtesy W. Lewicki)

of muscle juices. Not surprisingly, therefore, encouraged by Liebig, Australian farmers and landowners had been making their own extraction experiments (as well as ones on refrigeration that were to prove successful in 1879) at the very time that Giebert was approaching Liebig in Munich. It was this footnote that had caught Giebert's attention, as must Liebig's further kite also have done:

> It need hardly be pointed out specially, that those who may feel inclined to prepare extract of meat as an article of commerce, will

> entirely miss their aim, unless they most carefully and conscientiously seek to avoid the errors of those who have hitherto attempted it. Half an hour's boiling of the chopped meat with 8 or 10 times its weight of water, suffices to dissolve all the active ingredients. The decoction must, before it is evaporated, be most carefully cleansed of all fat (which would become rancid), and the evaporation must be conducted in the water bath. . . . The boiling of the meat in the first instance must be carried out in clean copper vessels, but for the evaporation of the soup, vessels of pure tin, or still better, of porcelain, should be employed. If the price of the extract should be found not to exceed about three shillings per pound, it would certainly become a most profitable commodity. In Giessen, without reckoning the cost of preparation [i.e., the cook's time] the extract of meat cannot be prepared for less than six shillings to seven shillings and sixpence per pound in meat alone.[24]

Giebert proposed that Liebig invest in the purchase of a *saladeros* on the outskirts of Villa Independencia (to be named Fray Bentos as the township expanded) on the eastern bank of the River Uruguay. Here a test extraction plant would be run. With Liebig's agreement, Giebert and several South American ranchers and the Antwerp entrepreneurs Cornelle David, Otto and George Gunther, and Josef Bennert founded the Société Fray Bentos Giebert et Cie, to prepare and sell an *Extractum carnis Liebig* on a small scale to test the market. By the end of 1864, some 50,000 pounds of extract worth £12,000 having been sold successfully on the world's markets via Amsterdam, Giebert concluded that expansion of the operation would be worthwhile; indeed, that there was a fortune to be made as King had predicted, even though it took a metric ton of cattle to produce just 25 kilograms of extract. The key to success was engineering a vacuum pan evaporation technology pioneered by the sugar refining industry. Boneless cattle cadavers were reduced to pulp by crushing in iron rollers revolved by a steam engine. The pulp was thrown into a vat of water and steamed for an hour before passing into a reservoir with a sieve. The extract then oozed into a second trough from which fats were skimmed off. (This prevented rancidity at the cost of nutritional value.) The resultant gravy was then poured into open vats heated by steam pipes and equipped with bellows to blast off the steam and to prevent condensation. After six to eight hours, the thick dark-brown and powerfully appetising extract was refiltered and allowed to cool before being packed into large tins.

Accordingly, in 1865 Giebert offered Liebig a directorship of the company, an annual salary of £1,000 and an immediate cash payment of

24. J. Liebig, *Familiar Letters on Chemistry* (1851), p. 426.

£5,000 if the company could use Liebig's name, fame, and reputation as a guarantee to raise sufficient capital on the London Stock Exchange to create the Liebig Extract of Meat Company. This was to take over the assets of the former exploratory company and to start with an authorised capital of £500,000 in £20 shares. The company's prospectus, published in 1865, anticipated an annual production of 1 million pounds of extract by 1868, which promised a profit to shareholders of 4 shillings in the pound, or £200,000 to the 450 original shareholders.

In 1866 Francis Clare Ford, the British consol in Uruguay, reported favourably on the Liebig process in a British parliamentary report:

> Its strength can be calculated from the fact that 33 lbs of meat are reduced to 1 lb of essence, which is sufficient to make broth for 128 men. A tin containing 1 lb of this extract is sold in London for 12s 6d. Eight small tins will hold the concentrated matter of an entire ox, at a price of 96s, and will make over 1,000 basins of soup, good strong soup; one teaspoonful to a large cup of water, and eaten alone or with the addition of a little bread, potato and salt, affords a good repast.[25]

This reminds us that Giebert's business acumen was in promoting the Liebig Company at a crucial moment when there were deep European discussion and concern over how its swelling populations could be adequately fed. The earlier emphasis on the medical qualities of the extract now shifted to its importance as a foodstuff. It is clear from the meat production statistics published by Critchell and Raymond in 1912 that until the introduction of frozen meat supplies in the 1880s, consumption of fresh meat in Britain overtook supply, with a consequent rise in prices, market shortages, and malnutrition amongst the poor.[26] Equally, it is clear that Australian graziers desperately needed to find new markets for meat if Eastern Australia, apart from its sheep, was not to become one vast knacker's yard devoted to tallow and leather production. This area had the populations shown in Table 8.1.

In a decade in which cattle plague (rinderpest) was rife and leading to the wholesale slaughter of thousands of head of cattle, when John Simon's co-workers (such as Edward Smith) were demonstrating that Lancashire cotton workers were able to afford less than one pound of meat per week, thoughts were inevitably turned to the possibility of importing fresh meat from British colonies. This was one of the functions of the food committee set up by Harry Chester of the Society of Arts in 1866. As early as 1853, in

25. *Report on the Methods Employed in the River Plate for Curing Meat, Parliamentary Papers,* 1866, LXXI [3747], 1–4.
26. J. T. Critchell and J. Raymond, *A History of the Frozen Meat Trade* (London, 1912; repr. 1969).

Table 8.1

Year	Human	Cattle	Sheep
1851	403,889	1,894,834	15,993,954
1861	1,153,973	3,846,554	20,135,286
1871	1,668,377	4,277,228	40,072,955
1881	2,252,617	8,010,991	65,078,341

a centenary address to the society, Chester (1806–68), who must have read Liebig, had drawn attention to the need to feed Britain's hungry mouths by developing some way of bringing meat from the colonies. Why, he asked, was Australia able to export the tallow and wool of her sheep but not her mutton? Another well-known social commentator, Andrew Wynter, put it more bluntly in an essay of 1861:

> But a vastly more important question than even the victualling of the navy with cheap and wholesome food is that of victualling the masses at home. What gives rise to the vast majority of disease in our hospitals? What is at the moment deteriorating the lower stratum of the population? – the want of a sufficient supply of nitrogenous food. Those who live by the wear and tear of their muscles are condemned by the present high price of meat to subsist upon food that cannot restore the power that is expended. In the income and expenditure of the human body, in short, they are living upon their capital, and of course sooner or later they must use themselves up. Bread is cheap, because free-trade pours the full sheaves of beautiful foreign lands into our eagerly-spread lap. Why should we not have meat too?[27]

In 1863 Sir William Trevelyan endowed the Society of Arts with a medal and a £70 (subsequently increased to £100) reward for any invention that would enable fresh meat to be exported from the colonies. (The prize was not awarded until 1884, when it was divided among five people.[28]) Before then, however, Trevelyan's offer succeeded in stimulating experiments in canning, refrigeration, and the vilely cruel injection of cattle with brine during slaughter (Morgan's process), and with meat alternatives such as fish, fungi, and horseflesh. It is interesting that in 1868 Liebig's pharmaceutical pupil Benjamin H. Paul lectured to the society on the possibilities of "artificial freezing and refrigeration." Liebig's extract,

27. Andrew Wynter, *Our Social Bees* (London, 1861), pp. 202–3.
28. Hudson and Luckhurst, *Royal Society of Arts*, pp. 300–2.

therefore, was easily advertisable as a solution to a vexing contemporary problem.

Liebig was no sleeping partner. He announced in 1866 that he was the director of the company's scientific department, where he would be joined by his former assistants Max Pettenkofer in Munich, the pharmacist Charles Finck in Antwerp, and Seekamp in Fray Bentos.[29] As Liebig's correspondence shows, he pursued his English contacts with the poet Charles Boner and the agriculturist John Mechi to persuade Florence Nightingale, as well as English landowners, to endorse the product for use in hospital, army, naval, prison, and workhouse dietaries, and as an incentive to landowners to expand their beef-farming activities. Nor was Liebig averse to puffing the extract repeatedly in the *Annalen*.[30] During the Abyssinian expedition of 1868, the British army used the extract as a substitute for fresh meat, and during the Franco-Prussian war, Ebswurst, a soup tablet combining dried vegetables with meat extract, was supplied to Prussian troops.[31] In 1872 Liebig obtained unsolicited independent testimonials from the African explorers Georg Schweinfurth (1836–1925) and Gerhard Rohlfs (1831–96), extolling the value of the extract for the serious traveller. (Rohlfs also sent Liebig samples of *Cannabis ind.*, wondering why it affected the brain and nerves; there is no evidence that Liebig made any tests with the plant.) In 1869 Liebig was awarded the Society of Arts' Albert Medal "for distinguished merit in promoting Arts, Manufactures and Commerce" – its sixth recipient and second foreigner (after Napoleon III!).[32]

In practice it took the Liebig Company several years to achieve the promised level of profitability because revolutions in South America, cholera outbreaks in the local workforce, problems with factory design and the breeding of suitable cattle, and European import and customs' duties all took their toll of investment. Within half a century, however, but after Liebig's death, it had become one of the largest cattle-farming enterprises in the world and renowned for two other beef products: Fray

29. Anon., "Extract of Meat," *Chemical News* 14(1866), 289.

30. J. Liebig, "Ueber den Werth des Fleischextraktes für Haushaltungen," *Ann. Chem.* 146(1868), 133–40. For more details, see Mark R. Finlay, "Quackery and cookery. Liebig's extract of meat," *Bulletin History of Medicine* 66(1992), 404–18; and his "Early marketing of the theory of nutrition; the science and culture of Liebig's extract of meat," in H. Kamminga and A. Cunningham, eds., *The Science and Culture of Nutrition* (Cambridge, 1995).

31. Anon., "Extract of Meat," *Chemical News* 17(1868), 19. See J. C. Crossley and R. Greenhill, "The River Plate beef trade," in D. C. M. Platt, ed., *Business Imperialism 1840–1930* (Oxford, 1977), pp. 284–336.

32. *Journal Royal Society of Arts* 17 (11 June 1869), 557.

Bentos Corned Beef, which was launched in 1879, and Oxo.[33] The latter, initially a cheaper inferior extract with ground beef added, was developed in consultation with the English chemist Henry Roscoe. Launched in 1900 it reached its familiar cubic solid form only in 1911 when it was cleverly advertised as the "penny a cube." Further profits accrued to the company from the sale of hides, the production of tallow, and, what Liebig would have approved, the sale of cattle dung as fertilizer. Liebig had even developed a process by which unused portions of the animal carcass could be turned into pig food.[34] By the 1890s, twenty years after Liebig's death, his company was worth £1.5 million per annum and was offering a 23.5 percent dividend. By then the refrigerated beef trade (first developed by Australian interests) had begun, and Fray Bentos was locationally disadvantaged by its distance from the railway networks of Uruguay. The company therefore bought up a rival extract factory that had been developed by Emil Kemmerich at Santa Elena and another at Colon, where corned beef was contained within a trapezoidal tin by 1908. Corned beef was to surpass the extract as the company's most profitable product.

Not surpisingly, competitors had quickly entered the field, led by the Australian Robert Tooth, who in 1864 had also tried to interest Liebig in a deal that he had struck with the London pharmacy of Allen & Hanbury, in which he would extract meat at Yeagaree, New South Wales, and distribute it through the pharmacy. Liebig was prevented from making a commercial deal with Tooth only when Giebert in 1865 offered him better financial returns provided he gave the Liebig Company his exclusive attention. Allen & Hansbury's were upset in 1866 when the Society of Arts' Food Committee seemed to imply that their *extractum carnis* was inferior to Liebig's and that the Liebig Company was trying to hold a monopoly of an article that Liebig had made public knowledge in 1847. All was well, however, when analyses by Liebig's pupil Thudichum led the committee to endorse all extracts for middle-class cooking, for use in hotels and restaurants, and for the nutriment of soldiers, sailors, and hospital patients.

One of the functions of the many Liebig Company agencies that were set up in the major European cities, the United States, Mexico, South America, and Australia was not merely to improve marketing and sales but to keep a wary eye on the opposition and if necessary to sue for damages. None of these court cases succeeded, however, on the grounds that the name "Liebig's Extract of Meat" had become general before the

33. C. Scarborough, *About Oxo. In Its Jubilee Year* (London, 1965).

34. A. Voelcker, "Annual Report of the Consulting Chemist for 1875," *J. Roy. Agri. Soc. of England* 12(1876), 298.

formation of any commercial company. "Liebig" was, therefore, common property and not a trademark. When Giebert died in 1874, a year after Liebig, he was briefly succeeded by the Fray Bentos works' doctor Emil Kemmerich, who had been appointed on the strength of physiological work he had done in 1869 as a student at Bonn under the physiologist Pflüger in which he had argued, despite accumulating evidence that the extract had little nutritional value, that it possessed valuable qualities as a nerve tonic. He had also caused the company momentary panic by suggesting that potassium salts (which were contained in the extract in significant percentages) could cause death – as it had done to dogs fed exclusively on the extract. The notion that the extract was a nerve tonic proved a godsend to Liebig, who was having a hard time dismissing the impressive experimental work of his former Munich pupil Karl Voit (1831–1908). As we saw in Chapter 7, as professor of physiology at Munich, Voit established a research programme on nutrition and metabolism. By 1867, confirming the work of Fick, Wislicenus, and Frankland, Voit showed that work, or energy, was not primarily due to muscular exertion.[35] It followed, as Voit made explicit, that meat extract, even if it did "feed" muscles (which he doubted), was not of any nutritional significance. Indeed, it was a waste of money as a food. Liebig himself spent the last years of his life fighting this conclusion, slowly moving towards Kemmerich's claim that despite possessing little or no nutritional worth, like tea and coffee it had an important effect upon the nervous system.

The definitive proof that the extract had no nutritional value, which was given prominent publicity in Britain by Edward Smith and others, proved embarrassing to the company, the chief selling point of which had hitherto been the scientific reputation of Liebig. Eventually, a few months before his death in 1873, in a letter to Josef Bennert, the Liebig Company's German agent at Antwerp, Max Pettenkofer, advised the company to promote the product as a stimulant and flavouring agent – a luxury like tea, coffee, wine, and beer, rather than as a food. The extract, Pettenkofer suggested, was a condiment as important as salt and pepper. In one of his final actions as co-editor of the *Annalen,* Liebig ordered Pettenkofer's twenty-three-page report to be printed in its entirety.[36] Thereafter, until just after World War II, nutritionists seem to have been unanimous that meat extracts had only a stimulative and flavouring function in diet. Then, between 1944 and 1951, it was shown that extracts contain significant

35. A. Fick and J. A. Wislicenus, "On the origin of muscular power," *Phil. Mag.* 31(1866), 485–503; E. Frankland, "On the origin of muscular power," *Phil. Mag.* 32(1866), 182–99; C. Voit, "Untersuchungen über den Stoffverbrauch des normalen Menschen," *Zeit. Biol.* 2(1866), 459–73.

36. M. von Pettenkofer, "Ueber Nahrungsmittel," *Ann. Chem.* 167(1873), 271–92.

amounts of the vitamin B2 group (riboflavin and nicotinic acid), important in growth and the maintenance of skin tone.

After a difference of opinion, Kemmerich resigned, only to found a rival company that also used the Liebig label. Despite lawsuits, he was able to continue marketing. Another rival, "Johnson's Fluid Beef," was an extract developed in Canada by a Scots-Canadian J. Lawson Johnson in 1877. In 1886 he successfully relaunched the product, very cleverly, as "Bovril." To the Victorians "Vril" meant the vital fluid that had been developed by the future men of Edward Bulwer-Lytton's science-fiction romance *The Coming Race* (1871).[37] As medical and pharmaceutical reports showed, Bovril had a nutritional edge over Liebig's extract insofar as ground lean meat was added to Johnson's extract before potting. In Australia, besides Tooth's enterprise, Liebig's former pupil C. G. Tindal began an extract works at Ramormie, New South Wales, in 1866. By 1912 it had grown into the Sydney Meat Preserving Company and the North Queensland Meat Export Company. In New Zealand, Richard Hellaby, a pioneer of meat refrigeration technology, used waste carcasses to make "Bovo," which was eventually sold to the Bovril Company in England. Not to be outdone, the North American meat industry retaliated with "Armour's (Chicago) Extract of Beef."

Repeated lawsuits tended to show that in the eyes of the law the Liebig Company had no exclusive right to Liebig's name, even though he was an interested party. Hence the great stress that the Liebig Company laid in its publicity towards the end of the century that only pots bearing Liebig's familiar signature on the blue labels were the authentic Liebig Extract. To this day the product is still marketed on the Continent in white pots bearing Liebig's signature. In Britain the company renamed itself "Lemco" from its initials. In 1968, in the intense expansion of the commercial food marketing system after World War II, Lemco was bought by the Brooke Bond Tea Company (Brooke Bond Liebig became the manufacturing title, with Brooke Bond Oxo as its distributing agency). In the 1980s this company in turn was absorbed by the giant Unilever interests. Today, therefore, Liebig's name retains an association with food only on the Continent as a tonic and as a flavouring for food and soup.

Although Liebig himself did not make the attempt, it would seem that Liebig's commercial success in marketing meat extract encouraged others to develop different soup flavourings. In the 1870s and 1880s Julius Maggi in Switzerland and Carl Knorr in Heilbronn began the manufac-

37. E. Bulwer-Lytton, *The Coming Race* (London, 1871). Both Playfair and Frankland were retained by Johnson's firm as quality controllers. See D. J. Jeremy, ed., *Dictionary of Business Biography*, 6 vols. (London, 1984–6), vol. 3, pp. 516–21, for J. L. Johnson.

ture of dried "instant" soups using vegetable extracts. Their portability, ease of preparation with hot water, as well as the expansion of the canned soup trade and bottled sauces pioneered in London by Crosse & Blackwell, signalled a revolution in urban mass food consumption.[38] The chemist was now to be required in the food factory as taste and flavour consultant as well as quality surveyor.

One important effect of international competition between meat extracts was a strong emphasis upon quality control. During Liebig's own lifetime and that of his sons Georg and Hermann, control had been achieved by personal analysis of sample batches by the Liebigs or Max Pettenkofer as the batches reached Antwerp and before they were marketed. This is a curious reminder of one of the early episodes in Liebig's career, when he joined Geiger as co-editor of the *Annalen der Pharmacie* in order to test analytically the claims for new pharmaceutical remedies and discoveries submitted by readers. By the early 1900s, Lemco employed several dozens of analytical chemists at its London factory to monitor production. The same must have been true of Bovril and other competitors. In 1867 Erlenmeyer revealed to Liebig that he had been asked to analyse a rival "Liebig Extract" being shipped from Buenos Aires by the German firm of Belte & Huebler.[39] Liebig warned him off; no doubt many other European chemists had similar offers. Although national legislation, such as the 1875 Public Health Act in Britain, had made it necessary for manufacturers to be more stringent concerning the purity and quality of their products, it is true to say that Liebig's insistence that the contract with his company involve the analytical monitoring of the product when it reached Europe was not only the shape of things to come but illustrates the pervading impact of Liebig's scientific exploration of food.

Cookery Books

As we have seen, Liebig's extract was first marketed in a small way as a food for invalids; indeed, a "Liebig" is sometimes mentioned as a restorative in Victorian novels when the hero or heroine lies dangerously ill (e.g., Charlotte M. Yonge, *Magnum Bonum,* 1879, which itself hinges upon a miraculous drug that a dying doctor has devised). Apart from extensive

38. H. J. Teuteberg, *Die Rolle des Fleischextrakts* (Stuttgart, 1990).

39. E. Heuser, ed., *Justus von Liebig und Emil Erlenmeyer in ihren Briefen* (Mannheim, 1988), pp. 22–23. See E. Cross, "Chemistry of meat products," in M. D. Curwen, ed., *Chemistry in Commerce,* vol. 1 (London, c. 1936), pp. 103–12, 211–17.

and often imaginative advertising, owners of proprietory foods at the end of the nineteenth century began to find that sales could be stimulated through the issue of recipes. During the 1870s, the Liebig Company issued an *Improved and Economic Cookery* with more than 100 recipes devised by the German cookery writer Henriette Davidis, most of which used the extract as a flavouring.[40] Quite possibly the idea did not originate with the Liebig Company, for there is some evidence that Allen & Hansbury's pamphlet, *True Extract of Meat* (c. 1868), contained a few soup recipes. Davidis took her work for the company seriously and freely confided her concerns to Liebig when she found that Bennert, the German distributor at Cologne, was selling incompletely filled jars of the extract. She was also concerned that the price of the extract should be reduced so that it could become a genuinely working-class commodity. On a lighter note, she sent Liebig recipes for pumpernickel bread to aid his recovery from illness in 1871. Convinced of the extract's value, she also included extract recipes in the revised editions of her important cookery books, *Beruf für Jungfrau* (1871) and the sixteenth edition of *Kochbuch* (1871). For good measure, she confirmed via a linguist friend that extract should take the male article (*der*) because its Latin form was *extractum*. In her *Kraftkuche* (1870), Davidis drew attention, no doubt at Liebig's behest, to:

> The irreplaceable value of Liebig's extract of meat for feeding typhus patients, for hospitals, sickrooms and armies, is well known. Similarly, the extract is important as a great help for hotel, station and other restaurants, banquets, charitable institutions, as well as for travellers, especially on board ships; for the sick, infirm and convalescent, however, no other restorative than Liebig's extract is so swiftly and securely effective. (p. 13)

In the same decade the company also commisioned the popular aristocratic Austrian cookery writer Katherina Prato (1818–99) (pseudonym Katherina von Scheigger) to produce a recipe book for the Austro-Hungarian region.[41] Presumably pleased by the reception of these German-language cookery books, which must have been approved by Liebig before his death, the company commissioned other nationally known cookery writers. In Britain, for example, Liebig Company's *Practical*

40. H. Davidis, *Kraftkuche*. See letters from Davidis to Liebig 1870–71, Staatsbibliothek, München. Elizabeth Driver, *A Bibliography of Cookery Books Published in Britain 1875–1914* (London, 1989), p. 526, cites a third edition (1887) and a fourth edition (1888). See also Friedrich Orend, "Henrietta Davidis und Liebig," in *Beruf der Jungfrau* (Oberhausen, 1990), and Teuteberg, *Die Rolle des Fleishchextrakts*.

41. K. Prato, *Die praktische Verwerthung Kochrecepte*. Prato was known for her *Kochbuch für Officers-essagen* (Grätz, 1879), which remained in print into the twentieth century.

Cookery Book was commissioned to the leading cookery writer of the 1880s and 1890s, Mrs Hannah M. Young (1858–1949).[42]

Young's energy and entrepreneurial abilities are evident in her own self-promotional advertising activities. Her many cookery books are filled with advertisments for her lectures and for her kitchen equipment and grocery products. Born in Birmingham, her father, Cornelius Young, made and sold gas-cookers in Warrington in Lancashire for the stove manufacturers Fletcher Russell & Co. Hannah began her career as her father's cookery demonstrator, developing and adapting recipes for use with the new gas cooking technology. Her *Domestic Cookery, with Special Reference to Cooking by Gas,* first appeared in 1886 and went through many editions. Significantly, she used Liebig's extract in several recipes, including Polish Stew, Spanish Stew, Stuffed Onions, Pork Cutlets, and Fillets of Beef with Tomatoes.

Advertisements in this book and her *Choice Cookery* (1888) promoted not only gas stoves but also a range of water heaters, broilers, and coal-effect gas fires; kitchenwares such as steamers, moulds, and forcing bags; fine leaf gelatine; superior baking powders; and "Flor-Ador Food" (apparently a substitute for cornflour, arrowroot, and tapioca). All these special items could be ordered directly from Young in Warrington, or later at the Waverley Temperance Hotel in Chester, which she evidently managed in the 1890s, and after 1904 from her home in Harston, Cambridge. With such a range of entrepreneurial activities, like Isabella Beeton, Young outstripped the fame of her husband. Unusually, she retained her maiden name when she married Dr. William Riding of Omskirk. To retain Victorian proprieties, it was Dr. Riding who altered his name to Young.

It would appear that Hannah Young must have been one of the earliest British recipe writers to exploit the availability of new manufactured foodstuffs; hence it is not hard to see why she should have been chosen by Liebig and Company to compile their English recipe book in 1893. This was her third cookery book. In the "Preface" to this handsome shilling compilation, she described cookery as "the culinary art" whose study "affords much pleasure and profit." Liebig would have approved. She noted the spread of cookery education during the previous decade. Teachers who had learned their skills at the National School of Cookery, founded in 1873, were passing on their knowledge as cookery entered the British school curriculum. Fortunately, she said, winding her readers up towards the advertising pitch, there were new products available that reduced the labour and drudgery of cooking, and "the Liebig Company's

42. Mrs H. M. Young, *Liebig Company's Practical Cookery Book* (London, 1893). For Young's many other cookery books, see Driver, *A Bibliography of Cookery Books.*

Extract of Meat holds foremost place in the field." In the "Preface" she wrote:

> The true object of cooking [is] to maintain and enhance the nutritive value of the food cooked and [to] secure its digestibility . . . to render it as palatable as possible and to please the senses of sight and smell.

Liebig's extract, she emphasized, assisted the cook to be varied and economical.

Young's recipes covered the range of her two previous cookery books in style. Every savoury dish uses some of the Liebig extract, including the fish recipes. Only the sweets escaped the extract, which included "Liebig sandwiches" and "Liebig on toast." All this was accompanied by the delightful illustrations reprinted from one of the company's German cookbooks. The result was decidedly kitsch, as when two "water babies" brandish saucepans, one standing upon a giant lobster, the other ready to assist: The recipe underneath is for Lobster Soup. There were also visual puns. "Liebig Stimulant or Nightcap" had a man wearing a nightcap and scarf, and the title page for the "Miscellanous" selection of recipes has the portrait busts of five very different men, fat and thin, in various styles of dress and fashions in facial hair. All this was bound in a maroon-coloured, finely lettered cloth cover on the front of which a grandly dressed young lady took the lid from a steaming copper pan while a young maid took a jar of Liebig's Extract from the shelf. The kitchen walls and stove were tiled and on the wall hung a pendulum clock. Other furnishings included some rococo shelves bearing blue and white storage jars, a set of spice drawers next to a wrought- or cast-iron bracket with a glass lamp. There was also a rococo water fountain and a basin in the corner. On the floor was a soup tureen with a generously displayed pile of assorted vegetables in a marmite (a French casserole-style pot). The vegetables overflowed into a gold rococo border at the bottom of the cover and included a shell-shaped frame enclosing a jar of Liebig's Extract. Finally, above a window overlooking a lake, mountains, and a stream, was the scrolled title *Liebig and Company's Practical Cookery Book*. The general effect was a combination of Dutch interior and Renaissance painting and above all of bourgeoisie luxury. Nor did the back cover escape the unknown German designer's attention. Two jars of Liebig's Extract were surrounded by plentiful vegetables, game, and a lobster. At the base of this arrangement were three scrolled texts that described the extract's many uses. The whole was again framed with a fine rococo border.[43]

43. My thanks to Valerie Mars for information and comments on Young and Parloa.

Such luxurious binding contrasted dramatically with the covers of the Liebig cookery book issued to Americans in the same year. Again the company chose a leading cookery writer Maria Parloa (1843–1909), rather than simply capitalising on Henriette Davidis's reputation amongst the large German immigrant population. (Davidis's cookery books had been issued in both German and English for American users.) Parloa, the founder of two cookery schools and the author of *The Appledore Cookbook* (Boston, 1872), *Miss Parloa's New Cookbook* (Boston, 1881), and *Miss Parloa's Young Housekeeper* (Boston, 1893), published *One Hundred Ways to Use Liebig Company's Extract of Meat: A Guide for American Housewives* in 1893. Although of the same size and quality as Hannah Young's book, it carried fewer recipes. The language was also a good deal less formal than Young's, beginning disingenuously like a storybook with the question, "Have you ever thought of the history of the little cream-colored jars so common in American households?" She then explained how the extract, "a boon to thousands of families," had been turned into a growing enterprise and with what care the product was examined at every stage. Its uses in the kitchen and as a reviver of exhausted travellers and weakened invalids were stressed. The recipes, for the most part, were not very different in style from those in Young's book. There is Julienne Soup and similar entrées and sauces. A few of the dishes could be said to be truly American, such as Creole Soup, Oysters à la Baltimore, and Clam Soup. Unlike the English text, there were, however, also instructions that incorporated tinned foods; Walled Salmon, for example, used a tin of salmon, the "wall" being a potato border. There were no sweet dishes, so every dish included the use of extract.

All these recipe books, as well as Davidis's original model, were clearly marketed with consumers in mind. The dishes described were from a conventional repertoire with no extremes of luxury or overelaborate preparation. This production of appropriate books, as Hans Teuteberg has stressed, is very much the ancestor of modern commercial recipes in respect of a known and well-understood middle-or-the-road public.[44] Liebig's Extract of Meat, like its modern cookery book equivalents of Stork Margarine or Be-Ro Flour, did not cater to the most recherché tastes.

The Liebig Company also exploited the advertising possibilities of household calendars, and from 1872 it began to include with the jars "*Liebigbilder,*" cards similar to cigarette or baseball cards, starting with pictures of the production of the extract and its distribution, as well as highlights of Baron Liebig's career. The many series of these cards, whose

44. Teuteberg, *Die Rolle des Fleischextrachts.*

production and distribution ceased only in 1940, covered such topics as flowers, fruits, music and opera, and native peoples. It was the forerunner of the cigarette card and was soon copied by other commercial companies as a certain way of ensuring that children cajoled their mothers to purchase the product so that their Liebig albums, like their stamp collections, could be complete. These *Bilder* are now collectors' items.

Bread, Coffee, and Milk

Liebig's interest in food extended well beyond his best-known commercial activity with meat extract. When his American pupil Eben Norton Horsford (1818–93) studied at Giessen in 1844, Liebig encouraged him to investigate the nitrogen content of vegetable foods such as wheat, rye, oats, corn, and clover. This interest remained with Horsford after he returned to America, where in 1847 he was appointed Rumford Professor of Science as Applied to the Arts at Harvard. When the businessman Abbot Lawrence donated funds for the establishment of a science school at Harvard, Horsford modelled the teaching and laboratory exercises on the Giessen prototype. In 1856 he patented a new baking powder containing acid phosphate and opened the Rumford Chemical Works in Providence, Rhode Island, for its manufacture. This replaced the yeast fermentation process traditionally used in bread making.[45]

Horsford's commercial success – he made a fortune – undoubtedly stimulated Liebig to interest himself in the chemistry of flour and of bread making. In 1868, stirred by news of famine in East Prussia (Poland), Liebig announced a new method of making white bread that received wide publicity in Europe, Britain, and the United States.[46] Liebig drew attention to the fact that bread making had hardly altered since biblical times insofar as "the chemical method of making bread" (i.e., using the Horsford type of baking powders instead of yeast) had failed to catch on in Germany because good results depended on close attention to the recipe. Black bread, which Liebig strongly advocated for its nutritional wholesomeness, was also frowned upon by the workingclasses who (as in Britain) chose the refined white flour preferred by the better-off (who undoubtedly preferred it because it was less likely to produce watery stools).[47]

45. E. N. Horsford, *The Theory and Art of Bread Making* (Providence, R.I., 1869).
46. For translations, see *Every Saturday* 7(1869) and *Lancet* 30 May 1868, p. 705. From *Augsburger Allgemeine Zeitung*, where his chemical letters had first appeared a quarter of a century earlier.
47. John Burnett, *Plenty and Want* (London, 1966; reprint 1979).

Recognizing the difficulty of changing public taste, therefore, Liebig's research problem was how to make a nutritious white loaf without including bran. In two essays on bread and its digestibility published in the *Augsburger Allgemeine Zeitung* in 1868, Liebig had shown that there was a serious loss of nutritional value when flour was refined.[48] As in washing fresh meat with water, refining led to a loss of nutritive phosphate salts. He claimed that a thousand parts of wheat and rye grains were composed of 8.94 and 5.65 grains respectively of phosphoric acid, whereas these proportions were reduced in white flour to between 2 and 3 grains only. Overall, he estimated, white flour contained 15.5 parts less of nutritive salts and between 6 and 7 parts less of phosphoric acid than wholemeal flour. Similar losses were reported for rye flour. The lack of calcium and magnesium phosphates in flour was serious, reported Liebig, because these salts, as animal feeding experiments conducted by Dr. Haubner of Dresden and Dr. Roloff of Halle showed, had important functions in skeletal growth and the maintenance of bones. Liebig also suspected, erroneously, that a deficiency of phosphates in the diet caused scurvy. Instead of bran, therefore, the nutritional quality of white flour could be raised by adding phosphates, and he noted that this was precisely what Horsford's development of phosphate baking powder had achieved. Moreover, its economical significance was that "with the same quantity of flour, a greater number of men may be satisfied and nourished," for the effect was as if wheat fields had "produced from one seventh to one eighth more wheat." Horsford's invention, Liebig declared, was "one of the weightiest and most beneficient . . . which have been made in recent times."[49]

Liebig had spent eight months experimenting with Horsford's baking powder (a mixture of calcium and magnesium phosphates with sodium bicarbonate).[50] The release of carbon dioxide in the oven puffed up the dough and made it porous during baking. Although this practice was justified by taste and cost, Liebig pointed out that potassium bicarbonate, rather than sodium bicarbonate, ought to be used because it was potassium rather than sodium salts that white flour lacked. (In fact, Horsford had already patented the use of potassium bicarbonate in bread preparation in 1864.) Liebig's ingenious solution was to use the cheaper potassium chloride instead of common salt in mixing the dough. It had

48. J. Liebig, "Über Wohlgeschmack und leichte Verdaulichkeit des Kleienbrodes," *Augsburger Allgemeine Zeitung*, 1868, Nr. 6; "Über Pumpernickel," idem 1868, Nr. 11.
49. J. Liebig, "A new method of making bread," *Every Saturday* 7(1869), 618–20.
50. P. R. Jones, "Liebig, Horsford and baking powder," *Ambix* 40(1993), 65–74.

been rendered cheap by the discovery of the vast layers of potassium salts in Strassfurt. Its use would reduce the price of bread, Liebig reasoned.

Liebig then calculated the proportions of potassium chloride needed in the revised baking powder for 100 pounds of flour, adding that the best results came from intimate sifting and mixing before water was added. The result was bread with a beautiful appearance, somewhat heavier than the common baker's loaf. The fact that an apparent 10 to 15 percent more loaf was made by this method compensated for the increased cost of baking powder over yeast. Such bread, he declared, "made without fermentation is daily consumed in my house and is preferred by my family and guests." Making bread rolls also became one of the highlights of the popular lectures Liebig delivered in Munich.

Liebig noted that in the United States, Horsford and others marketed a ready-mixed "self-rising flour" to which housewives had readily taken. Both Horsford and Liebig recognised the implications that "the new process of bread-making would be adopted by bakers," in which case the chief obstacle to "the industrial prosecution of the baker's art" would fall to the ground. When bread was made on an industrial scale, three or four men with an oven and a kneading machine could make hundreds of loaves daily, just as in the Portsmouth dockyards three men turned out 20,000 ships' biscuits a day.

> For an army in the field and for making bread on shipboard this new method of making bread appears to me of especial significance, and it is very desirable that the superintendents of prisons and poorhouses should collect observations in relation to the nutritive value of bread made with this baking powder.[51]

Eager to see baking powder (*Backpulver*) adopted in Germany, Liebig encouraged two of his former pupils, Ludwig Clamor Marquart (1804–81) and Carl Zimmer, who had set up chemical manufacturies at Bonn and Mannheim, to manufacture the substance in bulk. Although he directed that labels should bear the words "prepared according to the special directions devised by Baron von Liebig," aware of his commitment to the extract, he did not allow his signature to appear on the packets when the baking powder was finally launched in 1869.[52] Insofar as its retail price was higher than yeast, both Liebig and his manufacturers realised that the key to commercial success lay in the replacement of home

51. *Every Saturday* 7 (1869), p. 620.

52. Horsford did, however, issue trade cards in America for his phosphoric acid baking powder that carried an endorsement by Liebig for its nutritious quality. I thank Dr. James Bohning for drawing my attention to this.

bread-making and old-fashioned batch bakers by industrialised bakeries. As with the extract, therefore, much emphasis was placed upon the powder's convenience for armies and navies and in the superior nutritional value of the bread produced. Ironically, although the manufacture of baking powder proved financially successful in the United States and in Britain, sales were so sluggish in Germany that Zimmer abandoned manufacture in 1870. Two years later on his retirement, Marquart's sons also decided not to continue making baking powder, and the small commissions that Marquart and Zimmer had paid Liebig ceased. Unlike Horsford, therefore, Liebig did not become a millionaire through bread.

Earlier, in 1854, Liebig had investigated ways of improving the making of rye bread whose traditional sour taste had stimulated the poorer classes to emulate their betters by buying refined flour from which, in Liebig's view, a less nutritious white loaf was baked. Although British and Continental bakers had traditionally used copper sulphate or alum additives to diminish acidity, there was a thin line between the use of such "improvers" and the outright adulteration of flour for purposes of cheating customers. Second-grade flour, or damp flour whose gluten content had become immiscible, was often adulterated in this fashion. In the report on Liebig's process for making bread in the *Allgemeine Augsburger Zeitung* for June 1854, Liebig made it clear that he thought the addition of alum to flour was injurious to health and deserved police attention. Liebig proposed instead the use of lime water (calcium hydroxide) when mixing flour and yeast into dough. The result was, as he told the Glasgow meeting of the British Association in 1855, "a beautiful. sour-free, moist, elastic, nonvesicular, crusty loaf with a superb taste."[53] He took pains to point out the physiological advantages of his home-baked wholemeal loaf, drawing upon the experiences of the Munich doctor Benecke with the lack of calcium phosphate in scrofulous (i.e., rickety, tubercular) children. Liebig supposed that the calcium of the lime water combined with the free phosphoric acid in the flour to produce the calcium phosphate (bone earth) important in bone formation. Not only was the new loaf more nutritious, he claimed, but it would also be more economical to prepare. Without the addition of limewater, 19 pounds of flour baked 24 to 25 pounds of bread, whereas when 5 pounds of lime water was made into a dough with 19 pounds of flour, 26 to 28 pounds of bread was formed.

In 1856 Liebig contracted with the first baker in Germany, Sokeland (1806–84), "who, with the help of objective, clear, scientific intelligence,

53. J. Liebig, "Verbesserung und Entsäuerung des Roggenbrodes," *J. prakt. Chemie* 63(1854), 313–14; reported *BAAS Reports* 1855, p. 64.

comprehended the essentials of bread preparation and knew how to apply these to advantage in baking."[54] Nevertheless, although "Liebig's Bavarian Brown Bread" recipe was included in the posthumous revision of Elizabeth Acton's *Modern Cookery* in 1865, it is unclear whether Liebig's process was adopted by bakers on any scale. Since 1940, British bread has been fortified with chalk, initially on the grounds that expectant mothers and growing children need to supplement their diets with calcium to promote bone growth. On the other hand, in England Liebig's baking powder rivalled M'Dougall's baking powder in the 1890s, following the plug that was given to it by Heinrich Debus at the British Association in 1869.

If the quality of bread might be improved, what of the other staple of European diet, coffee? During 1864, while in the middle of planning the launch of his Extract of Meat and reconciling himself to the fact that his plans for the disposal of London's sewage would be ignored, Liebig investigated the best way of making a cup of coffee, at the same time developing a recipe, or formula, for humanised milk. Liebig had learned from a Captain Pfeufer (probably the son of the Liebigs' family doctor), who was a sanitary commissioner in the Bavarian army, that during the first Schleswig-Holstein and the last Italian campaigns, "coffee had very materially contributed to the general health of the German and French soldier." If a marching army was given coffee as a beverage, Liebig asserted, fatigue was diminished, and this claim was supported by evidence from the men who worked on the exhausting trading caravans in Central America. Rather in the manner of Bach's wonderful secular coffee cantata, Liebig sang coffee's praises in the resultant essay that appeared in *Dingler's polytechnisches Journal* in 1866 and simultaneously in *Chemical News, Popular Science Review,* and *Every Saturday.*[55] Because marching armies could not be burdened with the machines necessary for roasting and grinding coffee beans, Liebig had wondered whether an extract of coffee could be developed to parallel his success with Extract of Meat. Unfortunately, as Liebig quickly discovered, oxygen badly affected the taste and flavour of ground coffee. Moreover, if a cup of hot coffee was slowly evaporated, the remaining black mass was largely insoluble and produced a foul-tasting brew when reboiled with water.

In failing to solve the problem of coffee extract (which was not solved until the introduction of freeze-drying in 1940: the rapid freezing of ground roasted coffee beans followed by drying in vacuo), Liebig was led

54. Jones, "Liebig, Horsford and baking powder," p. 65.

55. J. Liebig, "Cup of coffee," *Every Saturday* 1(1866), 135–7, from *Dingler polytechnisches Journal,* 3(1866), 466–7.

to investigate the best method of brewing coffee. No chemistry was involved, only purely empirical investigation and tasting.

Germans drank more coffee than tea, whose preparation the English had perfected. Whereas tea aided stomach digestion, coffee stimulated peristalsis in the lower and upper intestines. "Therefore, the German man of letters, more accustomed to a sitting life, looks on a cup of coffee, without milk, and assisted by a cigar, as a very acceptable method of assisting certain organic processes."[56] Liebig stressed that the coffee berries must be well chosen and that impurities and black mouldy berries must be eradicated. The remaining good beans should be roasted only until they turned pale brown; any further heating brought about serious loss of the volatile caffeine. Roasted berries, if not for immediate use, were easily penetrated by oxygen, with loss of flavour; protection was achieved by mixing the hot berries with sugar, which rendered them impervious during the cooling process. Although sugar attracted moisture, storage in a dry place preserved the beans until they were wanted for grinding and use. This solved the problem of fresh coffee for marching armies.

Liebig then investigated and compared the three ways of making coffee: filtration, infusion, and boiling. Filtration, he discovered, caused the ground coffee to come into contact with air, whose oxygen attacked and spoiled the coffee's aromatic principles. Instead of dissolving some 20 percent of the coffee, filtration dissolved only 11 to 15 percent. Infusion, on the other hand, in which ground coffee was added to boiling water and allowed to stand for ten minutes, gave an aromatic coffee but one that contained little extract. Finally, adding the coffee to cold water, followed by boiling, produced a coffee rich in extract but having little aroma unless the boiling was of short duration.

Faults having been found with all three methods, Liebig's recipe was a compromise. Three quarters of the coffee to be used was boiled in cold water for ten to fifteen minutes, following which the remaining quarter of coffee was flung into the pot. The coffee was then allowed to stand for five minutes. The dregs could be filtered off, but Liebig thought this unnecessary. In this method, the boiling produced a strong coffee, and the additional coffee gave aroma to the beverage. Although this more time-consuming method may have made a more flavoursome drink, there is no evidence that his contemporaries took it to heart, despite the wide publicity his article received and its appearance in domestic science textbooks such as James Knight's *Food and Its Functions* (1895).

Matters were very different for Liebig's solution to the production of a humanised milk in 1865. Although Liebig was probably the first chemist

56. Liebig, "Cup of coffee," p. 135.

to try to produce an artificial milk, others, such as the Lancastrian chemist Edward Frankland, who had worked in Giessen in 1849, successfully solved the problem at about the same time. In both cases, the recipe was a practical outcome of feeding a baby whose mother was unable to suckle. Historically in such cases, when a family could afford it, a wet nurse (usually a poor but fit mother whose baby had died or whose illegitimate baby had been given away) was employed. However, by the 1860s, the middle classes, much more aware of public health and hygiene, were withdrawing from the use of wet-nursing. They were then faced with the problem of weaning their children. Cow's milk was frequently found to upset the baby's stomach, and similar problems were found with the use of cereals and wheys.[57]

Faced with the fact in 1865 that his German wife could not suckle their first child Percy, Frankland developed a modified cow's milk that was successful, and he reported his results to the Manchester Literary and Philosophical Society that year. Subsequently, after further investigations, he developed the milk commercially as "Frankland's Milk." It remained on sale into the early twentieth century when it was rapidly replaced by the dried milk developed by Glaxo and other pharmaceutical firms.[58]

Liebig's "neue Suppe für Kinder" was similarly developed in the summer of 1864, not for his wife but for his Nanny Thiersch's grandchildren.[59] When Liebig's publisher and friend Vieweg first heard of this, he misunderstood it to be an invalid soup for children instead of a food for suckling infants. The most influential (though in fact inaccurate) analysis of human milk was that reported by the Berlin physiological chemist J. F. Simon (1807–43) in his doctoral thesis of 1838. His analysis received wide publicity through his authoritative *Animal Chemistry*.[60] According to Simon, human milk contained 3.43 percent casein (in modern terms, 1.3 percent protein) and 4.82 percent of milk sugar (in modern terms, 7 percent of lactose). Liebig accepted that all mammalian milks contained a unique nitrogeneous base, the protein casein, which was also "the chief constituent of the leguminosae," or "vegetable casein." As we have seen, this fact was one of the foundations of his metabolic theory.

> How admirably simple, after we have acquired a knowledge of this relation between plants and animals, appears to us the process of

57. See T. B. Mepham, "Humanizing milk," *Medical History* 37(1993), 225–49.
58. C. A. Russell, *Edward Frankland* (Cambridge, 1996), pp. 484–5.
59. J. Liebig, "Eine neue Suppe für Kinde," *Ann. Chem.* 133(1865), 374–83; expanded as *Suppe für Säuglinge* (Braunschweig, 1865).
60. J. F. Simon, *Physiologische und pathologische Anthropochemie* (Berlin, 1842); trans. as *Animal Chemistry* (London, 1846).

Table 8.2. *Human and Cow's Milk (1992)*

	Human %	Cow %
Water	87.0	87.0
Protein	1.1	3.5
Fat	3.8	3.7
Lactose	6.8	4.9
Minerals	0.2	0.7

> formation of the animal body, the origins of its blood and its organs! The vegetable substances, which serve for the production of blood, contain already the chief constituent of blood, ready formed with all its elements.[61]

It followed that in preparing a substitute for human milk, vegetable casein from farinaceous materials would be suitable.

Using the analysis of human milk published by his pupil Haidlen in 1843, which merely confirmed Simon's erroneous formula, Liebig supposed that the ratio of plastic (flesh-forming) to respiratory (heat-conferring) aliments in human milk was 10:38. Liebig's recipe for humanised milk accordingly consisted in mixing ten parts of cow's milk with one part of wheaten flour, one part of malten flour (chosen in order to enrich the wheat starch with sugar), and some potassium bicarbonate to reduce the milk's overall acidity, as well as providing potassium for flesh. Because of its use of malt flour, the milk was often known in Britain as Liebig's Malt Extract. Although it preserved the then-supposed relative composition of human milk, Liebig's formulation was twice its strength – or, as he put it, "double the concentration of women's milk." Babies would need less of it because it was so rich! From the twentieth-century perspective, however, such a formulation carried dangers. It was overweighted with carbohydrates and wildly underestimated the significance of fat, besides being seriously deficient in key amino acids and vitamins, especially vitamin C. (See Table 8.2.)

As with the extract of meat, Liebig exploited all his contacts and publicity methods to make the new product well known. For example, he publicised the milk's recommendation by the director of Munich's Maternity Hospital, Dr. C. Hecker, and by the Munich physician Dr. Ludwig

61. J. Liebig, *Familiar Letters on Chemistry*, 4th ed. (London, 1859), p. 375.

Walther.[62] The formulation was reported by Liebig simultaneously in the *Annalen* and *Popular Science Review,* and the *Annalen* article was offprinted by Vieweg in 1866 as a twenty-page pamphlet, *Suppe für Säulinge.* A second enlarged edition, containing medical testimonials, was published in the same year, with English and French translations following in 1867. In Britain, Liebig arranged for the milk to be sold by the London pharmacy of Savory & Moore, and he allowed an English patent to be taken out by his translator, the philanthropist Baroness Elise von Lersner-Ebersburg.[63] Ebersburg, according to Drummond, licensed the patent to "Liebig's Registered Concentrated Milk Company Ltd," whose premises were in Tichborne Street off the Regent Quadrant. The company offered Liebig's malt milk in liquid form at 6d per quart, which was only a little more expensive than cow's milk. When sales did not come up to expectation, a cheaper dried form of the milk containing pea, instead of wheat, flour was sold. This would have had to be made up with cow's milk in the household. It was this dried milk that Lersner-Ebersburg, who had lost two of her own children because of her inability to breast feed, employed in her Lersner-Ebersburg British Institute for foundlings in Bethnel Green:

> I founded it in memory of two dear little babes of mine, whom the Lord took from me after a very short illness. I began with one district, but no sooner did the good results of this happy mission become known, when applications for opening neighbouring districts grew so numerous, and the good effects so marvellous, that it would have seemed cruel to deny this boon of life to so helpless a portion of humanity. During the period from February, 1870, to February, 1871, 8,256.5 lbs of malted food extract were distributed to twelve eastern [Bethnel Green] districts.[64]

The milk also received an accolade in *The Lancet* from Britain's leading food analyst Arthur Hassall:

> It appears to me that the great merit of Liebig's preparation consists in the use of malt as a constitutent of the food. This, from the diastase contained in it, exercises, when the fluid food or soup is properly prepared, a most remarkable influence upon the starch, quickly transforming it into dextrine and sugar, so that in the course of a few

62. J. Liebig, "Nachtrag zu meiner Suppe für Säuglinge," *Ann. Chem.* 138 (1866), 93–4, 95–6.

63. E. von Lersner-Ebersburg, trans. *Food for Infants,* 2nd. ed. (London, 1869). Under the name Lersner, she published *Children's Gifts and Mother's Duties* (London, 1865); and as Countess Ebersburg, *Six out of Ten* (London, 1877), which advocated Liebig's milk. See British Patent 3142 (29 November 1866) and 250 (30 January 1867).

64. Ebersburg as cited in C. H. F. Routh, *Infant Feeding* (3rd ed., London, 1879), p. 186.

> minutes the food, from being thick and sugarless, becomes comparatively thin and very sweet.[65]

Like Lersner-Ebersberg, who saw herself as an angel of mercy, Liebig was not averse to seeing himself as a saviour of infants. He wrote to the Hessen politician Reinhard Dalwigk in January 1865:

> I take the opportunity of sending you two offprints concerning the Kindersuppe which is now getting worldwide publicity. One may well ask what significance Kindersuppe can have for the world. I asked myself this right from the start and now, a year later after composing my articles, the milk has brought many needy children back to health and life, and it has exposed to view the huge mortality of suckling infants in England, France and many parts of Germany.[66]

Liebig's exposure of infant mortality stimulated an infant-food industry. In Switzerland, Henri Nestlé, using Liebig's language of plastic and respiratory aliments, developed a dried milk food from "good Swiss milk and bread, cooked after a new method of my invention, mixed in proportion, scientifically correct, so as to form a food which leaves nothing to be desired."[67] Simultaneously, the English pharmacists Gustav Mellin and James Horlick simplified Liebig's formula and soon found extensive markets in the United States, as well as Europe, for their "Mellin's Food for Infants and Invalids" and for "Horlick's Malted Milk." Like the Liebig Extract of Meat Company, these and other rival firms advertised widely in the popular press and women's magazines. In place of cookery books, they offered booklets on infant care and feeding that usually incorporated simple expositions of Liebig's nutritional theory, even though it was out of date by the 1870s.

Although baby-food products such as Liebig's and Frankland's preparations had a middle-class sale, it was the condensed milks that captured the working-class mother. That there was some confusion here no doubt occurred because Liebig's humanised milk involved the use of a malt extract that was evidently easily confused with the malty flavour of condensed milk. Despite his denial in 1871 that he was the inventor of condensed milk,[68] Liebig's name was often improperly – though usefully – associated with the burgeoning condensed and evaporated milk industry that also took off in the 1860s. It appears that his real association with these products was to advise the English agent of the Anglo-Swiss Con-

65. A. Hassall, "Liebig's extract of meat," *Lancet* 29 July 1866.

66. Letter in Staatsbibliothek, München.

67. Quoted in Rima D. Apple, *Mothers and Medicine. A Social History of Infant Feeding* (Madison, Wis., 1987), p. 9.

68. J. Liebig, "Erklarung," *Ann. Chem.* 158(1871), 136.

densed Milk Company (later Nestlés) to add beet sugar to the product to ensure its preservation.[69] Liebig himself did not patent this procedure.

Attempts to preserve milk by canning and evaporation were stimulated by the increased need to supply fresh milk to urban areas whose dairies and cowsheds were increasingly being abolished by nuisance inspectors and by-laws. In 1856 an American, Gail Borden, patented a method for condensing milk, and ten years later in 1865 he formed the Anglo-Swiss Condensed Milk Company, using cheap Swiss milk. Employing many of the same techniques as the Liebig Company and publicly endorsed by Liebig in company advertising, the Anglo-Swiss Company established outlets in most European countries and soon found itself in competition.[70] At this date, desiccated, or dried, cow's milk could be manufactured only from skimmed milk. Though harmless as a basis for the chocolate or malted drinks such as "Horlicks," being deficient in fats they were decidedly dangerous when used extensively for weaning. The problem of drying whole milk was not solved until the early 1900s when the Hatmaker process laid the foundation for the baby-food industry of Glaxo – which ironically was initially marketed by Brands, one of the many pharamceutical and grocery companies that had been established in the late 1860s as a purveyor of meat extracts and beef teas.[71]

Finally in this saga of Liebig's association with the food industry, it is important to notice Liebig's indirect connection with the development of the Australian wine industry. In the mid-1830s, James King, whom Liebig met in 1855 at the Glasgow meeting of the British Association while King was on furlough, settled at Irrawang in the Hunter Valley of New South Wales. Here he began to grow grapes. In October 1850, he sent samples of the wines to Liebig, who analysed them and reported favourably that they were up to the standard of the best French and German wines. By 1855, King was producing 2,500 gallons of wine per annum and selling his vintages throughout New South Wales. In 1855 he received medals at the Paris Exhibition, and meeting Liebig in London, he received a second endorsement from the German chemist. A year later King received the Royal Society of Arts' silver medal for his wine exports.[72] Unfortunately, the discovery of Australian gold temporarily destroyed the labour market for vinoculture, and it was not until the 1870s that Australian wines again were exported to Europe. King, the pioneer, had died in 1858.

69. *Journal Royal Society of Arts* 16(1867–68), 76.

70. A. Willard, "The American milk-condensing factories and condensed milk manufacture," *J. Roy. Agri. Soc. of England* 8(1872), 103–57.

71. R. P. T. Davenport-Hines and J. Shinn, *Glaxo, a History to 1962* (Cambridge, 1992).

72. J. King, "On the growth of wines in New South Wales," *Journal Royal Society of Arts* 4(1855–56), 575–78. See Jack and Liston, "Scottish Immigrant," 92–106.

Hofmann was in no doubt of Liebig's role in the industrialization of food:

> Who but knows that it was from Liebig's mouth that our housewives first learnt how to render the full nutritive value of meat available, or how to prepare a broth for invalids, combining the maximum of nourishing effect with the highest degree of digestibility? Who has but heard that, having carefully studied the nature of woman's milk, he was led to compound his infant food as a substitute for mother's milk, thus becoming the benefactor even of future generations? Who, lastly, is ignorant that it is owing to Liebig's researches in physiological chemistry, that the superabundant animal nourishment of the more thinly-populated quarters of the globe has been rendered accessible to the overcrowded populations of the opposite moiety; and that a grand commercial movement, uniting, as it were, by new bonds the two hemispheres, has been created by the organisation of a food industry already colossal, and tending still to expand with incalculable advantage to the inhabitants of Europe, thus lifted beyond the sharp pressure of deficient nourishment, and secured in the abundant supplies of those invigorating food constituents upon which the bodily and mental energies are alike dependent for their development?[73]

Liebig's work on the chemistry of food and his subsequent exploitation of commercial possibilities were theory-driven. Health, strength, and happiness would surely follow the application of science to cookery, either by improving the cook's efficiency or (as with his and Horsford's recipes for bread-making using self-raising flour instead of fermentation) removing domestic drudgery through the industrialization of the process. At the same time, we must recognize the employment opportunities that Liebig was opening up for chemists to act as quality controllers and as gatekeepers of quality, and for the making of money from science. The domestication of nature for chemists was an extension of employment opportunities, for the patenting of recipes, and for the making of fortunes. Hence the irony that in highlighting the fact that cookery was chemistry, Liebig was preparing the way for the disappearance of chemistry from the kitchen into the factory. The lower-class lad who had learned French in a ducal kitchen, who became a baron in 1845, was never a Baron Hardup of the fairy story but a very rich baron indeed when he died in 1873. The company he founded survives as part of the multinational Unilever Company.

73. A. W. Hofmann, *The Lifework of Liebig* (London, 1875), pp. 36–7.

9

Liebig and London: The Chemistry of Sewage

> Your nation, Britain, has caused the grave depletion of its fields' fertility, by waste of its excretion. As a famous German chemist I issue the precaution that this cabin creates havoc of Noah's Flood proportion. Each pull of the chain is like a Noah's Flood taking in its cataracts so much nutritious good. What gets flushed away in these gallons of flood water in terms of drowned potential amounts to wholesale slaughter.[1]

As we saw in Chapter 6, although a scientific and theoretical approach to practical farming in Great Britain had been heralded by Humphry Davy in 1812, chemists' lack of knowledge of the role of nitrogen in plant growth and of the chemical processes in which inorganic materials entered into plants meant that the crucial question of the action of the soil and its constituents had yet to be solved in 1840. Liebig's *Agricultural Chemistry* did not provide a complete answer to all aspects of plant nourishment; it did, however, make an important point about soil exhaustion and about the potential value of human excrement as a manure in recycling the elements of human and plant foods:

> It must be admitted as a principle of agriculture that those substances which have been removed from a soil must be completely restored to it, and whether this restoration be effected by means of excrements, ashes or bones, is in great measure a matter of indifference.[2]

Liebig's interest in the potential value of human excrement was evidently a source of amusement amongst his British students. William Francis reported to his father in 1841 of "a fine caricature" that was circulating amongst the Giessen students. This portrayed Liebig "easing himself in a basin held by Playfair," whose analysis of dungs appeared in the second edition of *Agricultural Chemistry*.[3]

1. Liebig's speech from Tony Harrison, *Square Rounds* (London, 1992), p. 13.
2. Liebig, *Die organische Chemie in ihrer Anwendung auf Agricultur* (Braunschweig, 1840), p. 167; *Agricultural Chemistry* (London, 1840), p. 176.
3. William Francis to Richard Taylor, 26 November 1841, Taylor & Francis papers, St

When Edwin Chadwick published his momentous *Report on the Sanitary Condition of the Labouring Population of Great Britain* in 1842, he was therefore able to cite Liebig approvingly as a chemical authority who supported his vision of cities in which human sewage and wastes were piped away from houses in water-flushed sewers, to be collected at some distant point and there recycled back onto the land. Here was an arterial-venous circulation of pure and foul water that could be of economic benefit to urban and country-dwellers alike. In 1845 Chadwick formed the Metropolitan Sewage Manure Company (part of his Towns Improvement Company) to work such a system commercially; however, it never actually proceeded to business because of the way capital and engineering expertise was attracted to railway speculation during the 1840s.

Such recycling ideas were not new in 1840. For example, between 1827 and 1850, the apocalyptic English artist John Martin (1789–1854) repeatedly published grandiose but practical schemes for supplying London with pure water and for using a series of intercepting sewers on each side of the Thames to remove human and animal wastes and to prevent them from polluting the river. Fed into huge receptacles, the ordures would then be distributed by covered boats on canals to the agricultural districts around the city. Martin, like Liebig later, cited the Chinese, Japanese, Flemings, and French as exemplars of such husbandry.[4]

By the mid-1850s the many streams and rivers that flowed into the urban territory of the Thames had been covered over and were serving as sewers and drains. The growing popularity of the water closet amongst the upper and middle classes had had the unfortunate effect of turning the Thames into a giant sewer – a *cloaca maxima* – that at low tide produced a foul, noisome stench. In 1855 Michael Faraday complained publicly about the appalling state of the river, and the Royal Society of Arts, led by

Bride's Printing Library, London. Besides Playfair, another English pupil, John Rogers, also examined faeces in 1848. Liebig consulted Hofmann over Rogers's financial prospects when he began to court Liebig's daughter Agnes. See Hofmann to Liebig, 18 April 1848, in W. H. Brock, ed., *Justus von Liebig und August Wilhelm Hofmann in ihren Briefen* (Weinheim, 1984), p. 78.

4. Thomas Balston, *John Martin 1789–1854. His Life and Works* (London, 1947), Chap. 15. Martin inspired others such as Robert Angus Smith (a pupil of Liebig's), *Manchester Lit. & Phil. Memoirs* 12(1855), 155–75, and the physiologist Marshall Hall, who recommended that sewage be loaded onto closed railway trucks. See *Lancet* 27 September 1856, p. 363. For Chadwick's experiments with liquid sewage hosed from canal barges, see R. A. Lewis, *Edwin Chadwick and the Public Health Movement 1832–1854* (London, 1952), pp. 122–3. More generally, see Graeme Davison, "The city as a natural system," in D. Fraser and A. Sutcliffe, eds., *The Pursuit of Urban History* (London, 1983), pp. 349–71; Nicholas Goddard, "Nineteenth-century recycling. The Victorians and the agricultural utilisation of sewage," *History Today* June 1981, pp. 32–6.

Liebig adopted the English fashion of exchanging photographic *carte de visite* with his many visitors. This photograph from the 1860s was presented to Professor Marchi on Liebig's visit to Florence 6 August 1867. (Courtesy W. Lewicki)

Sir John Lawes, debated the issue.[5] The Society became the nation's chief forum for debate concerning the disposal and utilization of sewage. In May 1863, for example, Liebig's former student J. L. W. Thudichum seriously proposed collecting solid and liquid excretions in a specially adapted closet since, after all, in men at least, "nature separates these two excretions by propelling them in directions that diverge at an angle of 40°, more or less."[6] Even so, nothing was done; in 1856, known popularly as

5. J. B. Lawes, "The sewage of London," *J. Roy. Agri. Soc. of England* 3(1854–5), 263–84, and discussion pp. 311–25.
6. J. L. W. Thudichum, "On an improved mode of collecting excrementitious matter, with a view to its application to the benefit of agriculture and the relief of local taxation," *J. Roy. Agri. Soc.* 11(1862–3), 440–52, 13 May 1863.

"the year of the great stench," Parliament was forced to postpone business because of smells, and nearly £900 a week was spent on disinfecting the river but to little avail. As is well known, it was through experiments on the deodorization of Carlisle's sewage with carbolic acid (phenol) in the late 1850s that the surgeon Joseph Lister was led to try it as a surgical antiseptic in 1865 – one of the few success stories of sewage utilization.

In 1856, the newly created Metropolitan Board of Works (MBW) for London decided to solve the problem of London's drains once and for all. Its plan was to intercept sewage before it passed into the Thames and to pipe it some seventeen miles downriver to Barking Creek on the north bank and Crossness on the south bank. At both outfalls the sewage would be discharged into the Thames twice a day whenever the river was at ebb tide. This was the impressive sewerage and Thames embankment engineering system that Joseph Bazelgette built between 1858 and 1865 and that remains essentially London's sewerage system today.[7] That the MBW intended the Barking and Crossness outfalls to be only temporary expedients is clear from the fact that it advertised for tenders to treat as fertilizer the sewage that ended up at these stations. There were some eight contenders for this privilege in 1860, but because of parliamentary inquiries into the value of sewage as a fertilizer, in which both Graham and Hofmann were involved, nothing was done immediately. Indeed, nothing was done for thirty years.

The expense that water-borne sewage schemes cost London and other urban ratepayers (Bazelgette's system cost the MBW £4.6 million) inevitably made the possibility of selling sewage to farmers an extremely attractive proposition. Could the expenses of sewerage systems be offset against sewage sales?

The question was not an easy one for agriculturists to answer. Farmers had taken enthusiastically to the so-called artificial fertilizers of the 1840s – to the guano imported from Peru, the coprolites and superphosphates of Lawes. Such fertilizers were not cheap; guano, for example, cost about £10 per ton. Hence, if sewage were cheaper and as effective, farmers would be attracted to the idea of using town sewage. But what was the pounds, shillings, and pence value of human sewage?

Estimates varied enormously from 0.5d per ton to 2d per ton. Liebig was to suggest 4d before dropping in 1864 to 2.5d. So, for example, a farmer using 5,000 tons of human excrement per annum might have to

7. Bazelgette's scheme was based upon several earlier plans, including the *Report of Messrs Alton, Simpson and Blackwell on Main Drainage* (London, 1857) and the *Reports of Hawksley, Bidder and Bazelgette* (London, 1858). See Leslie B. Wood, *The Restoration of the Tidal Thames* (Bristol and Boston, 1982).

pay a town rate of £10 or £41 or £52 per annum. In practice, such estimates (Liebig's included) were inherently absurd because metropolitan sewage was diluted by thousands of gallons of water; even if farmers had been willing to pay 2d a ton in the spring when they needed fertilizer, they would have refused to pay even 0.5d a ton if forced to take it in untreated liquid form day and night and all the year round. A sewerage system cannot be switched off during the rainy season.

How then could sewage be used profitably? Would the farmer be wise to make legal contracts to use town sewage as urban ratepayers were suggesting? Clearly, the sewage question raised many issues, and since massive capital investment was also involved, the subject had to be treated cautiously and the very best advice sought. This was where Liebig came in.

There is no German pamphlet version of Liebig's *Letters on the Subject of the Utilization of the Metropolitan Sewage,* which was published by the Corporation of London in 1865.[8] It is therefore a unique publication in Liebig's bibliography, because all his other books and pamphlets are represented in both German and English editions. Although Volhard commented on the *Sewage Letters* in his remarkably thorough biography of Liebig in 1909, it has been ignored by later scholars.[9] The *Sewage Letters* raises the obvious questions: What on earth was a German chemist who lived in Munich doing in writing to the Lord Mayor of a foreign capital city about sewage disposal? How did Liebig come to write two such letters? What were his aims and intentions? and What effects, if any, did their publication have?

Liebig's Views on Excrement

Although Bazelgette's sewerage scheme, which was completed in 1865, cleared central London of sewage, the tidal waterway to the east of Barking and Crossness became a vast sewer. It had never been Bazelgette's intention that the sewer outfalls at Barking and Crossness were to be other

8. Liebig may have hoped that Vieweg would publish a translation. See Liebig to Vieweg, 18 October 1864 and 23 January 1865, in M. and W. Schneider, eds., *Justus von Liebig Briefe an Vieweg* (Braunschweig, 1986), pp. 389, 393. A number of German agricultural journals did, in fact, translate parts of Liebig's letters, as did Julius von Hoftendoff in *Ueber Zusammensetzung, den Werth und die Benutzung des stadtischen Cloakendungers von J. B. Lawes und Dr J. H. Gilbert* (Glogau[=Clogow, Poland], 1867). I am grateful to Dr E. Heuser for this information.
9. Volhard, *Liebig,* vol. ii, pp. 48–71.

than temporary. The problem was that no one could agree on how best to recycle the sewage once it was seventeen miles downriver.

We must bear in mind that, despite his move from Giessen to Munich in 1852, Liebig was in a restless state of mind during the mid-1850s. There is abundant evidence that he would willingly have abandoned Munich for a position in Britain had one been forthcoming. In correspondence with his Glaswegian friend, the calico printer Walter Crum in 1856 and 1857, Liebig mentioned the possibility of emigrating to America to farm. Crum responded by trying to get Liebig elected to the Regius Chair of Agriculture at the University of Edinburgh. Failing in this, Crum then attempted to persuade the millionaire chemist James ("Paraffin") Young to set up a Society for the Improvement of Agriculture that would have employed Liebig as its consultant chemist. But nothing came of this scheme either.

Again, in a letter to Michael Faraday, Liebig said in 1856:

> I am dominated by the desire to establish a school of practical farming for the education of teachers of practical agriculture. It seems to me that there is no other way of showing the application of scientific principles. It must be done on a large scale, and I am confident to succeed [*sic*]. I think I could do for agriculture what I have done 30 years ago for the practical education of experimental chemists. All my friends tell me that it is folly to give up the most brilliant position which a man of science has ever held, but I am tired of lecturing . . . I am sick of my schoolmastership and all my happiness depends to get [i.e. on getting] rid of it.[10]

And as we have seen, Liebig went on to regret that Hofmann and Graham had raised the agricultural testimonial to him in 1852 in the form of silverplate instead of persuading Parliament to award him a state pension. That, he told Faraday,

> might have given me full liberty to resign my professorship. By this supply [i.e., in this way] I should be in a position to spend 3–4 months in Scotland or England, and to devote all my powers to agricultural questions. . . . But all that is to[o] late.

This depression, or ennui, which reached its peak at the time of the death of King Maximilian in 1864, does much to explain Liebig's reentry into the world of commerce at this time (Chapters 5 and 8). As to the source of the depression, we must bear in mind that during the 1850s he was in the middle of the bitter dispute with Lawes and Gilbert, as well as being

10. Crum to Liebig, 30 September 1856 and 24 February 1857, Liebigiana, Munich; Liebig to Faraday, 17 September 1856, in *The Selected Correspondence of Michael Faraday*, ed. L. P. Williams (Cambridge, 1971), vol. 2, pp. 851–52.

frustrated by the ban on his right to reply in the pages of the Royal Agricultural Society's *Journal.* He had been forced to climb down and admit that soils had the power of absorbing nitrogen as ammonia, as J. T. Way had demonstrated in 1850. On the other hand, this discovery also pointed to the fact that sewage could be "purified" by passing it through soil – though Way, Lawes, and Voelcker (the three principal agricultural scientists in mid-Victorian Britain) all agreed that the utilization of town sewage as manure would be uneconomic.[11]

This aside suggests that not only was Liebig keen to come to Britain to restore the reputation he had lost through Pusey's and Lawes's attacks on his practical credibility but that he was eager to find an agricultural context in which, once and for all, his theoretical teachings could be put to the test. If, to all intents and purposes, the British agricultural establishment was more persuaded by Lawes's and Gilbert's field trials than by Liebig's theory, perhaps the sewage question would persuade them to think again. Although such tests and demonstrations could also be made in Germany (as, indeed, they were), because Britain was the leading agricultural, as well as industrial and urban, nation in the world, Liebig looked to Britain as the exemplar. Where Britain led, other nations would follow. By the same token, London being the world's leading capital city, what London decided to do with its sewage, other towns throughout the world would copy.

Liebig's involvement with London sewage began innocently enough in December 1858 when a certain unknown Charles Lewell wrote to Liebig in German from London to offer the potential services of one of his friends as a translator. Lewell added that he had been following Liebig's agricultural writings attentively, for they were extremely relevant to the row that was brewing in London over the value of sewage. Was it possible, he asked casually, to devise a method of deodorizing sewage and selling it as a manure? If Liebig had an article on the subject, Lewell could use it to enlighten the "dum Aldermen" of the city.[12]

Liebig replied early in January 1859 that his Irish pupil John Blyth had just finished a new translation of Liebig's lectures on practical agriculture, the *Naturwissenschaftliche Briefe über die moderne Landwirtschaft* (Leipzig, 1859). He added that he had also just finished lecturing in Munich on the value of sewage,

> and I am firmly of the opinion that if England wishes to remain an agricultural country she must use as manure the nightsoil and similar

11. J. T. Way, *J. Roy. Agri. Soc.* 15(1855), 135; J. B. Lawes *J. Roy. Agri. Soc.* 23(1862), 462; J. C. A. Voelcker, *J. Roy. Agri. Soc.* 1(1865), 231.
12. Lewell to Liebig, 23 December 1858, Liebigiana, Munich.

> residues produced in large cities. The necessity would be increased in the event of war with America, when supplies of guano would cease.[13]

If Lewell had any interest in the welfare of his country, Liebig suggested, he should strive to convert British agriculturists to this essentially Malthusian view.

Lewell responded by sending Liebig's private letter to *The Times* (9 January 1859). This was a foretaste of the bitter thoughts expressed in the *Agricultural Letters* when they appeared in English in June 1859. Liebig wrote angrily of Britain's "spoilation system" of agriculture.

> In the year 1855–1856 above 10 million hundredweights of guano were imported [from Chile], of which the greater portion remained in England. In the course of half a century above 60 million hundredweights of bones have been imported into that country; yet all this mass of manure is not worth mentioning when considered in relation to the arable surface of Great Britain, and it is but as a drop when compared to the sea of human excrements carried by rivers to the sea.[14]

The *Agricultural Letters* strongly recommended that European agriculturists adopt the Chinese practice of returning human and animal wastes to the soil in exchange for every sack of corn or hundredweight of rape, turnips, and potatoes sold in the market. Was it not significant, Liebig warned, that the Chinese civilization, unlike Roman civilization, had existed for thousands of years?

Liebig's English warnings and recommendations were taken up in *The Times* by John Mechi (1802–80). The son of an Italian immigrant, Mechi had made a fortune by inventing a strop for sharpening razors. In 1841 this amiable, hearty, hospitable, and genial man had bought a farm at Tiptree in Essex, where he carried out farming on scientific principles – including manuring with human excrement from a reservoir 30 feet in diameter and 20 feet deep, and using a pair of force pumps capable of discharging 100 gallons per minute. Although he overreached himself and went bankrupt in the 1870s, during the previous decade his "model farm" attracted some 400 to 500 visitors a year; he was also one of the decade's most prolific writers on agriculture. He became England's principal advocate for spreading sewage on land by a steam-engine–driven hose and sprinkler system.

13. Liebig to Lewell, in *The Times*, 9 January 1859.
14. J. Liebig, *Letters on Modern Agriculture* (London, 1859), Letter 11, p. 222. The point had been made earlier, if less fiercely, in *Agricultural Chemistry*, 3rd ed. (1843), p. 165.

As early as 1853, in a typically lively and robust lecture to the Royal Society of Arts, Mechi had extolled the value of human manure:

> I venture to predict, that the people of this country will soon connect ample water supply, cleanliness, and health, with the idea of ample and cheap physical supplies – they will identify the well-washed contents of their [water] closets with rounds of beef, saddles of mutton, big leaves, and rich milk. . . . We know by our great chemists, that our sewers contain the elements of our food – of, in fact, our very selves – and that to waste them, as we now do, is a cruel robbery on the welfare and happiness of our people.[15]

Using Liebig's *Agricultural Chemistry* to great effect and calling Liebig "the Isaac Newton of agricultural science" (which must have pleased Liebig), Mechi warned readers of *The Times* in November 1859 of the dangers of soil exhaustion and of the Englishman's "false delicacy" over human excreta. In an editorial the same day (7 November 1859), *The Times* took up "The Great Metropolitan Drainage Question" and emphasized Mechi's Liebigian point that now ratepayers had paid handsomely for Bazelgette's *cloaca maxima,* which took sewage out to sea, ratepayers were throwing their money away. This issue must be rethought, it pontificated.

Mechi's letter to *The Times* was seen by Liebig. Probably Mechi sent him a copy to strike up a correspondence, for there is no evidence of earlier correspondence or friendship, or that Liebig had met Mechi socially during his final visit to England in 1855. Liebig replied with a very long letter, which Mechi promptly sent to *The Times* (23 December 1859). In this important essay, part of which also appeared in German in the *Augsburger Allgemeine Zeitung,* Liebig accused farmers of naiveté in maintaining that tillage and good weather were sufficient to guarantee a good harvest.[16] Such people were like the French socialist Charles Fourier, who planned to feed the inhabitants of his utopian *phalanges* on eggs. By providing a couple of thousand hens each capable of laying 365 eggs a year, there would be sufficient thousands of eggs left over to sell to the English at a profit. What Fourier overlooked, said Liebig, was that "in order to lay an egg [each chicken] must eat an amount of corn equal in weight." Liebig then proceeded to give a vivid Malthusian sermon. If men

15. J. J. Mechi, "Third paper on British agriculture, with some account of his own operations at Tiptree Hill Farm," *J. Roy. Agri. Soc.* 2(1853–4), 65–72, lecture of 16 December 1853. See also the "Fourth Paper," *J. Roy. Agri. Soc.* 3(1854–5), 49–58, read 6 December 1854, where "our excrement is literally our food."
16. *The Times* 23 December 1859, p. 6; *Augsburger Allgemeine Zeitung,* Nr. 342 (8 December 1859).

wished to avoid having to balance excess population by wars, revolutions, plagues, famines, and emigration, they must return to the soil what was extracted:

> If the British people do not take pains to secure the natural conditions of the permanent fertility of their land – if they allow these conditions, as hitherto, to be squandered – their fields and meadows will at no distant time cease to yield their returns of corn and meat.[17]

This message received wide publicity; Mechi was so enraptured by its "truth and eloquence [which] entitles you to the gratitude of the whole human race," that he had copies printed and sent to 500 different newspaper editors throughout the country.[18] An important consequence of this publicity was that Liebig was invited and apparently agreed to become consultant chemist to a company that the engineer George Sheppard was forming to supply sewage to farms.[19] Was this the opportunity he had been looking for to settle in Britain? Continental newspapers actually reported that Liebig had resigned his Munich chair for a London consultancy.[20] Although Sheppard's company was awarded a contract to dispose of Croydon's sewage, such tantalizingly brief evidence as remains suggests that, although Liebig accepted, he withdrew from the company after less than six months. Most likely he objected to the way Sheppard was using his name to solicit shareholders. In fact, by 1865 the Croydon scheme had collapsed, and any hopes that Sheppard had entertained that its success would lead his company to the franchise for the disposal of London's sewage also disappeared. Others in the market were only slightly more successful.

In 1861 two entrepreneurs William Napier and William Hope approached the Metropolitan Board of Works with a dramatic scheme of sewage irrigation, basing it upon the way Edinburgh had successfully disposed of its sewage for over sixty years by using it to manure the meadows of Craigentinny to the east of the city. For an estimated capital cost of £2 million they proposed to reclaim from the sea Maplin Sands and Dengie Mud Flats off the coast of Essex by conducting sewage from the Board's northern outfall at Barking Creek in a 44-mile-long canal. The project was clearly indebted to Hope's experience as a military engineer in the Crimean War, in which he had won the Victoria Cross. At the end of its journey, the sewage would irrigate the sands and transform them into a

17. *The Times* 23 December 1859, p. 6.
18. Mechi to Liebig, 9 January 1860, Liebigiana, Munich.
19. Sheppard to Liebig, 29 December 1859, Liebigiana, Munich.
20. As reported by *British Medical Journal* 11(1864), 630.

vast dairy farm and market garden for London's citizens. Farmers with land adjacent to the sewage conduit would also be able to tap the sewage for their fields; the rents from this use, Napier and Hope estimated, would produce £2 million per annum. The finance for the project was to be raised through land improvement and land security companies to which Napier and Hope were attached through their business as agents for the international finance bank, Crédit Mobilier. The citizens of London were not being asked to pay for the capital works. During 1862, after much indecision and squabbling, the Metropolitan Board of Works granted Napier and Hope a fifty-year concession – providing that they were able to obtain the necessary parliamentary permission to reclaim the Maplin Sands.

The only serious alternative scheme offered to the MBW was that of an Irish solicitor Thomas Ellis. He planned to pump 700,000 tons of London's sewage to a height of 400 feet and contain it within enormous reservoirs on Hampstead Heath and Shooter's Hill. The sewage would then be gravity-piped to farms within a 30-mile radius of London, where it would be hosed and jetted onto the land by farmers in the manner practised by Mechi at Tiptree. This fantastic and ill-thought-out scheme would have cost £3.5 million and have attracted very large annual running costs. In retrospect, too, we can see that it would have inhibited the twentieth-century development of metropolitan London.

Nevertheless, it was strongly supported by an Irish doctor and Member of Parliament, Dr. John Brady, who in 1862 struck up a correspondence with Liebig that was to badly affect Liebig's judgement. Both Ellis himself and Brady his chief advocate became convinced that John Bennet Lawes and other fertilizer manufacturers and investors, as well as those involved in the lucrative importation of guano from South America, were conspiring to blacken the reputations of those who were arguing for the agricultural usefulness of town sewage. However, far from joining ranks with Napier and Hope, who were also proposing to use sewage, Ellis and Brady seized every opportunity to ridicule the Maplin Sands project and to accuse Napier and Hope of duplicity. The debate was not, therefore, ideological – not the equivalent, say, of today's ecologists arguing for recycling as against the power and might of chemical companies; rather, it was a straightforward squabble between two rival business schemes.

In a series of long letters to three London newspapers in 1863, Ellis waged verbal war, pressing Liebig into service not only against the artificial fertilizer lobby (which Liebig, to his discredit, also privately accused of self-interest) but also against Napier and Hope.[21] And because Liebig

21. Thomas Ellis, *The Sewage of the Metropolis and How to Utilise It* (London, 1863).

detested Lawes so much, he fell for Ellis's argument and with it opposition to the Napier–Hope irrigation plan that rationally he ought to have supported. Persuaded that Lawes and his ilk were conspiring against the use of human sewage for agricultural purposes, Liebig wrote:

> The manufacturers [of artificial fertilisers] are a very stupid set of people, because the application of sewage to agricultural purposes must necessarily increase tenfold their trade.[22]

And Liebig noted that his new book, *The Natural Laws of Husbandry,* published in that year, would strongly advocate sewage farming.[23]

It is odd that Liebig was willing to support such an expensive system. Despite the active support of the chairman of a select committee of the House of Commons, Lord Robert Montagu, who was obsessed with a belief in the corruption of the Metropolitan Board of Works, Ellis was neither able to raise sufficient capital nor to obtain the franchise. Liebig repeated his accusation that artificial fertilizer manufacturers were "inimical to the utilization of sewage" in a further letter to Mechi. This too was sent to *The Times* by Mechi, who also publicised it at the Royal Society of Arts where, in a sour debate, Gilbert sprang to Lawes's defence. Such an accusation, Gilbert retorted,

> may, perhaps, be naturally enough conceived and propagated by those who are directly interested in keeping the public ignorant on this question; but that it should be echoed by Baron Liebig,when himself putting forward estimates and calculations which, if based upon well-established facts would most strikingly confirm the [lower] estimates and views of those he sets to calumniate, is highly discreditable; though it will be well understood by all acquainted with recent agricultural discussions having no reference to the sewage question.[24]

It was, of course, inevitable that as the country's leading agriculturists, both Lawes and Gilbert should have been involved in the sewage question. Lawes was a leading member of a select committee on the sewage question, and he also acted as a consultant for some interesting experiments on the disposal of sewage by irrigation that were carried out on the meadows

Liebig's letter to Ellis of 22 June 1863 is at p. 23. Ellis's newspaper letters were published 7, 11, and 14 September 1863. For the plans of Ellis and others, see Metropolitan Board of Works, *Report of the Main Drainage Committee on the Tenders for the Metropolitan Sewage,* 20 July 1863.

22. Liebig to Ellis, Liebigiana, Munich.
23. J. Liebig, *The Natural Laws of Husbandry* (London, 1863), Chap. 7.
24. J. J. Mechi, "Utilization of sewage," *J. Roy. Agri. Soc.* 11(1862–3), 655–7, Liebig's letter of 21 August 1863; Gilbert's reply, 4 September 1863, *J. Roy. Agri. Soc.* pp. 685–6. Note also Andrew Wynter, *Subtle Brains and Lisson Fingers, being some chisel marks of our industrial and scientific progress* (London, 1863), pp. 105–25.

around Rugby.[25] It was equally inevitable that Napier and Hope should have retained a number of agricultural chemists, including Thomas Way, as consultants.[26]

In 1864 Liebig's polemic against Lawes and Gilbert had reached its climax in the *Journal of the Royal Agricultural Society* with his equivocal defence of the mineral theory. In the same year, Napier and Hope also addressed several letters to the chairman of the MBW, John Thwaites. In these they reviewed the development of sewage schemes and emphasized the advantages of their proposal – notably that it would use dilute sewage, would not be dependent upon the good will of farmers in winter or bad weather to siphon off the sewage, and had the sound financial backing of land improvement and land securities companies.

At this stage the City of London's governing council, the Court of Common Council (which included John Mechi amongst its aldermen) began to take an active interest in the profitability of sewage. Would their ratepayers be getting value for money if Parliament granted Napier and Hope, rather than Ellis, the North Thames sewage concession? On 21 July 1864, their oddly named Coal, Corn and Finance Committee set up a subcommittee on sewage disposal that took evidence from Brady and another MP who had been an active spokesman for sewage utilisation, the previously mentioned Lord Montagu. Knowing that they had to give persuasive evidence to the City burghers, both men appealed to Liebig for evidence in support of the Ellis scheme, despite the fact that Napier and Hope had already been notionally granted the franchise. What did "the greatest living authority upon the matter" think?[27]

This was the origin of Liebig's first English letter, 4 October 1864, on sewage utilization. In it "the great chieftain of chemistry"[28] warned Montagu against assuming that human sewage was a unique panacea; lacking the phosphorus from bones, field rotation with sheep fed upon turnips would still be necessary. (He was to revise his views on phosphorus within the next few months.) More ominously, he observed that "the agriculturist must be made aware that, by the introduction of sewage, his whole system of farming undergoes a change, and . . . he has to make an apprenticeship to learn to apply it rightly and economically, in order to benefit and not to injure the fields." Liebig feared that millions might be spent on

25. J. B. Lawes, "Utilisation of town sewage," *J. Roy. Agri. Soc.* 24(1863), 65–90.

26. T. Way to Gilbert, 3 March 1865; Napier to Gilbert the same day, Rothamsted archives.

27. Brady to Liebig, 13 September 1864; Montagu to Liebig, 17 September 1864. Liebigiana, Munich.

28. *Lancet* 19 November 1864, p. 582; cf. "Liebig is a man of the century, every sentence that issues from his pen is the bearer of originality and sterling reflection," *Lancet* 9 July 1842, p. 485.

a scheme; farmers might then misapply the sewage and became so disgruntled that "the great and important example of England should thus be lost to Europe."[29]

Liebig's other main point, which was bound to please advocates of the Ellis scheme, was a reapplication of his patent-manure scheme of twenty years earlier,

> that for each crop the composition of sewage ought to be corrected according to the nature of the soil; by adding those ingredients which are wanting in sewage, and which the plants to be grown require in the largest proportion.[30]

Although Hope could (and did) deny that his scheme ruled out the possibility of adjusting the chemical quality of sewage to suit the soil and its crops, Liebig's statement made it easy for Brady and Montagu to persuade the Court of Common Council that the full potential value of London's sewage – they reckoned £0.25 million per annum from 260 million tons of sewage – would not be forthcoming from Napier and Hope's scheme.

Hope did reply forcibly to this argument and to the clear bias of Brady and Montagu, but we can pass over this. For meanwhile, Montagu had persuaded Liebig to send a detailed evaluation of the potential value of London's sewage to the Lord Mayor of London. "The sewage question," Liebig told Wöhler (23 January 1865),

> is of such importance and so momentous in its consequences that I look forward to the outcome with some anxiety. For twenty years I have taken the trouble to bring people's attention to the utilisation of sewage for agricultural purposes, and the time has now come when it will have to be decided whether the nation has an understanding of its future welfare. The example of England will be telling. Sewage is worth some two million [sic] pounds sterling. I have made a quantitative analysis of water in which fish, potatoes, cauliflowers and white cabbage have been cooked, and arrived at the almost unbelievable result that in London's drainage water nearly a million pounds of potassium and 281,000 pounds of phosphoric acid flows into the sewage. The material is as weak as water and scarcely looks dirty, so great is the dilution. The difficulty will be to bring the sewage water to the fields. Another problem is the dilution of profits since under these conditions the fertile materials can only be distributed by machines and pumps.[31]

29. J. von Liebig, *Letters on the Subject of the Utilisation of the Metropolitan Sewage Addressed to the Lord Mayor of London* (London, 1865), pp. 7–9.
30. *Ibid.*
31. A. W. Hofmann, ed., *Aus Justus Liebig's und Friedrich Wöhler's Briefwechsel*, 2 vols. (Braunschweig, 1888), vol. 2, p. 176.

It is unclear why Liebig believed that the great dilution of sewage ruled out irrigation in favour of pumps and hoses. By doing so he was in effect ignoring (or rather rejecting) twenty years of British experience to the contrary at Craigentinny, Rugby, and Croydon.[32] But this opinion, and a great deal more, was contained in Liebig's letter to the Lord Mayor on 19 January 1865. In it, Liebig took some pains to emphasize that the monetary value he placed upon London's sewage – an extraordinary £2.5 million – was only an estimate and that this was necessarily so because of the variable amounts of phosphoric acid, potash, and ammonia excreted by Londoners and their horses and cows. If it were worth so much, Napier and Hope would be purloining profits that rightly belonged to Londoners. The letter also reviewed Liebig's own experiments on the absorbent powers of soil (here he again did public penance for his earlier mistake in introducing insoluble manures in 1845) and gave evidence that whatever the dilution of sewage, it had fruitful fertilizing effects after absorption. The advantage of dilute sewage was its manoeuvrability by gravity or force pump; unlike the existing agricultural use of night soil, it was easily distributed.

This letter had precisely the effect that Brady and Montagu had wanted, for the Court of Common Council decided in February 1865 to petition Parliament for the rejection on engineering, agricultural, and commercial grounds of the Napier–Hope scheme. Liebig had persuaded them that Londoners would get more value from their sewage if it was used for the occasional small dressing of soil rather than if it was used to flood a large tract of land, as Napier and Hope intended. The reservoir on Hampstead Heath or some other hilly neigbourhood was the ideal solution.

Hope responded immediately, accusing Liebig of blind prejudice:

> The Baron as well as the Corporation, seems to think that the greater he can prove the value of sewage to be, the more completely he smashes our scheme, so he sets himself manfully to the task of proving that the sewage of London is a veritable stream of gold.[33]

But, Hope observed, Liebig had never personally analysed real London sewage water. If it really was worth the amount Liebig claimed, then surely this would work to the advantage of the Napier–Hope scheme. What Liebig had failed to consider was the British climate. Not all of London's sewage could possibly be used in a small dressing scheme such

32. John C. Morton, "London's sewage from the agricultural point of view," *J. Roy. Agri. Soc.* 13(1863–64), 185–93, 202–8, read 3 February 1864 with discussion.

33. W. Hope and W. Napier, *The Sewage of the Metropolis. A Letter to John Thwaites* (London, 13 February 1865), p. 6. From his later committed writings, it is obvious that Hope was the writer in the partnership.

as Ellis or Mechi had in mind. Unlike water, the stream of sewage could not be turned off, and the reservoir possibility was ridiculous.

This was fair criticism and made points that clearly worried Brady in his further correspondence with Liebig, who, goaded by Brady and Montagu, now composed a second letter to the Lord Mayor. In it he thoroughly condemned the Napier–Hope project as hopelessly impracticable. Grass, he declared confidently, would never grow on the sea sand of Maplin. It grew abundantly only in the meadows of Craigentinny, Edinburgh, because that city had fewer water closets; consequently, most of that city's sewage consisted of animal excreta and street mud that created an alluvial soil when mixed with the Craigentinny sands. The Napier–Hope plan was

> baseless, for the land to be experimented on does not as yet exist, being covered at high water by the sea. All the calculations, therefore, as to crops, returns and percentage of capital, are absolutely fabulous. It appears to me like a soap-bubble, glistening with bright colours, but inside hollow and empty. There is not the slightest doubt that every penny expended in that frivolous undertaking would not only be a squandering of an enormous amount of money, but before long would be looked upon as a national calamity.[34]

Liebig was being outrageously unfair. He had not been to Edinburgh for ten years, and he had not tried to grow plants in sand irrigated with sewage – a technique that is today an important feature of market gardening. And even if Napier and Hope were a couple of financial rogues, their scheme was in principle viable, as Melbourne's later experience was to show. Whether it could ever have been profitable is a different matter; although again Melbourne's experience is suggestive. The 10,851-hectare sewage farm at Werribee, Victoria, was established in 1897 and still handles West Melbourne's sewage and continues to make a profit for the city. An activated sewage sludge treatment for East Melbourne's sewage was opened at Carrum in 1975.[35] But Liebig was convinced *a priori* that the scheme would fail, and that such a catastrophic failure would be the eternal damnation of his hopes and aspirations for the utilization of sewage agriculturally. For where Britain led, other nations would follow. As he told Wöhler:

34. Liebig, *Letters on Sewage*, p. 41.
35. A. H. Croxford, *Melbourne, Australia, Wastewater System. A Case Study*, Paper 78–2576 given to the American Society of Agricultural Engineers, 18–20 December 1978 (copy Melbourne University Agriculture Library). Berlin, to Volhard's delight, had also developed such a scheme by 1909. See Volhard, vol. ii, p. 65.

> The case is now before Parliament upon whose decision either for the acceptance of these plans or their rejection, everything depends. Upon this all my hopes for the future of agriculture through the employment of sewage by the state are grounded. The English alone possess the financial resources, energy and ability not to be afraid of the [Ellis] piping plan. And so I hold it to be my duty to eliminate the harmful plans with all my energy and to forward what appears to be appropriate. It is, indeed, this that is now the point of my life.[36]

Liebig's spirited opposition to the Napier–Hope plan so delighted the Corporation of London that they recorded their personal thanks to him with an illuminated vellum address – displayed today in the Liebig Museum in Giessen.[37]

The "sewage question" now became rather unpleasant and took on an *ad hominem* form. Hope replied to Liebig's accusations, protesting at Liebig's "audacious misrepresentation, . . . monstrous contradictions and . . . pedantic rubbish," and even raised a doubt whether the letters to the Lord Mayor were authentically Liebig's. All this appeared in a privately published pamphlet.[38] In an angry private letter to Liebig, Hope complained that the Munich professor had no business intervening in Hope's affairs; that he had cost them (Napier and Hope) time and money in delaying the scheme, that Liebig had been thoroughly misled by hostile sources (which was probably true), and that some of Liebig's facts were incorrect. Hope appealed to Liebig to come to England so that he could be crossexamined in a legal way before the House of Lords (whose Committee was about to examine Napier and Hope's private Parliamentary bill to reclaim the Essex coastline). If Liebig declined to come, Hope threatened ominously, "We shall do our very utmost to destroy your reputation, and that we shall do so with every chance of success."[39]

Liebig, now realizing that he was dealing with fairly ruthless businessmen, replied pugnaciously:

> I leave it to the public before whom this correspondence will be laid, to judge how far passion and intimidation are to prevail against hon-

36. Hofmann, *Liebig's und Wöhler's Briefwechsel,* vol. 2, p. 177.
37. *Lancet,* 4 March 1865, p. 245. Hope later queried whether this was a legal use of ratepayers' money!
38. W. Hope, *The Sewage of the Metropolis. A Letter to John Thwaites Chairman of the Metropolitan Board of Works. Being a Comparative Analysis of Baron Liebig's Three Letters* (London, 8 March 1865), p. 39. See also Hope's *A Letter to John Thwaites, Esq., Chairman of the Metropolitan Board of Works* (London, 29 April 1864), p. 23; *The Sewage of the Metropolis* (London, 14 November 1864), p. 12.
39. Hope to Liebig, 7 April 1865, Liebigiana, Staatsbibliothek, München.

Vellum scroll presented to Liebig by the Lord Mayor of the City of London on 23 January 1865 as a mark of gratitude for his reports on the utilisation of London sewage. Now in the Liebig-Museum Giessen.

> esty and truth. A scientific reputation which could be so easily destroyed would not be worth defending.[40]

There is, indeed, some evidence that Hope did subsequently seize upon every opportunity to ridicule the German chemist's reservations concerning the scheme. For, in June 1865 – by which date eight British towns were using sewage as manure – Napier and Hope's private bill for the reclamation of Maplin Sands received Royal assent and the Metropolitan Board of Works granted them the concession for the disposal of the sewage from the northern side of the Thames.

40. Liebig to Hope, appended to *ibid.*

Frustrated and disappointed by this turn of events, Liebig abandoned the sewage question and found solace in publicising his Extract of Meat and other dietary foodstuffs. As he told his English friend, the poet Charles Boner, who had probably helped him with the English of the *Sewage Letters:*

> I have put the sewage question out of my mind and have written to Dr Brady to say that I have no wish to make a second reply [to Hope's letter]. I have had enough with one confrontation. As to the franchise for the southern side of the Thames, I am indifferent. All that I wish is that the undertakers of the scheme will have the farmers' support: there are no Maplin Sands![41]

And he told his Saxon agricultural friend Theodor Reuning:

> My many exertions in England to employ sewage in a useful way have, for the time being, proved unavailing.[42]

Indeed, although English friends continued to send him information about the fate of London's sewage, Liebig said nothing more about the matter.

The Aftermath

That was not the end of the matter for Napier and Hope but only the beginning. In 1865 they began to raise money for their project using the recently invented idea of a corporate investment (i.e., finance company) for the necessary capital. Hope was the manager of the English branch of the Crédit Mobilier, the International Financial Society; Napier, the second son of the ninth Lord Napier, was a managing director of another of its associated companies, the Lands Improvement Company. One of the principal financiers involved in the International Financial Society was the banker Anthony Gibbs, who had made his fortune from holding the monopoly on the European import of Peruvian guano. Because Gibbs had lost the guano contract in the early 1860s, he was naturally attracted by Napier and Hope's plan to make money from using London sewage as a fertilizer.

Unfortunately for the schemers, because of the bad publicity that Liebig's dismissal of the scheme had received in the City of London, unlike the Liebig Meat Company launched the same year, the financial public would not buy shares in the International Financial Society's Essex Reclamation Company when the market was opened in July 1865. Conse-

41. Liebig to Boner, 29 June 1865, Liebigiana, Munich.
42. Quoted Volhard, vol. ii, p. 398.

quently, almost the entire working capital of £2 million had to be raised by the International Financial Society.[43] This proved impossible because of the financial crash of British money markets in 1866. (The same financial crash also prevented Thomas Ellis from ever raising capital for his scheme, which was now looking towards the franchise for the southern outfall.) Minor engineering work on the Maplin Sands project was abandoned in 1868, and in 1870 the Essex Reclamation Company went into liquidation. Hope also lost his managerial post with the International Financial Society after bitter disagreements. Even so, in a witty lecture to the Royal Society of Arts, he was able to claim to have made his case against Liebig.[44] Finally, in 1871, the Metropolitan Board of Works, angered that nothing had been done, cancelled Napier and Hope's sewage concession and made them forfeit the £25,000 the two speculators had had to deposit as a security. Ironically, this was the sole profit that Londoners ever made from sewage.[45] The only money that Napier and Hope recouped, after investors had been repaid, was £2,600.

Napier was doubtless a rogue and a villain. He was reputed to have swindled a widow of her savings, and on at least two occasions he had to flee to Belgium to escape creditors. But William Hope, although a ruthless businessman with a silver tongue who was prepared to cut corners when necessary, was undoubtedly genuinely committed to the cause of sewage utilization. It was all the more unfortunate, therefore, that he and Liebig found themselves at odds. In 1866, at his own expense, Hope arranged for some land near Barking Creek to be covered with 4,000 tons of Maplin sand and irrigated with London sewage. He was then able to challenge Liebig by growing and marketing cow's milk, celery, radishes, carrots, and rape. Even so, the venture lost money. Undeterred, in 1869, Hope bought Breton's Farm in Essex, which he irrigated with sewage from the town of Romford, for which he paid £600 per annum.[46] This venture also was not successful financially, for the experiment ended in a lawsuit between Hope and the Romford Sewage Board. This farm is now the site of the University of East London. Despite his failures, Hope was clearly widely respected for his practical knowledge of sewage. Along with Liebig's pupils Gilbert, Paul, Smith, and Williamson, he served on the British Associa-

43. For financial details, see P. L. Cottrell, *Investment Banking in England 1856–1881. A Case Study of International Financial Society,* 2 vols. (New York and London, 1985).

44. W. Hope, "On the use and abuse of town sewage," *J. Roy. Agri. Soc.* 18(1869–70), 298–308, lecture of 23 February 1870. Also Hope, *Food Manufacture versus River Pollution. A Letter Addressed to the Newspaper Press of England* (London, 1875), 50 pp.

45. See David Owen, *The Government of Victorian London 1855–1889* (Cambridge, Mass. and London, 1982), Chap. 3.

46. *Lancet,* 3 September 1870, p. 348.

tion's Committee on the treatment and utilization of sewage between 1868 and 1875.[47]

The story of South Thames sewage disposal was also one of failure. In 1867 the franchise was granted to the Native Guano Company, which used alum, blood, and clay (hence the "ABC method") to precipitate sewage sludge, which was then sold as a solid fertilizer. Unfortunately, sales never matched expenses and the process had to be abandoned in 1873.[48]

By then, mainly as a consequence of the work on river pollution done by the chemist Edward Frankland, it was understood (as Liebig had suggested in the 1860s) that the absorption of faecal organic matter depended upon the nature of the soils that sewer water passed through before reaching a river. This paved the way for the development of the filtration and chemical treatment of sewage at outfalls, or "sewage works." Even so, in the case of London, nothing was done until 1890, despite accusations that many passengers had been poisoned by sewage when their excursion boat *Princess Alice* capsized outside the Barking outfall in 1878.[49]

Since 1890, London's sewage has been chemically and bacteriologically treated, and some 40 percent of the sewage sludge is spread on land as a low-grade fertilizer. The remaining 60 percent is conveyed by barge down the Thames and dumped into Barrow Deep – ironically only a few miles east of Maplin Sands, which is currently an army gunnery range. Environmental agencies continue to criticize London for persisting with this method of disposal, and the situation is likely to change very shortly. Liebig would be pleased to know that in several English counties, including Leicestershire, treated sewage is recycled and sprayed onto the land, and that methane gas produced as a by-product of such treatment plants is often used to power the sewage works.

There was widespread interest in and experimentation with the irrigation of town sewage during Liebig's lifetime, but the problems were manifold. Once piped water and the water closet had become general by the 1860s,

47. *BAAS Reports*, 1869, and W. Hope, *Sewage Irrigation. A Lecture delivered to the Ratepayers of West Derby* (London, 1871), which has details of Hope's farm. For a picture, see Herbert J. Little, "Sewage farming," *J. Roy. Agri. Soc.* 7(1871), 389–420 at p. 401.

48. J.C.A. Voelcker showed that the ABC method was worthless, *J. Roy. Agri. Soc.* 6(1870), 415–24.

49. Wood, *Restoration of Tidal Thames*, (1982), and J. C. Wylie, *Fertility from Town Wastes* (London, 1955).

huge quantities of sewage in a highly diluted form had to be dealt with at outfalls; transport costs were high in relation to any manurial value that such liquid sewage possessed. Moreover, given the British climate, where rainfall exceeds evaporation for at least nine months of the year, it was difficult to persuade farmers to take on liquid sewage all the year round, however wet the conditions. There were doubts, too, especially in Liebig's mind, whether a field irrigated with liquid sewage would ever sustain more than rye grass. Hence agriculturists like Liebig, Mechi, and Ellis, who had been brought up on the husbandry of night soil manuring (as in the Third World today), opted for intermittent hosing and spraying of liquid sewage rather than the wholesale flooding of fields that Edinburgh's citizens practised and that Napier and Hope's plan would have entailed. In addition, by the time of Liebig's death, sewage irrigationists were poised to contend with the "nuisance" image of sewage as well as with the potential dangers from pathogens in untreated sewage that bacteriologists were beginning to reveal. This point had been made to Liebig by the barrister James Manning, who had received advice from Liebig in 1860 on patenting a solid fertilizer made from human excrement.[50] Moreover, Liebig's model of disease was difficult to reconcile with an image of sewage as beneficent.

By the 1870s, enthusiasts of sewage irrigation had become less optimistic, for their sanguine estimates of agricultural profits had not been realized, heavy losses had been incurred during the wet seasons of the early 1870s, and above all, untreated sewage was found to be far too weak in ammonia for its very heavy cost of distribution. By 1880, therefore, leading agricultural chemists, such as Voelcker, were of the opinion that sewage could be profitably irrigated only if local conditions were extremely favorable; otherwise, it was a nuisance best got rid of by dispersal at sea.[51]

Nevertheless, by then, some 100 British local authorities had "sewage farms" for sewage disposal. But what finally killed off all these schemes was the relentless growth of urban population and housing, which made it impossible to keep adequate acreages of arable land aside for this purpose. With the firm establishment of bacteriology, there was also a local consumer resistance to crops grown from untreated sewage. Sewage farms were therefore gradually turned into sewage works, and although some

50. James A. Manning, *The Utilization of Sewage: Being a Reply to Baron Liebig's Letter to Robert Montagu* (London, 1864). Manning owned the Wolverhampton Manure (Manning's Patents) Company as well as practising law at the Inner Temple.

51. J. C. A. Voelcker, *J. Roy. Agri. Soc.* 14(1878), 830–34.

proportion of sewage sludge was returned to the land, the costs of working the sewage again confirmed that town sewage was not "the cheap and miraculous fertilizer that its protagonists" had imagined.[52]

Although Liebig was incorrect in placing a high profitability upon town sewage, his intervention in the London sewage question in the last decade of his life vividly demonstrates his total commitment to the concept of recycling, to the law of the minimum, to the abolition of soil exhaustion, and to his deep moral concern that the laws of nature should be exploited for the relief of man's estate. Above all, the intervention shows Liebig's belief that European husbandry would never become rational and scientific unless British agriculture showed and led the way. The story illustrates both the extent and the limitations of Liebig's powers of persuasion over British opinion. Like *Agricultural Chemistry* at the midpoint of his career, the *Kloakenbriefe* at the other end is a landmark in Liebig's astonishingly important contribution to agriculture in Germany, Great Britain, and throughout the world.

52. G. E. Fussell, "Sewage irrigation farms in the nineteenth century," *Agriculture* 64 (1957–58), 138–41 at p. 138.

10

Populariser of Science: Chemical Letters

> My father's *Chemical Letters* belong to our national literature and will remain, as he himself noted, a memorial to his intellect and to the stage to which he brought his science. [Georg von Liebig, 1878][1]

Free from telephones and e-mail computer systems and aided by a remarkably cheap and efficient European postal service, the nineteenth-century scientist relied a good deal on letter writing for the exchange of opinion, gossip, and the latest technical information in advance of its reportage in the more leisurely produced scientific press. Moreover, in the absence of the regular meeting opportunities that conferences and air travel have provided the twentieth-century scientist, Liebig's generation forged their friendships through correspondence. Although Liebig and Wöhler did collaborative work together and went on holiday together a few times, we may estimate that they met only a few dozen times during their lifetime. On the other hand, they exchanged more than 1,000 letters.[2] Although an unknown number of the many thousands of letters that Liebig wrote and received was obviously destroyed, a surprisingly large number of letters received by Liebig have survived. On a conservative estimate, there are some 1,500 items of correspondence at the Liebig Museum in Giessen and well over 6,000 letters in the Bayerische Staatsbibliothek in Munich. The latter collection, labelled Liebigiana, was patiently amassed by Liebig's son Georg and his grandson Julius Carrière and formed the basis for Volhard's huge biography in 1909. Further material was added to this archive in 1974, and it now occupies fifteen metres of shelving. And since Liebig wrote to practically every contemporary scientist of note, as well as to lesser mortals, most scientific archives throughout Europe contain rep-

1. J. Liebig, *Chemische Briefe,* Foreword to 7th ed. by Georg von Liebig (Leipzig/Heidelberg, 1878), p. xvi.
2. A much-edited sample of these letters was published by Wöhler's daughter with Hofmann's help. See A. W. Hofmann, ed., *Aus. Liebig's und Wöhler Briefwechsel,* 2 vols. (Braunschweig, 1888). A complete edition is currently in preparation by Christoph Meinel.

resentative examples of his letters. They are distinctive in style – what Caneva has described aptly as "a casual [German] prose,"[3] and that his translators often found difficult.[4]

Such collections of correspondence are of enormous value to historians and they have served their purpose in the writing of this biography. Letters not only illuminate the activities and interests of the respective correspondents, but they are of value for what the theologian John Henry Newman described as "arriving at the inside of things." Through their letters we eavesdrop upon a conversation. As Newman said, "When a [person] is himself the speaker, he interprets his own action. His words are the index of his own life, as far as that life can be known to man."[5] What is interesting and almost unique about Liebig is that he also addressed the impersonal general public through the medium of letters. Of course, they were letters only in a literary sense and might more properly be classified as popular essays on chemistry. These "familiar letters," as their English versions were described, came to form an important aspect of Liebig's literary activity after 1840 and were responsible for much of his worldwide fame and adulation – for they were translated not just into English but also into French, Italian, Spanish, Danish, Swedish, Dutch, Polish, and Russian. A Japanese translation was made as recently as 1952, though for historical purposes rather than as a belated aid to the public understanding of chemistry. The great German philologist Jacob Grimm was in no doubt of the significance of Liebig's *Chemical Letters* when he was compiling his dictionary of the German language in 1854. "Chemistry is a gibberish of Latin and German; but in Liebig's hands it becomes a powerful language." The significance of Grimm's choice of the word "powerful" (*sprachgewaltig*) will become apparent later.[6]

The essay has a long tradition in the history of chemistry, beginning in the sixteenth century with Libavius's *Rerum Chymicarium Epistolica* (1595) with its 100 letters on all aspects of chemistry,[7] and continuing in the seventeenth century with Robert Boyle's effusive reports on colours and many other subjects, and proceeding notably through Richard Watson's *Chemical Essays* (Cambridge, 1781–88) to the twentieth century,

3. Kenneth L. Caneva, *Robert Mayer and the Conservation of Energy* (Princeton, 1993), p. 353, note 78. Caneva's study incoporates a meticulous analysis of Liebig's "mutually contradictory" views on vitalism.
4. E.g., William Gregory to Liebig, 9 May 1851, Liebigiana, Staatsbibliothek, München.
5. J. H. Newman, *Apologia Pro Vita Sua* (London, 1864).
6. "Die Chemie kauderwelscht in Latein und Deutsch, aber in Liebig's munde wird sie sprachgewaltig," J. Grimm, ed., *Deutsche Wörterbuch* (Leipzig, 1854).
7. Andreas Libavius, *Rerum Chymicarium Epistolica* (Frankfurt, 1595).

with the thought-provoking writings of Erwin Chargaff and Primo Levi.[8] In this literary tradition, outsiders are introduced to the meaning, significance, and relevance of chemistry as an essential part of natural philosophy or of applied science. In Boyle's case the essay form was used to argue and to demonstrate that the black art of chemistry, relieved of its spurious elements, forms, and qualities, was worthy of the serious attention of natural philosophers both for the evidence it provided for the corpuscular philosophy and (more discretely) for the malleability and transmutability of matter. For Richard Watson, chemistry was the essential science for those of "the first Rank, Fortune and Ability," for its cultivation by the aristocracy offered "the most certain means of bringing to their utmost perfection, the manufactures of their country." For Watson, therefore, essays about chemical manufactures and chemical manipulations were useful for Cambridge University's poll men (who were "unhappily deprived of one of the strongest incentives to intellectual exertion – narrowness of Fortune") to prepare themselves "for becoming, at a proper age, intelligent Legislators of their Country, [inspiring them] with such a taste for Husbandry, as might constitute the chief felicity of their future lives."[9]

Sixty years later, Liebig's best-selling *Chemische Briefe* became, like the writings of Goethe, Schiller, and Humboldt, part of German literature – though the collection's role in the dissemination of the chemical word in many other languages has yet to be properly analysed. What was the impact, for example, of distributing 60,000 copies of the first English edition in tabloid newspaper form for 40 cents a copy in the United States?[10] Liebig's essays, which began simply as informative newspaper articles, grew in book publication into vehicles for arguing a point of view and for paying off old scores; but above all they served the gatekeeping function of demonstrating that chemistry was essential knowledge for understanding plant and human physiology and for solving problems of health and nutrition, as well as saving the world from Malthusian agricultural damnation and misery:

> They were . . . written for the especial purpose of exciting the attention of governments, and an enlightened public, to the necessity of establishing Schools of Chemistry, and of promoting, by every means,

8. E. Chargaff, *Serious Questions* (Boston, 1986). Note also Chargaff's title, *Essays on Nucleic Acids* (Amsterdam, 1963); P. Levi, *Il sistema periodico* (1975), *The Periodic Table* (London, 1985).
9. R. Watson, *Chemical Essays,* 3rd ed., 4 vols. (Cambridge, 1788), Preface.
10. Liebig makes this claim in the "Preface" to the sixth German edition of the letters. Two editions were also printed by D. Appleton of New York in 1843.

> the study of a science so intimately connected with the arts, pursuits, and social well-being of modern civilised nations. For my own part, I do not scruple to avow the conviction, that ere long, a knowledge of the principal truths of Chemistry will be expected in every educated man, and that it will be as necessary to the Statesman and Political Economist, and the Practical Agriculturist, as it is already indispensible to the Physician and the Manufacturer.[11]

As with each of Liebig's books, patronage and dedications proved important. He spent much time and effort with Hofmann, for example, getting the wording right for his dedication of the third English edition to Sir James Clark in 1851.[12] The original German and French editions were dedicated to Dumas in a spirit of reconciliation and admiration, after Dumas broke the ice in 1850 by conferring the Legion d'Honneur upon Liebig in the name of the Republic at a meeting they both attended in Lille. Liebig and Dumas had shared a love of science for a quarter of a century, Liebig declared, but whereas the whole world and not just the German people knew of Dumas's work and discoveries, no one except Liebig himself appreciated better Dumas's hard work and genius. Liebig knew how to flatter! "If the roads by which we endeavoured to attain the goal were often different, in the proximity of that goal we always met in order to shake hands with each other."[13] The fourth enlarged German edition of 1859 was dedicated to King Maximilian II in appreciation of the royal patronage of the Munich circle of men of arts and science. The Italian translations carried no dedications from Liebig, though interestingly, Bruni, the translator of the first edition in 1844, dedicated the book to Amadeo Avogadro for his expert knowledge of the work of German experimental physicists.

A detailed study of the publishing history of these letters has yet to be made and would prove quite complex. Consider, for example, the fact that in Letter 1 of the first English edition, only the first paragraph remains in the third edition (1851); the remainder of the essay is shifted to Letter 11, with ten new letters filling the gap. In each case, extensive revisions and expansions have been made between the first and the third editions. Moreover, because the translators were different, the actual English text differs slightly (and occasionally significantly) from the first to the third

11. Preface, *Familiar Letters on Chemistry* (London, 1843), pp. v–vi.
12. Liebig to Hofmann, 26 April 1851, in W. H. Brock, ed., *Justus von Liebig und August Wilhelm Hofmann in ihren Briefen* (Weinheim, 1984), p. 107; and 19 April 1851 in E. Heuser and R. Zott, eds., *Justus von Liebig und August Wilhelm Hofmann in ihren Briefen* (Mannheim, 1988), p. 21.
13. W. Shenstone, *Justus von Liebig* (London, 1895), pp. 51–52.

editions.[14] Once Liebig had moved to Munich, where one of his expected tasks was to deliver popular lectures on chemistry, he had further opportunities of publishing material in the *Allgemeine Zeitung* that he subsequently reprinted as Letters in the fourth definitive edition of 1859. This edition added one further letter to the thirty-five of the third edition. Whereas the first English edition was a slim octavo volume of only 179 pages, the fourth and final English edition of 1859 was a quarto volume of 536 pages. The series continued, however. Fuelled by the controversy over agriculture, Liebig published a large number of polemical and advisory essays on agricultural theory and practice in the *Allgemeine Zeitung* during the 1850s. He gathered fourteen of these together in 1859 as *Naturwissenschaftliche Briefe über die moderne Landwirtschaft,* the English translation by John Blyth, *Letters on Modern Agriculture,* appearing simultaneously. When Winter called for a fifth "cheap" edition of the original *Chemische Briefe* in 1865, the two sets were amalgamated in one volume containing altogether fifty letters. Neither this nor the posthumous two-volume edition of 1878 was reissued in English.

Liebig's essays were a direct consequence of the appearance of his *Agricultural Chemistry* and his derogatory comments on the state of chemistry in Prussia. One particularly impressed reader of the former was the newspaper proprietor and publisher Johann Georg Cotta (1796–1863). Cotta's publishing house had been founded in Tübingen in 1640 and had been developed by a succession of scholarly family members. Georg's father Johann F. Cotta von Cottendorf (1764–1832) had founded the important newspaper *Augsburger Allgemeine Zeitung* in 1798 and published the literary works of Schiller, Goethe, Herder, Schelling, Humboldt, and Jean Paul. In 1810 the firm relocated in Stuttgart, where in 1824 it introduced the first steam-powered printing press into Bavaria. At the same time, Karl Tauchnitz at Leipzig was developing stereotyping and cheaper binding methods, so that by the 1840s, as in Britain, it was possible to market cheap books for the common reader.[15] A member of the aristocracy and a man of considerable wealth, Johann F. Cotta owned large estates in Württemberg, all of which were inherited by Georg. One of the principal aims of Cotta's publishing group was to make the middle classes more aware of their culture and to look forward to political change in which the barriers between them and the aristocracy would be loos-

14. Gregory, in preparing the third English edition, commented that he had had to correct Gardner's original translation. Gregory to Liebig, 27 December 1850, Liebigiana, Staatsbibliothek, München.
15. Robert E. Cazden, *A Social History of the German Book Trade in America to the Civil War* (Columbia, S.C., 1984), p. 67.

A Liebig visiting card presented to Elizabeth Hofmann of Dresden on 23 September 1872. (Courtesy W. Lewicki)

ened. This could be achieved, Cotta believed, through the development of German science and industry.[16]

As a practising farmer and landowner, Georg Cotta was naturally interested in what Liebig had to say about agriculture. Deeply impressed, he wrote to Liebig in April 1841 and concluded with an invitation:

> I would be particularly delighted if you would express your opinions on scientific questions, your own research, and the practical implications which they have for the mother country, etc. from time to time in my Allgemeine Zeitung or quarterly journal.[17]

16. For a good discussion, see Timothy Lenoir, "Laboratories, medicine and public life," in A. Cunningham and P. Williams, eds., *The Laboratory Revolution in Medicine* (Cambridge, 1992), pp. 24–65.
17. A. Kleinert, *Justus von Liebig "Hochwohlgeborner Freyherr." Die Briefe an Georg von*

The quarterly journal mentioned here was the *Deutsche Vieteljahrschrift* for essays on politics, science, medicine, and religion that Cotta had started in 1837 on the model of the English quarterlies. In 1846 Liebig contributed a seventy-page essay on the relationship between physiology, pathology, chemistry, and physics to this quarterly journal. This essay, as we saw in Chapter 7, was separately printed as the second volume of the third edition of *Animal Chemistry.*

Liebig responded with alacrity to Cotta's invitation, saying that he had long admired Cotta's firm for its promotion of German literature. He even claimed that his polemical essay on the state of Prussian chemistry had been conceived originally for Cotta's campaigning newspaper; however, in view of his long-standing relationship with Vieweg, he had had to decide against publishing with Cotta. Liebig declared himself happy to publicise his ideas for the reform of agriculture in Cotta's newspaper, and as we saw in Chapter 6, he recommended that Petzholdt should be commissioned to write articles on reform.[18] Given how noticeably little interest German intellectuals took in scientific affairs and of the effects of science on "the pulse of the circulating blood of the national economy," he believed that it was vital that the needs of science, life, and industry should be brought more prominently before the nation. He agreed, therefore, that from time to time he would write informative articles for placement in one of Cotta's several newspapers and journals. These would, he thought, examine "the position of chemistry, its development and its practical applications." Cotta's invitation, therefore, offered him a unique public opportunity to act as the gatekeeper of chemistry. No fee was mentioned.

In his answer to Cotta, Liebig referred to "essays" (*Aufsätze*) rather than "letters," but when he next communicated with Cotta in August 1841, he described the five articles he had enclosed as *Chemische Briefe.* They had been written, he said, for the benefit of pure chemists, physicists, manufacturers, and industrialists (*Techniker*). If they were not good enough for the *Allgemeine Zeitung,* perhaps Cotta would like to insert them in his *Vieteljahrschrift?* Cotta was delighted, and the five letters were duly published anonymously in the *Allgemeine Zeitung* during September 1841. By August 1843, when the collected English edition of the letters appeared, a dozen of the letters had been published in the German newspaper; the English book version contained sixteen letters, some of which were published by Cotta during 1844 prior to the appearance of the first

Cotta und die anonymen Beiträge zur Augsburger Allgemeinen Zeitung (Mannheim, 1979), p. 1.

18. *Ibid.,* pp. 2–3. For a translation, see P. Munday, *Sturm and Dung* (Ph.D., Cornell University, 1990), pp. 211–12.

German edition of the letters in July of that year. By May 1844 Cotta had published twenty-one letters in his newspaper, all of which appeared in the Italian edition of May 1844, which had been made by G. D. Bruni of Torin. The German edition, for which Anton Winter of Heidelberg had to pay a royalty,[19] included a further five letters, making twenty-six in all. According to *The Lancet* in 1843, echoing Liebig's English "Preface," the newspaper articles had already led to the establishment of new chemistry chairs at Göttingen and Würtzburg for the pursuit of "new lines of research in physiology, medicine and agriculture."[20] It is interesting to note that, as with his *Annalen der Chemie,* Liebig insisted that his *Letters* be set in Roman type. Later in the century, however, printers began to reintroduce Fraktur as a symbol of cultural and political nationalism, despite protests from the scientific community. Later German editions of the *Chemische Briefe* all appeared in Gothic typefaces.

Who translated the letters into English is unclear. It is usually assumed that it was John Gardner, an English apothecary who had visited Giessen in 1843 and who is described as the work's editor on the title page. Since Gardner is known to have translated Liebig's lectures on organic chemistry for *The Lancet* in 1843, this is not an unreasonable assumption. On the other hand, in his English "Preface," Liebig explicitly thanks his former pupil Dr. Ernst Dieffenbach (1811–55) for suggesting that his newspaper articles should be published in collective English form. In the same "Preface," Liebig thanked Gardner only for revising the manuscript and correcting the galley proofs. It seems far more likely, therefore, that it was Dieffenbach who made the original translations, which Gardner then lightly corrected. For Gardner, the *Familiar Letters on Chemistry* were useful propaganda for his and Bullock's campaign to set up a school of chemistry in London.

As we saw in Chapter 5, Dieffenbach was one of Liebig's most interesting German pupils. The son of a Giessen professor of theology, he had studied medicine there from 1828 to 1833, during which time he also attended Liebig's chemistry classes. As a politically active student, he was unfortunately forced into exile before he could qualify, as a result of activities during the outbreak of revolutionary activities in 1833. Following qualification at Zurich in 1835, he moved to London, where he earned his living teaching at Guy's Hospital and sending articles about English affairs to the *Allgemeine Zeitung,* or translating English medical books and articles into German. His translation work brought him into contact

19. Liebig to Vieweg, 21 June 1854, in M. and W. Schneider, eds., *Justus von Liebig Briefe an Vieweg* (Braunschweig, 1986), pp. 273–4.

20. *Lancet* 7 October 1843, p. 31.

with the Geological Society, including members like Lyell and Darwin. The latter's wonderful account of the *Beagle* voyage (1839) was to be translated by Dieffenbach in 1844. It was Liebig who arranged publication with Vieweg. Meanwhile, in 1839, Dieffenbach, inspired by Darwin's voyage, became ship's surgeon and naturalist on *HMS Troy*, which was to explore New Zealand on behalf of the New Zealand Company. Dieffenbach's *Travels in New Zealand,* which was published in 1845, was the first scientific account of the island's flora, fauna, and geology. As we have seen, Liebig also used Dieffenbach at this time to promote his artificial fertilizers in Britain. Having been pardoned for his political offences in 1837, Dieffenbach returned to Germany in 1848 to edit a Hessian newspaper. During the final years of Dieffenbach's short life, Liebig was able to see that he was appointed professor of geology at Giessen.[21]

As happened frequently in Liebig's relationships with his scientific publishers, Walton, Winter, and Vieweg, Liebig soon seized the initiative with Cotta to suggest other publishing projects. In January 1842, when the serialization of his letters had scarcely started, he suggested that Cotta should publish a German verse translation of Petrarch's sonnets that had been made by two of Liebig's Darmstadt friends, the jurists Karl Ludwig Kekulé (1802–43) and Ludwig von Biegeleben (1812–72). In case Cotta doubted the ability of a mere chemist to be a good judge of poetry, Liebig mentioned his intimate past friendship with Platen, who "had played a significant role in my life and who had a decisive influence on my mind's bent (*Geiste Richtung*)." Perhaps the knowledge that he would be attending the centenary celebrations of the founding of the University of Erlangen in 1843 had brought Platen back to mind.

Liebig's powerful plea was successful, and the German Petrarch (which Biegeleben referred to as "Liebig's fosterchild") duly appeared in 1844. Clearly, years of practical chemistry had not entirely dimmed Liebig's cultural sensibilities. Indeed, in 1865, he was similarly to persuade Vieweg to undertake several literary projects, including a translation of Brillat-Savarin's *Physiologie de goût* (1825). The beauty of having a German version of this witty compendium of the art of dining was that Liebig was able to incorporate reprints of his papers on meat extract, infant's milk, invalid food, and recipes for bread and coffee making.[22]

21. *Hessische Biographien,* vol. 2 (Darmstadt, 1927; reprint, Wiesbaden, 1973), pp. 146–50; *An Encyclopaedia of New Zealand,* 3 vols. (Wellington, 1966), vol. 1, p. 472; Gerda E. Bell, *Ernest Dieffenbach* (Palmerston North, 1976).

22. *Physiologie des Geschmacks oder Physiologische Anleitung zum Studium der Tafelgenüsse . . .* von Brillat-Savarin, trans. by Carl Vogt (Braunschweig, 1865), 5th ed. 1888, pp. 394–421. See Schneider, *Briefe an Vieweg,* 25 August and 18 November 1865.

Although Liebig had to offer the German edition of his letters to Vieweg, he was able to exploit the fact that English sales had been so good (the print runs were 3,000 copies in August 1843 and another 3,000 in February 1844) to encourage Cotta to publish similar letters connected with other scientific disciplines. Insofar as Liebig's primary interest was now physiological chemistry, not surprisingly, physiological letters were an obvious subject. Again Liebig was able to deliver patronage to a friend; this time to another of his former German pupils Carl Vogt. Like Dieffenbach, Vogt had been forced to seek political refuge in Switzerland from where in 1844 he had migrated to Paris. Liebig not only found him work with Cotta as the newspaper's French scientific correspondent but also got him to begin a series of physiological letters that were closely modelled on his own popular essays on chemical matters. Cotta published Vogt's essays in book form in 1847.[23]

Winter's first German edition (1844) of the *Chemische Briefe* had contained twenty-six letters; seven more letters written for Cotta's newspaper in 1851 were added to the third edition in 1851, which was completely recast into English by William Gregory. The fourth edition (1859), the principal edition, added five general letters plus fourteen more on agriculture, all of which had appeared in Cotta's paper between 1857 and 1858. The new material for the English edition of 1859 was translated by Liebig's Irish pupil John Blyth. As Kleinert has discovered, Liebig wrote many other ephemeral news items for Cotta's newspaper that were not suited to book publication. These ranged from an obituary of Ludwig Schleiermacher, the physicist son of Liebig's former ducal patron in Darmstadt, to reports on the state of the University of Giessen, the cholera, and his recipes for quinoidine, bread, and coffee. One of his last pieces of writing, a long commentary on the future of the University of Munich, was serialised in 1873.[24]

Given the nature of Cotta's original invitation, the first letter was not surprisingly a much-toned-down version of the "Zustand" polemic of 1840. In it, Liebig again strove to convince readers that chemistry was a mature and independent science, offering a "most powerful means towards the attainment of a higher mental cultivation," and profitable both in a utilitarian sense as well as in the philosophical sense of enhancing man's feeling of awe towards the wonder of creation. It was from organic chemistry that "the laws of life, the science of physiology, will be devel-

23. C. Vogt, *Physiologische Briefe für Gebildete aller Stände* (Stuttgart, 1845–47). The letters had appeared in *Allgemeine Zeitung* from 1845 to 1847. See F. Gregory, *Scientific Materialism in Nineteenth-Century Germany* (Dordrecht, 1977), Chap. 3; and Kleinert, *Justus von Liebig*, pp. 8–9.

24. Kleinert, *Justus von Liebig*, pp. 52–61.

oped." Having extolled chemistry's independence, Liebig then explained how modern chemistry differed from the more speculative approach of the scholastics and early natural philosophers. Two letters on the history of chemistry (inspired by his colleague Herman Kopp's important *History,* which had appeared in 1844[25]) were added to the third edition in 1851. Like Kopp, Liebig divided the development of chemistry into an alchemical period, a confused phlogistic period in the eighteenth century, and the contemporary period of experimental chemistry. This approach enabled him to return to the theme of the "Zustand" article. The older German natural philosophy had been "a dead tree, which bore the finest leaves and and most beautiful powers, but no fruit. With an infinite expenditure of mind and sagacity, only pictures were created." In contrast, today's chemical philosophers explained something "by seeking out the causes which have preceded it. To those which are manifest to the senses, he gives the name of *conditions;* and that of *forces* to those which are not thus manifest."[26]

Having established chemistry's history and empirical independence, in Letter 5 Liebig began a simple exegesis of chemistry, beginning with combination and decomposition, chemical affinity, analysis, and synthesis. In the seven successive letters, he sketched in an elegant, popular account of chemical ideas, experiments, techniques, and apparatus, including the atomic theory and some of the basic principles of organic chemistry. In Letter 11, Liebig described the present state of chemical industry; this section included his much quoted and misattributed aphorisms on soap production and sulphuric acid production as being measures of modern civilization. He stressed, too, the importance of Britain's soda (alkali) industry and the country's dependence upon the import of sulphur. A remarkable twelfth essay drew on Faraday's electrical work to compare the merits of electrical and steam power, before he discussed the sugar-beet and gas lighting industries. Liebig was convinced that the electric motor would never be an economic source of power like the steam engine. Zinc might be abundant in Germany and produce about four times as much power during electrolysis as when it was burned; but it remained a stubborn fact that when coal was burned, it produced six times as much power as a pound of zinc, which, in any case, required coal to smelt it from its ores. All this was a prelude to a discussion of Mayer's ideas on the conservation of force, which led to the crucial idea that food was the fuel

25. H. Kopp, *Geschichte der Chemie,* 4 vols (Braunschweig, 1844–47). Liebig was also indebted to his son-in-law Moritz Carrière, whose *Die philosophische Weltanschauung der Reformationszeit* had appeared in 1847.

26. J. Liebig, *Familiar Letters on Chemistry,* 4th ed. (London, 1859), p. 26.

of animals for their nutrition and animal heat. This letter was based on one of his first popular lectures given in Munich in March 1858.[27]

Liebig next turned to the phenomena of isomerism and allotropy (mentioning the work of Schönbein on ozone) as important peculiarities of organic and inorganic chemistry that were clearly caused by the physical arrangements of atoms, as well as being indicators of the probable role of electric polarity at the moment of chemical change. He referred to the recent work of his former English pupil Benjamin Brodie, who was making some interesting experiments and speculations on chemical reactivity.[28] Following some further thoughts on the roles of heat, light, electricity, and mechanical motion on chemical affinity, Liebig began a popular account of the state of organic chemistry in his seventeenth letter. Three preliminary letters (17–19) concentrated on fermentation and ferments (such as pepsin) and the role of oxygen in putrefaction (eremacausis). A practical essay on wine and beer making and the preservation of food (20–21) reincorporated Liebig's theory of disease that he had first announced in 1840 in *Agricultural Chemistry.* Letters 22 to 23, which examined the relationship between chemistry and physiology, effectively recapitulated in popular form the ideas and convictions of *Animal Chemistry,* though they also represented Liebig's most explicit support for vitalism and opposition towards materialism and the spontaneous generation of life. The materialists were,

> total strangers to all investigations connected with chemical and physiological forces. . . . They are the speculations of amateurs, who assume the right, after a very slight acquaintances with the investigations of natural phenomena, of expounding to an ignorant and credulous public the wonders of creation and of life, and of setting forth to them what progress has been attained in the highest departments of science. And such ignorant and presumptuous dreamers are listened to more readily than the most philosophic inquirer.[29]

It was for this section that Liebig was roundly condemmed by Moleschott.[30]

At this juncture, Letter 24 on the phenomenon of so-called spontaneous combustion was interjected both to illustrate man's infinite gullibility and presumably to offset any materialistic claim that because human beings were simply a conglomeration of chemicals, their spontaneous combus-

27. Liebig's notes are reproduced in facsimile in Volhard, vol. i, pp. 439–56.
28. B. C. Brodie, "On the condition of certain elements at the moment of chemical change," *Phil. Trans.* (1850), 750–804.
29. Liebig, *Familiar Letters on Chemistry,* 4th ed. 1859, Letter 23.
30. J. Moleschott, *Der Kreislauf des Lebens; physiologische Antworten auf Liebigs Chemische Briefe* (Mainz, 1852; 2nd ed. 1855); see Gregory, *Materialism,* Chap. 3.

tion was theoretically possible. Although Liebig drew upon a longer account he had published in 1850,[31] this chapter forms one of the highlights of *Familiar Letters on Chemistry.* On 13 July 1847, Count Friedrich von Görlitz (b. 1795) had returned to his Darmstadt home to find his wife had burned to death. Apart from some charred furniture close to the body, which showed no signs of foul play, the rest of the room and house was undamaged. Some of the countess's possessions were, however, found to be missing. Oddly, one of the witnesses to these events was Liebig's future student August Kekulé, whose parents' house was opposite. It was Kekulé's sister who had called her brother's attention to flames behind the count's upstairs blinds when she had been staring out of the window.

Some older biographies of Kekulé attributed his decision to become a chemist to hearing Liebig give evidence at the trial. In fact, after matriculating at Giessen in the winter semester of 1847–48, with the intention of becoming an architect, Kekulé had already begun to attend Liebig's lectures in the summer of 1848. He spent the winter semester 1848–49 at the Gewerbeschule in Darmstadt studying mathematics and science before his family reluctantly allowed him to abandon architecture completely for chemistry. He rejoined Liebig at Giessen in the summer of 1849; the Görlitz trial did not take place until March 1850.

At first the count himself had fallen under suspicion, but he had a good alibi. Instead, a valet named Johann Stauff was arrested on suspicion of robbery, with the additional suspicion of a murderous attack, even though the count's family doctor and several of his medical colleagues argued that it was a clear case of spontaneous combustion. Stauff was therefore freed, but the revolutionary events of 1848 led to a press campaign suggesting that there had been a cover-up by the aristocratic elite and that both the count and countess were alcoholics. The Hessen medical faculty thereupon dismissed spontaneous combustion as unscientific and demanded an exhumation in August 1848. But because the body had been so badly destroyed, post mortem examination proved crucially indecisive as to whether or not there were signs of physical injury. And since one very prominent member of the Hessian medical faculty, the state medical officer Theodor Siebold, remained convinced that spontaneous combustion had occurred, the Darmstadt Court called on the city's most famous chemical son, Justus Liebig, to advise them of the scientific feasibility of this method of death. Liebig's Giessen physiological colleague Theodor

31. J. Liebig, *Zur Beurteilung der Selbstverbrennung des menschlichen Körpers* (Heidelberg, 1850); there was no English translation. Note the *Times* report of the trial 18 April 1850 and *London Medical Gazette,* 11(1850), 899–903, 944–9.

Bischoff was also asked to help in the inquiries, as was Liebig's friend Emanuel Merck, Darmstadt's leading apothecary.[32]

Liebig's and Bischoff's approach was twofold. On the one hand, they were legalistic, examining some forty case reports of spontaneous combustion. All these they found flawed insofar as testimony had never been subjected to cross-examination or the critical analysis of such beliefs as that heavy drinkers accumulated combustible hydrogen phosphide in their bodies. Most of these cases could be simply explained in terms of the drunken carelessness and to the proximity of candles or gas lights to heavy curtaining and bed draperies. These historical cases, and not the Görlitz trial itself, formed the basis of Liebig's Letter 24. Bischoff's and Liebig's principal approach, however, was experimental; in a series of macabre experiments they simply showed that it was extremely difficult to ignite human flesh and ensure complete oxidation to cinders or to a fatty melt unless extraordinarily high temperatures were used. This made it extremely unlikely that a body could set itself alight internally.

All this convinced the court, as well as Siebold, that the Görlitz case was not explicable by spontaneous combustion. Since theft was involved, there was therefore a high degree of probability that the valet was guilty of murder – indeed, Stauff duly confessed and was sentenced to life imprisonment. Later, in the spirit of post-März liberalism, he was released and he emigrated to America with a common-law wife and several children. Intriguingly, despite the publicity given to Liebig's and Bischoff's forensic arguments in *The Times* and in the *London Medical Gazette* in 1850, and by Liebig himself in *Familiar Letters on Chemistry* in 1851, Dickens still dispatched Mr. Krook this way in *Bleak House* in 1853.[33] Given that Gregory was well disposed towards the occult, it is not surprising to find that he felt that Liebig had been too categorical in rejecting the possibility of spontaneous combustion, and he claimed support from his Edinburgh colleague, the toxicologist Sir Robert Christison.[34] Cases of "spontaneous combustion" are probably due to a candlewick effect. Clothing acts as a wick fed by burning human fat. Legs and arms do not burn so well and often remain as a macabre feature of the incident.

32. For helpful English orientation, see E. Gaskell, "More about spontaneous combustion," *Dickensian,* 69 (1973), 25–35; and for Kekulé's involvement, O. P. Krätz and C. Priessner, *Liebigs Experimentalvorlesung* (Weinheim, 1983), pp. 367–73. Michael Harrison, *Fire from Heaven* (London, 1976), is uncritical. See also J. L. Heilbron, "The Affair of the Countess Görlitz," *Proc. Amer. Phil. Soc.* 138(1994), 284–316.
33. Dickens was strongly rebuked by George Henry Lewes, who with George Eliot, visited Liebig in Munich in 1854. See Chapter 12.
34. Gregory to Liebig, 12 February 1851, Liebigiana, Staatsbibliothek, München.

Following this debunking of gullible medical men, the final eight letters of the fourth edition concerned agriculture and formed a popular recapitulation of *Agricultural Chemistry,* together with some digs at Lawes and Gilbert's attack on him in the *Journal of the Royal Agricultural Society* in 1850. In a curious reversal of his normal publication practice, three of these new letters (on the absorption of oxygen by blood, the influence of chemistry upon agriculture, and the relations between combustible foods and life processes) were inserted in *Annalen* after their appearance in the third edition of the *Letters.*[35] That Liebig was unwilling to concede much, if anything, to his English and German critics becomes especially clear in the fourth principal German edition of 1859, with its further eighteen agricultural letters that were included in a second volume, as well as being published independently as *Letters on Agriculture,* which he dedicated to King Maximilian II. Although he once remarked to Wöhler that "Praise is always acceptable, but reproof is of more use," there was little sign that criticism had altered his views.[36]

Blyth, Liebig's Irish pupil, who translated these "important and interesting letters" in 1859, drew attention to their political message (so poignant to an Irishman):

> The wants of an increasing population, and the danger of a possible stoppage, at any moment, of supplies drawn from foreign sources, makes us all feel a deep interest in the discovery of the means of producing more bread and meat on a given surface.[37]

Blyth also drew attention to the law of the minimum that Liebig made much more explicit in the eighth agricultural letter than he had done in previous editions of *Agricultural Chemistry:* If one particular nutrient were deficient, there could be no growth after it had been exhausted. Or, in Liebig's strange presentation:

$$P = F - R$$

> The P in this formula stands for Produce (corn, turnips, &c); the F stands for Food (phosphoric acid, potash, lime, ammonia, &c); the R stands for Resistance. . . . The amount of produce yielded by a field corresponds with, or is proportionate to the quantity of food in the soil (to the conditions for the production of the crop) *minus* the sum of the resisting forces which hinder the production of the crop from the elements of food present.[38]

35. *Ann. Chem.* 79(1851), 112–16, 116–23, 205–21, 358–69.
36. Liebig to Wöhler, 29 July 1857, in Hofmann, *Liebig's und Wöhler's Briefwechsel,* vol. 2, p. 45.
37. J. Liebig, *Letters on Modern Agriculture* (London, 1859), Blyth's "Preface," p. v.
38. *Ibid.,* p. 265.

These fifteen agricultural letters ranged over the conflicting needs of theory and praxis, the mineral theory, the nitrogen theory, soil absorption, manures, soil exhaustion, and the recommendation of sewage reclamation that was to lead to the London sewage debate already discussed in Chapter 9.

It was in the context of agriculture that Liebig made his final remarks on science teaching. In his fourteenth agricultural letters of 1859 he referred to the fact that the opening of agricultural institutes in Germany had made it imperative that chemistry be taught in schools. Maximilian, he noted, had wisely ordered that this be done in Bavaria. However, neither schools nor institutes could teach a practical art like farming; they could only teach principles:

> I formerly conducted at Giessen a school for practical chemistry, analysis, and other branches connected therewith, and thirty years' experience has taught me that nothing is to be gained by the combination of theoretical with practical instruction. A student of chemistry who attends the lecture-hall and the laboratory concurrently, positively defeats thereby the object of his stay at school, and misses the aim of his studies.[39]

On first reading, this quotation seems to contradict Liebig's own practice at Giessen, in which lectures were held concurrently with practical classes. Liebig's views on the distinction between pure and applied science had hardened considerably, and in 1859 he was making a plea for a sound secondary education in science before training in more practical, or vocational, skills took place:

> It is only after having gone through a complete course of theoretical instruction in the lecture-hall that the student can with advantage enter upon the practical part of chemistry; he must bring with him into the laboratory a thorough knowledge of the principles of the science, or he cannot possibly understand the practical operations. [For] if he is ignorant of these principles, he has no business in the laboratory.[40]

Although Liebig was not unwise to emphasise the importance of a general, liberal education before the technical part of an industrial pursuit was mastered, he was surely proved wrong in assuming that "an individual who may be thoroughly master of the technical part [of an industry] is altogether incapable of seizing upon any new fact that has not previously

39. *Ibid.*, p. 257.
40. *Ibid.*, p. 257.

presented itself to him; or of comprehending a scientific principle and its application."[41]

Overall, *Familiar Letters on Chemistry* represent the dogmatic views of a chemical colossus who was able to express the central significance of chemistry in society in simple, nontechnical, and elegant language. Wöhler's remarkable reaction to the enlarged principal edition of the *Letters* is worth quoting extensively, because even allowing for Wöhler's close friendship and admiration of Liebig, it sums up beautifully the contemporary sense of Liebig's achievement as a populariser:

> I am sitting in my small study on this winter's evening reading your *Chemical Letters* – I cannot tell you with what delight, and with what enlightenment. I had the thought flashing through my mind that you were sticking your neck out. Never before has the world been told so clearly what chemistry is, how it relates to physiological processes in living nature, on its relationships with medicine, agriculture, industry and commerce. These connections, so clearly set out so that a child could understand them, must be sufficient to make this book a classic. The influence that it must surely exert, or already has asserted, is limitless; thousands will feed from it and, standing on your shoulders, make use of the stimulating ideas in it. Everything in this discourse possesses a clarity, simplicity and characteristic of presentation that makes it appear to be done with the greatest of ease. And then, what studies, what troubles, what knowledge of manifold subjects are set before us! I have been continually amazed at your scholarship, your familiarity with things that you could only make your own after laborious study, and of which we others never bother to cultivate. It is a genuine philosophy of chemistry in the informative style of letters – a form, besides, which is always most successful in carrying significance. And what about the practical applications of the views it expresses? These will be as gold for the unfortunate majority of doctors and agriculturists of the present generation who understand it, but who stubbornly refuse to utilise it because their old brains no longer understand the fundamentals of science. However, the time cannot be far off when they will all see, as we chemists already do, that you have come up with and revealed the truth – a triumph which you will perhaps live to see. However, with so many old heads still about, I think we must simply give them up. They first need to study the fundamentals of chemistry, physics and plant physiology; but these subjects are no longer comprehensible to them.[42]

41. *Ibid.*, p. 258.
42. Wöhler to Liebig, 27 January 1859, in Hofmann, *Liebig's und Wöhler's Briefwechsel*, vol. 2, pp. 63–4. Volhard quotes this, vol. i, pp. 401–2.

Liebig's *Chemische Briefe,* or *Familiar Letters on Chemistry,* as the book was known in English, began as a series of popular essays that were commissioned by the editor of the *Augsburger Allgemeine Zeitung* between 1842 and 1843. Significantly for Liebig's British and American reputation and influence, the letters were first published collectively in an English translation in 1843. Enlarged Italian and German editions, which appeared in 1844, were followed by translations into French, Dutch, Swedish, Danish, Russian, Polish, and Spanish. By the time of the final German edition of 1865, the book had swollen into fifty essays that covered an astonishing range of pure and applied chemistry, especially chemistry's significance for agriculture and the chemistry of health and disease. The *Familiar Letters on Chemistry* were superb and influential popularizations of science and were indirectly a powerful suasive factor in encouraging students to take up careers in chemistry, to commit themselves to a gospel of civic virtue through science, and for governments to undertake financial and institutional responsibility for scientific teaching and research. Just before his death, Liebig could think of no better compliment to one of his former pupils than to compare their respective attempts at popularisation:

> The style of your book is excellent; it is really a masterpiece of simplicity and clearness. To most readers it may appear off-hand, as many have said of my chemical letters, but I am sure that you have expended much care and attention on the composition, as I did on mine; there is an art in simplicity and freshness of style which, however, must be kept out of sight.[43]

Conceived as a way of raising the profile of chemistry and its practitioners among government officials, industrialists, and the general public, Liebig's *Familiar Letters on Chemistry* proved an extremely effective work of popularization, not only in the Germanies but throughout Europe and America.

43. To an unknown pupil, quoted by Max Pettenkofer, "Liebig' s scientific achievement," *Contemporary Review* 29(1876–77), p. 875. Note Pettenkofer's popular lectures: *Beziehungen der Luft zu Kleidung, Wohnung und Boden* (Braunschweig, 1872; trans. *The Relations of the Air to the Clothes We Wear,* London, 1873), which makes him the most likely recipient.

11

Philosopher of Science: The Bacon Affair

> Out of the old discarded rags of science, Bacon patched together a new garment for his countrymen; and although it did not hide their nakedness, each one found it sat easily and looked well.[1]

One of the bizarre consequences of Liebig's love of Britain and of his polemic with Lawes and Gilbert over the mineral theory was his decision in 1862 to examine the life, work, and reputation of Francis Bacon (1561–1626). Once again he found himself the centre of controversy in both Germany and Britain, as philosophers and historians countermanded his fierce denunciation of Bacon's reputation. The controversy is of interest not only because of what it reveals about Bacon as a British icon and the role of Bacon in British culture but also for its exposure of Liebig's own ideas concerning scientific method and the philosophy of science. The following discussion will enable us to examine another simmering controversy to which Liebig was continually exposed after 1840, namely, his contribution to the debate concerning vitalism and his argument with those of his German scientific compatriots who espoused a full-blown materialistic philosophy in which living systems were reduced entirely to the properties and movements of atoms.

Liebig at the Court of Bavaria

If Francis Bacon praised King James I of England, Wales, Scotland, and Ireland as a philosopher king and owed his advancement entirely to him, there is some irony in the way Liebig, who had been ennobled by the duchy of Hessen Darmstadt in 1845, was "called" to Munich in 1852 to

1. J. Liebig, "Lord Bacon as natural philosopher," *Macmillan's Magazine* 8 (July and August 1863), 237–49, 257–67. This was an English abstract of *Ueber Francis Bacon von Verulam, und die Methode der Naturforschung* (Munich, 1863), reprinted in *Reden und Abhandlungen* (Leipzig and Heidelberg, 1874; repr. 1965) under the original 28 March 1863 lecture title, "Francis Bacon von Verulam und die Geschichte der Naturwissenschaften," pp. 220–54.

add lustre to the learned court that surrounded King Maximilian II (1811–64). Because Liebig became perpetual president of the Royal Bavarian Academy of Sciences in November 1858, it fell to him to pronounce a eulogy on his royal patron in 1864.[2] He was perfectly honest that on his arrival in Munich he had held no very high opinion of the academy, which he compared to a blind and lame old carthorse; but in the king's hands and those of the other scholars that Maximilian had brought to Munich, it had begun to thrive and to achieve a hefty output of annual publications.[3]

As we have seen, Liebig repeatedly received invitations to chairs of chemistry at universities like Heidelberg, Vienna, and the Royal College of Chemistry, and he always used them to bolster his conditions at Giessen. At the beginning of 1852, when he was forty-nine and at the height of his powers, he received an offer from Munich he could not refuse. As he told Hofmann (15 April 1852):

> I have decided to go to Munich. I sent off my definitive demands today; if they are accepted, I shall consider myself committed. The thought of being condemned for the rest of my life to give lectures in Giessen decided me. The position in Munich offers anything one wants to make of it; it is an honourable retreat. The people of Darmstadt are asses if they don't call you to Giessen [to replace me].[4]

Three months later he was able to reflect upon his motives more clearly:

> I returned yesterday from Munich having committed myself firmly and irrevocably. My lectures there begin in November and I shall leave Giessen in September. I hope I shall never regret this decision. What depresses me here and saps my strength is the practical course. I would have given it up here too in any event; I shall be 50 years old next year and if I still want to achieve something in science I must limit my activities. I could not do so here; if I were to close the laboratory today, no more foreigners would come tomorrow, half of my students would fall away. I have this contract here which no one will let me off. I can't let the rooms of the laboratory stand empty, nor is it possible to take on a second professor or to split it up since my own assistants would be constantly quarreling with the others. Munich offers a new sphere of activity. I have committed myself to six hours experimental chemistry in the winter semester, in the summer the general public, *nothing more, no practical students*! I shall let my students work in the laboratory, but according to my choice; I shall

2. J. Liebig, "Nach dem Tode von König Max," in *Reden* (Leipzig, 1874), pp. 330–31.
3. *Sitzungsberichte der mathematisch-physikalischen Classe der k.b. Akademie der Wissenschaften zu München.*
4. W. H. Brock, ed., *Justus von Liebig und August Wilhelm Hofmann in ihren Briefen* (Weinheim, 1984), p. 127.

> get 30,000 thalers for a completely new laboratory, 5,000 thalers for equipment, a salary of 2,250 thalers per year, 2,000 thalers for current expenditure and *no practical students*! Can one refuse a thing like that? Impossible.[5]

Who could blame Liebig? Any drop in salary was more than offset by the conditions of service offered in Bavaria. Ironically, he was to discover that even in Munich he was unable to avoid the mind-deadening examination of medical and pharmacy students.[6]

Max Pettenkofer (1818–1901), who had studied with Liebig at Giessen in 1844 and who was directly involved in the negotiations with Liebig, later noted that "to his honour" Liebig "did not embarrass the negotiations by exorbitant demands, but only stipulated that he should not have the conduct of as large a laboratory as at Giessen."[7] In fact, rather than a laboratory – though he got both – Liebig demanded and got a brand new lecture room, raked like a theatre and seating 300 people. Its focal point was a huge demonstration bench placed under a chimney mantel that ran the full width of the lecture hall. When any particularly dangerous gases were prepared, the experiments were performed under a large eight-sided glass bell that had direct access to the chimney through a downpipe. This auditorium was to be the model of all German university lecture theatres built from the 1860s onwards.[8] Liebig's demands suited King Maximilian II, who wanted Liebig as a scientific advisor who would be willing to propagate his science to the Bavarian people. Also, as befitting his grand status, Liebig was to be made president of the Bavarian Academy of Arts and Sciences, a body that had had a desultory history since its foundation in 1759 but that the king wished to put on a par with the learned academies in Vienna, Berlin, Paris, and London.

There had been no university at Munich before 1826 when King Ludwig moved the former Jesuit university of Ingolstadt (founded 1472) from the quarters it had occupied at Landshut since 1800, with the intention of expanding it into a great centre of freedom of learning and teaching such

5. *Ibid.*, p. 136.
6. "What wouldn't I give to shake off the mind-deadening examinations of the medical and pharmacy students here in Munich, but in Berlin the examinations are annihilating." Liebig to Hofmann, 14 November 1863, in Brock, *Liebig und Hofmann in ihren Briefen*, p. 197.
7. M. von Pettenkofer, "Liebig's scientific achievements," *Contemporary Review* 29(April 1877), 865–87 at p. 879, translated from his *Rede und Gedächtnis Justus von Liebig* (Braunschweig, 1876).
8. See W. Prandtl, "Das chemische Laboratorium der Bayerischen Akademie der Wissenschaften in München," *Chymia*, 2(1949), 81–97; and *Das chemische Laboratorium der königlichen Akademie der Wissenschaften in München*, ed. A. von Voit (Vieweg, 1859), with atlas. Voit was also commissioned to design laboratories at Erlangen.

as Wilhelm von Humboldt had established for Prussia at Berlin (f. 1809) and Bonn (f. 1818).[9] On the other hand, the Bavarian Academy had built its own small chemical laboratory in 1815 in the old botanical garden fronting on Arcisstrasse. Its "curator," a pharmacist named H. A. Vogel, was duly made a professor of chemistry and pharmacy at the new university in 1826. A similar fusion of academy and university had occurred in Berlin. When Vogel died in 1852, to be replaced by Liebig, it was obvious that the laboratory and lecturing facilities at Munich were grossly inferior to those Liebig had enjoyed at Giessen. Accordingly, with the help of the architect, Voit (the father of the physiologist Carl Voit), Liebig designed a new building. With ingenuity the former academy laboratory next door was transformed into a house for the Liebigs that was furnished to a high standard for domestic comfort and for entertaining the intelligentsia of Munich. The cupola skylight, which had been the former laboratory's only form of lighting, now became an impressive domestic feature:

> My wife and I have been here six weeks and in our house for the last ten days; for two days we had to live with difficulty in one room because of workmen. Our home is large, roomy and quite practical. There are nine rooms and in the middle a round salon measuring thirty feet from one end to the other and prettily lit from above by a domed skylight. . . . The new laboratory is still being created, but the lecture theatre will be ready today, and I intend to give my first lecture there tomorrow morning. I am quite anxious since the people here hold such false views of themselves and of my future operations. Because I have, for Munich's circumstances, an uncommonly high salary, the expectations made of me are correspondingly high. It will give me much joy to deliver my first lecture since it deals with the role that the study of chemistry has in the sciences.[10]

This was the first of a series of fortnightly public lectures similar to those conducted by Faraday and Tyndall at the Royal Institution in London.

Adjacent to the Liebigs' new home, and in the former Institute's garden, Voit designed the huge lecture theatre, 52 feet square. To one side of this he placed a laboratory with six or seven work places – its small size reflecting Liebig's nonengagement with teaching. The building housed a number of special rooms for gas analysis, quantitative analysis, and

9. Karl Alexander von Müller, *Die wissenschaftlichen Anstalten der Ludwig-Maximilians-Universität zu München* (Munich, 1926).

10. To Hofmann, 16 November 1852, in Brock, *Liebig und Hofmann in ihren Briefen*, p. 148. This inaugural lecture was "Ueber das Studium der Naturwissenschaften," *Reden* (1874), pp. 156–71.

distillation. When the American agriculturist Evan Pugh (1828–64) visited Liebig in Munich in August 1854 (he found Liebig in bed with so-called cholera), he noted:

> It would exhaust too much . . . space to give the minutiae of the lecture rooms. It is enough to say that *everything was complete,* convenient, and of the very best kind. . . . The students' laboratory is small, only adapted to admitting seven students. He has almost entirely withdrawn from the drudgery of giving instruction in analytical chemistry. He informs a student . . . that he does not obligate himself in any manner whatever to give the least attention to them [sic] or to interest himself in their investigations.[11]

In keeping with Munich's image as the new Athens, the gabled end of the lecture theatre was eventually dressed by four large allegorical murals by Löffler and Thiersch depicting female images of Chemistry and Agriculture that symbolized Liebig's work. These murals were executed in stereochromie, a silicate fresco technique to which both Liebig and Pettenkofer contributed and which greatly interested both King Maximilian and Prince Albert in England.[12] The buildings were totally destroyed by Allied bombs in 1945.

Although not explicitly mentioned by Liebig, there were undoubtedly political reasons for his departure from his homeland. His great patron Grand Duke Ludwig I had abdicated in 1830 to be succeeded by his son, who had continued to treat the university generously amidst growing political troubles that were caused by his imperviousness to change. When he died in 1848, however, his successor Ludwig III proved extremely reactionary to the events of that year. The possibility of a political alliance with Prussia, which did not come about in practice until 1866, would not have pleased Liebig. Despite Liebig's notionally Lutheran background, he had plenty of Catholic friends at Giessen, where there was a Catholic theology division of the university. In any case, his wife was a practising Catholic. Hence, a call to Catholic Bavaria by a king noted for his liberalism and love of the arts and sciences must have seemed a welcome oasis from possible future political disturbances in northern and central Germany. An attempt by Heinrich Ferber, the mayor of Giessen, to persuade the Hessen government to retain Liebig is a mark of how seriously the

11. Quoted by C. A. Browne, "European laboratory experiences of an early American agricultural chemist, Dr Evan Pugh (1828–64), *J. Chem. Educ.* 7(1930), 502.
12. See P. Barlow, *Proceedings Royal Institution* 1(1851–52), 424–26; Johann N. Fuchs, "Manufacture, properties and application of water glass, including a process of stereochromic painting," translated by order of Prince Albert. *J. Royal Society Arts* 7(1858–59), 521–6, 532–8.

townspeople viewed the likely effects on student numbers if Liebig departed.[13] Nor was Liebig left in any doubt that "the [Bavarian] king wishes to influence the agriculturists through me."[14]

During the Napoleonic Wars, Bavaria had cocked a snoot at its traditional Austrian enemy by siding with Napoleon. As a reward for its loyalty, Napoleon raised the Bavarian Elector Max-Joseph to the crown in 1806. As King Maximilian I, he began the process of turning Munich into a great capital city of the arts and sciences, a trend that was continued by his son Ludwig, who from 1825 rebuilt the city in Greek style. Ludwig, a somewhat unconventional ruler, was forced to abdicate following the 1848 revolution in favour of his son Maximilian II, who immediately showed the same benevolence towards the sciences that his father and grandfather had shown towards the arts.[15] In particular, he was keen to transform the moribund University of Munich into a centre of intellectual research and discovery, to modernize his country's agricultural system, and to provide industrial employment for people displaced from the land. To this end, to the dislike of ultramontane politicians, several "foreigners" like Liebig were called to Munich, where they formed their own distinctive, usually Protestant community. Their links with the royal court were close. Indeed, being of an intellectual cast of mind (it was said he would have preferred to have been born a professor rather than a king), Maximilian formed a dining circle modelled on that of the Weimar Court of Friedrich II, which met for weekly symposia in the Residenz Schloss whenever the king was not living outside the city in the mountain countryside he so loved.

Liebig's initial response to Munich was not entirely favorable:

> My lectures have been given in the new lecture hall which has all of its 280 places taken [i.e., subscribed], and it is beautiful, well-nigh luxurious. The acoustics are excellent. It seems after all that everyone is pleased with what I am offering. The king wishes me well personally and wants to make Munich into a great city. I doubt in the performance, for too much dung has accumulated in the Augean stables, and he seems no Hercules![16]

13. *Hessische Biographien,* Bd. I (Darmstadt, 1918; reprint Wiesbaden, 1973), p. 7. Liebig was given a large oil painting, "View of Giessen" (by Reiffenstein), as a farewell present. This is now displayed at the Liebig Museum in Giessen.
14. Liebig to Wöhler, 11 April and 25 June 1852, in A. W. Hofmann, ed., *Aus J. Liebig's und F. Wöhler's Briefwechsel,* 2 vols. (Braunschweig, 1888), vol. 1, pp. 379, 382.
15. F. Girard, *The Romance of Ludwig II of Bavaria* (London, 1899); and Benno Hubensteiner, *Bayerische Geschichte* (München, 1977).
16. Liebig to Hofmann, 3 December 1852, in Brock, *Liebig und Hofmann in ihren Briefen,* p. 149.

He was also a little upset by the fact that the government controlled doctors' appointments, so that Georg Liebig, who had qualified in Berlin, could not immediately practise in Munich. In the end, Georg decided to serve in British India for several years. In 1858, with his father's very considerable influence, he returned to Germany to practise as a spa and county court physician at Bad Reichenhall, to the south of Munich.

Liebig's life in Munich was therefore very different from the one he had led in Giessen for twenty-eight years. There, constant laboratory work and literary activity had taken their toll. His American pupil Eben Horsford remarked how young he still looked when he arrived at Giessen in 1844, only to note two years later how Liebig had aged. In Munich, life took on a new gaiety. Three of Liebig's children, Georg, Agnes, and Hermann, were already in their twenties, and although Agnes's health was a continual worry to the Liebigs, only the two younger girls, Nanny and Maria, were still being educated. With no laboratory classes to run, competent assistants to demonstrate at his lectures, and plenty of time to write during the day, Liebig's evenings could now be filled with theatre- and concert-going, dancing, conversation, and whist-playing.

His home became the centre for the north German expatriates whom Maximilian had called to Munich. We have a fine pen portrait of this sociable Liebig in the memoirs of Luise von Kobell, the daughter of the professor of mineralogy Franz von Kobell (1803–82). Following public lectures at the university, Frau Liebig would invite several members of the audience to tea afterwards, where

> Liebig supplemented his address by answering the questions and objections of his guests. He was master not only of his science, but he possessed also the knack of being able to express his thoughts clearly, so that with him as leader one cheerfully followed him into the wonderland of nature's laws and had confidence in him. Sometimes a learned discussion developed with Jolly, Pettenkofer, Bischoff, my father, etc., which was attractive to follow. Wanting to bring the scientific talk to an end, he would make a literary observation in the middle of it, which gave an opportunity for the discussion to move off in a different direction. Once he declared in the middle of an argument on soil conditions that the subject was boring the ladies, and Carrière wanted to have a lecture in the Appolosaal.[17]

She went on to describe attending whist parties held by the Liebigs three or four times a week. At Liebig's frequent dinner parties and dances, there were usually foreign guests and only the best wines were drunk. She

17. L. von Kobell, *Unter den vier ersten königen Bayerns nach Briefen und eigenen Erinnerungen*, 2 vols. (Munich, 1894), vol. 2, p. 28.

recalls that his sociability was so great that she was once wagered by Baron Völdernhoff that he could infiltrate a dinner party uninvited, which he succeeded in doing. Such high living must have been expensive to maintain, even on Liebig's generous salary and literary income. It goes a long way towards explaining why he became interested in making money commercially after he moved to Munich.

The Bacon Affair

Under the king's express command, Liebig returned to the agricultural matters that he had largely neglected since the debacle of his fertilizer in 1846. He thereby decided to take up the challenges to his mineral theory thrown down by, in particular, the English agriculturists Pusey and Lawes. It was while reflecting on their attacks and writings that he recalled the observations that he had once made to Faraday in 1844: "England is not the land of science." This generalization seemed as true in the 1860s. Why were the British, whom he loved, unique, as he believed, in this respect? Why did men like Pusey, Thompson, Acland, and Lawes all put the practical consequences of agriculture over and above scientific truth? The answer, he decided, lay in their peculiarly long-lived subscription to the philosophy of Francis Bacon.

In his early writings, indeed right up until the 1860s, Liebig had been happy to exploit utilitarian arguments for the value of experimental science against the supposed evils of speculative *Naturphilosophie,* or for the material benefits it would bring to governments that aided its institutionalisation. In 1840 Gregory could think of no greater compliment for Liebig's *Animal Chemistry* than to label it "your truly Baconian treatise."[18] And as late as 1856 Liebig had had no compunction about quoting Bacon *against* Lawes and Gilbert at the end of his lengthy Royal Agriculture Society rebuttal of their views. Hence, his accusations in 1863 that Bacon had been a liar, plagiarist, court sycophant, and naive methodologist represented a complete about-face.

Liebig had probably first come across a discussion of Bacon's method when reading John Stuart Mill's *System of Logic* (1843), a work that Liebig naturally admired because of the praise Mill had showered on Liebig's method of deduction in *Animal Chemistry:*

> The recent speculation of Liebig in organic chemistry shows some of the most remarkable examples since Newton of the explanation of

18. Gregory to Liebig, 25 September 1840, Liebigiana, Staatsbibliothek, München; see Brock and Stark, *Ambix* 37(1990), 136.

> laws of causation subsisting among complex phenomena, by resolving them into simple and more general laws.[19]

Liebig lost no time in persuading Vieweg to issue a German translation of Mill, from which he would have gathered details of Bacon's method of inductive enumeration.[20] Liebig was also much taken by the writings of the English historian Baron Macaulay (for example, he quoted Macaulay in the dedication of the third English edition of his *Chemical Letters*). Macaulay's *History of England* contained a section on the life and times of Francis Bacon. Macaulay had also written a long essay on Bacon in the *Edinburgh Review* in 1837 that condemned Bacon's personal character whilst extolling him as a prose writer.[21] Consequently, it slowly dawned on Liebig that not only was the English obsession with utility attributable to Bacon but the prevailing view of science among the educated landowning classes was that the investigation of the laws of nature had to proceed inductively. This, he concluded, must be why his own theoretical approach to agricultural questions was not understood by his chief English critics Pusey, Gilbert, and Lawes. (The fact that Gilbert was a trained chemist was always ignored by Liebig.)

Convinced that he had found the key to English mentality, he seems to have acquired the collected edition of Bacon's works edited by James Spedding and Robert Ellis between 1857 and 1859, and read through the entire seven volumes, including Latin essays, during the space of a few months in the spring of 1862. Having done this, he was the more puzzled why Bacon's name "still shines as a guiding star which, it is asserted, has shown us the right road and true aim of science."[22] Compared with his contemporaries Kepler, Galileo, Stevin, Gilbert, Harriot, and Harvey, Bacon had made no significant contribution to science; moreover, as Liebig showed, he had ignored, appeared indifferent to, or even pillaged many of

19. J. S. Mill, *System of Logic* (1843), pp. i, 562. *System der deductiven und inductiven Logik*, 2 vols. (Braunschweig, 1849). Whewell was surprised that Mill was so confident of the validity of Liebig's speculations.
20. M. and W. Schneider, eds., *Justus von Liebig Briefe an Vieweg* (Braunschweig, 1986), p. 216.
21. Macaulay was reviewing the sixteen volumes of Bacon's *Works* edited by Basil Montagu between 1825 and 1834. See *Edinburgh Review* July 1837, 1–104, and *Macaulay's Collected Essays*, vol. 3. Following both Macaulay's and Liebig's deaths, Macaulay's review was answered at extraordinary length by the Bacon scholar James Spedding in *Evenings with a Reviewer, or Macaulay and Bacon*, 2 vols. (London, 1881).
22. J. Liebig, "Lord Bacon as natural philosopher," *Macmillan's Magazine* 8 (July 1863), 237–49; (August 1863), 257–67. For a convenient analysis of modern views of Bacon, see Brian Vickers, ed., *English Science, Bacon to Newton* (Cambridge, 1987); and R. Yeo, "An idol of the market place: Baconianism in nineteenth-century Britain," *History of Science* 23(1985), 251–98.

the findings and views of these same contemporaries and denied the Copernican heliocentric astronomy.

By subjecting one of Bacon's factual works, the *Historia Naturalis,* or *Sylvia Sylvarum,* to detailed analysis, Liebig was able to claim that it was a ragbag of anecdotal information that had been subjected to no sort of critical or experimental investigation. When Bacon had proposed experiments – his "fructiferous" and "luminiferous" examinations of instances – they seemed to bear no resemblance to the controlled and designed experiments that would occur to a nineteenth-century investigator. By citing Bacon's recipe for making gold, Liebig believed he epitomized the whole man in his ignorance:

> All the means that he indicates to make gold are erroneous and deceptive and the axioms of which his theory is made mere fantastic conceits.[23]

Far from leading readers to the promised land, Bacon led them into a barren wasteland, the journey often proceeding by "unblushing plagiarism" from other writers, "vain self-praise and detraction of others' merits." All of Bacon's work was external, and nowhere did Liebig detect the inner joy and love of science that permeated the work of Kepler, Galileo, and Newton:

> Compared to those men, Bacon shows like a quack doctor, who, standing before his booth, tries to make his rivals appear as ignorant as possible; who vaunts his wondrous cures, and praises the remedies with which he promises to raise the dead and banish illness from the world; and, finally, hints that such services to humanity are not unworthy of recompense.[24]

Bacon's world was not as God had created it but was one "of illusion and deceit." This showed up most clearly in "the new instrument" of the inductive method that he had commended to natural philosophers as the way of attaining their goal while avoiding the speculations and fantasies of the ancients.

Liebig then gave an analysis of Bacon's only example of the inductive procedure, in the *Novum Organon,* of how the philosopher should investigate the phenomenon of heat. Bacon's tables of affirmative and negative instances of heat were, suggested Liebig, clearly those of a scribe who had simply extracted from books "every passage where the words warmth, warm, hot, heating, burns, and cold, cooking were met with."[25] How else to explain why sulphuric acid, which "burns" holes in clothing, was

23. Liebig, "Lord Bacon," p. 242.
24. *Ibid.*, p. 244.
25. *Ibid.*, p. 245.

tabulated below the wool and feathers that warm and the dung that smokes? Following this indexing, Bacon then supposed the philosopher would cast aside instances that were insufficiently palpable, satisfactory, or convincing, failing to notice that this could be achieved only on the basis of a preformed view of the phenomenon.

Bacon then sifted the instances and made further exclusions on the basis of "understanding," which seemed to Liebig quite arbitary, or because it would not fit the conclusion that was in aim, namely, that heat was motion. For Liebig, Bacon's method was intelligible only when it was remembered that Bacon had been a lawyer and that the "instances" were the equivalent of witnesses in a civil or criminal trial. Liebig's conclusion was sharp and devastating:

> Bacon promises to show us a road that shall lead to a solution of the highest questions on the nature of things; and, when we accompany him, he leads us round and round in a labyrinth, and is himself unable to get out. His inductive method leaves him perfectly helpless in determining the simplest conceptions; and at the end of a diffuse investigation we learn as much as we knew at the beginning. . . . In all his explanations it is invariably Bacon that is the speaker; he never allows the things themselves to say a word. To be their interpreter it must be necessary for him to understand their language; and this is just the thing of which he is ignorant.[26]

Fruitful scientific conceptions were usually savagely opposed by contemporaries; but Bacon's essays were never opposed, being "in unison with the popular views of the ignorant crowd." Bacon's philosophy required no hard preparation and fell in with the antiauthoritarian attitude and spirit of inquiry of his age. Moreover, in James I he had possessed a patron "who was vain in his learning, boastful of his knowledge and insatiable in his greed of praise."[27] To such a philosopher king the idea that the aim of intelligence was utility could not but appeal. Liebig quoted the two famous aphorisms from the *Novum Organon* that "The true and legitimate aim of science is no other than to enrich our lives by new *inventions* and discoveries," and "Our true office is to lay the foundation of man's *power* over nature, and to enlarge the boundaries of his *dominion.*" And he noted drily that the real aim of science, *truth*, was nowhere mentioned:

> The aim of science is neither invention, nor utility, nor power, nor dominion. Invention is the object of Art [technology], while that of

26. *Ibid.*, p. 249.
27. *Ibid.*, p. 259.

> Science is "to recognise the cause." The former finds, or finds out (invents) facts; the second explains them.[28]

So, whereas in the 1830s and 1840s Liebig had found it expedient to stress the material and spiritual utilitarian benefits of chemistry and to quote Bacon approvingly, he now urged that utility was not the main goal.

As Liebig had seen and experienced through bitter controversies over the role of chemistry in agriculture and physiology, and as his pupil Hofmann experienced directly in the financial management of the RCC in London, "failure to recognize . . . that the practical usefuless of science was neither direct, immediate, nor specifically predictable . . . could only harm the course of science because it led to expectations that could not be fulfilled and [therefore] to the questioning of science's ultimate utility."[29] In this switch of viewpoints, Liebig surely proved himself to be as "many-faced" as Francis Bacon had been to seventeenth-century natural philosophers. Liebig's British pupils absorbed both messages; the first wave of "civic scientists" used the rhetoric of utility to justify their expectations of patronage; the later institutionalized "academics" deployed the rhetoric of pure science as a necessary prior step before it bore fruit in application.[30]

What, then, was the appropriate method of science? Rather than the enumeration of instances, said Liebig, it is through the examination of a single instance; if this instance is explained satisfactorily, then by analogy all similar cases are expounded. Moreover,

> in science all investigation is deductive, or *a priori*. The experiment is but the aid to the process of thought, as an arithmetical operation is; and the thought, the idea, must always precede it – necessarily precede it – in every case where a result of importance is looked at.[31]

Bacon's approach was therefore wrong in principle because there was no such thing as a strictly empirical method of investigation. "An experiment not preceded by a theory – that is, by an idea – stands in the same relation to physical investigation as a child's rattle to music."[32]

Unfortunately, the influence of Bacon's methodological claims had struck a chord in the English mind: the landowning class that continued "to hold a sort of patronizing intercourse with science," and the practical man who, knowing little or nothing of the real nature of science, con-

28. *Ibid.*, p. 261.
29. Otto Sonntag, "Liebig and Francis Bacon and the utility of science," *Annals of Science* 31(1974), 373–86 at p. 382.
30. R. H. Kargon, *Science in Victorian Manchester* (Manchester 1977).
31. J. Liebig, "Lord Bacon," p. 263.
32. *Ibid.*, p. 263.

nected "scientific principles, with everything that was useless and impracticable." To illustrate his point, Leibig compared, in a long, double-columned footnote, one of Bacon's deductions with a quotation from Lawes and Gilbert's paper that had counterattacked him in 1855. Unable to get back at them through the pages of the *Journal of the Royal Agricultural Society,* he slandered them anew by labelling them Baconians and utilitarians who were unable to understand that power and influence were based upon mental culture.[33]

In offering an English translation of Liebig's views on Bacon, the editor of *Macmillan's Magazine,* David Masson, had referred to their "unpalatable" nature. He even consulted William Whewell on whether there should be a rebuttal.[34] But apart from a severe reaction in Germany from Christopher Sigwart (1830–1904), a professor of philosophy at Tübingen, who leaped to Bacon's defence as a logician, there was no English stir until the French translation of Liebig's book on Bacon appeared in 1866.[35] This translation was reviewed anonymously in *Fraser's Magazine* under the rhetorical title "Was Lord Bacon an Impostor?"[36] The scholarly and knowledgeable author may well have been *Fraser's* editor, the historian of Elizabethan England, Anthony Froude. Internal evidence suggests that the review was not by Spedding, who was beginning his twenty-year-long effort to reply to Macaulay.[37]

We need not examine the anonymous *Fraser's* reply in detail. Suffice to say that it was as much concerned with the Russian translator's ardent support of Liebig's dethronement of Bacon as with Liebig's text, which was condemned as being by a shoemaker venturing beyond his last. What Liebig and his translator Tchihatchef had failed to grasp was Bacon's role as the reformer of natural philosophy – a positive view that twentieth-century historians of science are happy to accept. In emphasizing the historical significance of Bacon, the reviewer failed to grasp the point at which Liebig was aiming – the continued use of Bacon's name to justify a particular and erroneous way of doing science. On the other hand, he easily saw that Liebig's real grievance was the rejection of his own mineral

33. *Ibid.*, p. 264.
34. Yeo, "Idol," p. 279.
35. J. Liebig, *Ueber Francis Bacon von Verulam und die Methode der Naturforschung* (München, 1863), was translated into French by the Russian traveller Pierre de Tchihatchef as *Lord Bacon* (Paris, 1866). Surprisingly, there were second French editions in 1877 and a third in 1894. Tchihatchef dedicated the translation to the geologist Sir Robert Murchison.
36. *Fraser's Magazine* (December 1866), 718–40.
37. See Spedding, *Evenings with a Reviewer.* The *Wellseley Index of Victorian Periodicals* is unable to identify the author.

theory by English agriculturists. This grievance was summarily rejected, the reviewer plainly siding with Lawes and Gilbert and even furnishing an obscure agricultural work of 1830 as having priority over Liebig's theory.[38] Liebig confessed to not having heard of the book or of its author.

The *Fraser's* attack was sufficiently provoking for Liebig to reply in April 1867. But Liebig was placed at a disadvantage, for his unknown critic was permitted to comment on his renewed claims in footnotes – the sort of treatment that Liebig himself had long handed out as editor of *Annalen.*[39] In this retort Liebig was explicit that what had troubled him before was that

> there existed in England a class of literary men perfectly unknown in France and Germany; a class formed by gentlemen of education, who, on the whole, are not addicted to professional studies, and who are generally devoid of professional knowledge either in philosophy or natural science, but who take part in, and frequently lead public opinion on, the most difficult and complicated scientific questions, as, for instance, on the value of Darwin's theory; on the cause of the fertility of the soil; on the Atlantic cable, on utilization of sewage, &c. Frequently these gentlemen are members of Parliament, and also have often a certain influence on the press; and thus, as in an instance I could mention [sewage?], do their best to prevent such views as clash with their preconceived views from being made known.[40]

This gentlemen clerisy, according to Liebig, had its own logic; to them an experiment meant something entirely different from the meaning it had for a scientist. Because he had identified the cause (the "demon") as Bacon, it did not follow that Liebig was hostile towards the English, as his hostile reviewer had hinted. In a remarkable and revealing passage, he reviewed his love of British culture:

> It is true, I was many years ago full of bitterness, and I may say full of hatred, against the above-mentioned class of scientific dilettanti; but I confess my wrong, for I love England; I love the land and the English. During my travels in England and in Scotland I learned much, and spent with my friends there the happiest days of my life. I dare say, there is no man on the Continent who admires and esteems Englishmen more than I do. My friends in Germany and France even reproach me, that in this respect I go rather too far; and the fact is, that my criticisms against England, attentively examined, will be found to be those of over-regard; *such reproof as a passionate but disappointed*

38. W. Grisenthwaite, *A New Theory of Agriculture* (London and Nottingham, 1830). A rare work, there are copies in the British Library.
39. J. Liebig, "Was Lord Bacon an impostor?" *Fraser's Magazine* 76(April 1867), 482–95.
40. *Ibid.*, p. 483.

> *lover addresses to his mistress.* What I might quote in favour of this statement is, doubtless, of no great weight, and I am almost ashamed to be obliged to allude to it; but at all events, slight proofs founded on facts are better than strong protestations in words. I am a friend of English literature, and I read almost more English than German works.[41]

He went on to note how he had arranged for his German publishers to translate Darwin's *Journal of Researches,* as well as the works of Mill and Buckle, of how he had tried to advise London's citizens on the disposal of their sewage, and of how imperceptibly England had changed for the better in the last thirty years: Sewage was now sewage and not euphemistically nightsoil; his views on agriculture were being treated more evenhandedly by the Royal Agricultural Society. Perhaps it would be the same with Bacon, he speculated.

In retrospect, the row over Liebig's demonstration that Bacon had had no influence on the reform and progress of science or its methods of inquiry seems slightly ridiculous, as does the effort Liebig put into establishing his view in Germany, France, and Britain. The long-term effect of these Baconian studies, however, was to interest him further in the history and philosophy of science. What were the sources of ideas that led to inventions and the progress of science if they were not produced by Bacon's method? Such reflections led to two further lectures to the Munich Academy in 1865 and 1866 – "Induction and Deduction" and "The Development of Ideas in Science."[42] Liebig claimed that Masson refused to publish the former in *Macmillan's Magazine* because of the offence he had given over Bacon; instead, it was published by the courageous and "more independent editor" of the *Cornhill Magazine,* George Smith, who later founded the *Dictionary of National Biography.* Liebig's *Fraser's* critic unkindly suggested that it was turned down by Masson because it contained nothing new and especially nothing that Bacon had not adumbrated in the *Novum Organon.*[43]

In fact, the interesting point about this essay was the emphasis Liebig laid upon the role of imagination in science, to which he attributed the real meaning of induction. "The reasoning of the chemist or natural philosopher can best be compared to the peculiar faculty of the composer, who

41. *Ibid.,* p. 484.
42. "Induction and deduction," read 28 March 1865, in *Reden,* pp. 296–309. English translations (by different hands) *Cornhill Magazine* 12 (September 1865), 296–305, and *Annual Reports Smithsonian Institution* 1870, pp. 258–67; "Die Entwickelung der Ideen in der Naturwissenschaft," read 25 July 1866, in *Reden* (1874), pp. 310–29, and offprinted as a pamphlet; English translation by Playfair, Edinburgh (1867).
43. *Fraser's Magazine* 76 (1867), 495, editorial footnote.

thinks in sounds."[44] When examined logically, the explanation of phenomena seemed to rest upon facts all linked together like a chain; however, many of these joints and linkages depended upon phenomena that did not occur naturally in nature but that had to be teased from nature experimentally or imagined inductively. If imagined, further experiments had to be devised to support the plausibility of the linkage in question.

By way of example, Liebig discussed the work of the Swiss chemist Friedrich Schönbein, with whom he had struck up a friendly correspondence since meeeting him for the first time in Munich in 1853.[45] (Schönbein had been at Erlangen at the same time as Liebig, but they had never met until 1853.) In 1840 Schönbein found that when an electric discharge was passed through air, its oxygen became an even more powerful oxidizing agent. It also possessed a peculiar "phosphoric smell," which was why he gave the new form of oxygen the name *ozone* (Greek, to smell). In 1844 he had found that if phosphorus was slowly burned in air, the same smell was produced. From this Schönbein had made the induction, or (as we would now say) he had imagined or formulated the hypothesis that burning phosphorus also electrolysed oxygen. It was then a matter of experimental test to verify the existence of ozone in the air surrounding burning phosphorus. Liebig's point was that *smell,* not understanding, had led to the joining together of two disparate pieces of information. Most inventions and technical processes had originated in this way, including photography, many of whose techniques were still not understood – that is, they were not deductively explicable from first principles.

Many people, Liebig continued, attributed discoveries to chance, as with Daguerre's discovery that mercury fixed images on plates covered with silver iodide, or Talbot's discovery that gallic acid would do the same for impregnated paper; but to Liebig these were inductions arrived at by imaginative reasoning, followed by systematic experiment. Such operations of the mind could not be examined logically; no syllogisms were involved, and often the starting ideas could prove quite wrong, the "correct" ones emerging only as a consequence of the investigation. "Working does everything, and . . . whatever theory stimulates working leads to discoveries."[46]

"Reason and fantasy," he concluded,

> are equally necessary for science; to each of them belongs a certain defined portion of all problems occurring in natural philosophy and

44. *Cornhill Magazine* 12 (1865), 299.
45. G. W. A. Kahlbaum and E. Thon, eds., *Justus von Liebig und Christian Friedrich Schönbein Briefwechsel 1853–1868* (Leipzig, 1900).
46. *Cornhill Magazine* 12 (1865), p. 305.

> chemistry, in medicine and political economy, history and philology, and each occupies a certain space in these respective domains. The portion over which fancy presides is wider and more extensive in the very ratio that the positive knowledge encompassed by the understanding is undefined and vague. What characterizes progress is, that, with the increase of knowledge, those ideas vanish, which had their origin in the imagination; and whereas, during the first period of science, fantasy has full sway, it afterwards subordinates itself to the understanding, and becomes its useful and willing servant.[47]

This servitude was chiefly accomplished through measurement and the quantification and mathematization of phenomena. "Imagination compares and discriminates, but does not measure; for to measure one must have a standard to measure by, and this is a product of the understanding." By shaping what was art (industry) into science and technology through the development of rules of understanding that could be acquired and taught, "What had first belonged only to an individual becomes the common property of all."[48]

This essay was an excellent distillation of his unacknowledged reading of Whewell and Mill, together with reflections upon his own forty years' experience as a chemist. As he said to Wöhler at the time, "I find from the history of science that to understand facts, one must already have some definite ideas in one's head and that one does not see what was not already thought of in spirit."[49] Although today we would add a social dimension to Liebig's portrayal of scientific method, his discussion and especially his stress upon the role of imagination in science (the latter was to be echoed by Tyndall, as well as by the chemists Kekulé and van't Hoff later in the century) seem very modern. In fact, Liebig's perspective was shared by more of his contemporaries than he realized.[50]

Not surprisingly, Liebig followed up his anti-Baconian stance with an essay on the history of science. On one level this was merely a further demonstration that Bacon had not developed any new method of scientific investigation; on the contrary, science had developed continuously from the time of the Greeks through reflection upon practical experiences. In a startlingly modern way, Liebig denied that any fact or occurrence was simply waiting to be gathered by an investigator; rather, facts themselves were mental constructs, or, as we would say now, all observation is

47. *Ibid.*, p. 305.
48. *Ibid.*, p. 305.
49. Hofmann, *Liebig's und Wöhler's Briefwechsel*, vol. 2, p. 193.
50. See Jonathan Smith, *Fact and Feeling: Baconianism, Science and the Nineteenth-Century Literary Imagination* (Madison, Wis., 1994). Professor D. M. Knight kindly drew my attention to this book.

theory-laden. For Liebig, the development of such factual ideas was possible only because of the prior development of mechanical inventions and devices. The further development of astronomy, for example, had been dependent upon the telescope, which itself was dependent upon the prior development of colourless glass, and whose further development depended upon the manufacture of flint glass and achromatic lenses. Similarly, the advance of chemical analysis had been beholden to metallurgical assay, mineral chemistry had been dependent upon the art of the pharmacist or of the chemical tradesman, organic chemistry had emerged from medicine, and contemporary ideas of heat and light, he claimed, had been forged through the invention of the steam engine and photography.

Although Liebig's claim that scientific ideas were dependent upon prior technical developments and inventions was tendentious, it was an interesting thesis. The tone of the lecture, however, was a good deal more radical than this outline suggests, for in commenting on the replacement of Greek slaves by machinery, Liebig drew the essentially political conclusion that "freedom, that is the loosening of all fetters which hinder mankind's progress to allow man to fully exploit the powers that God has endowed him with, is the foundation and most important of all the conditions for the advancement of mankind in civilization and mental culture."[51] It followed that if the advance of mankind was solely dependent on the conceptualization of the particular experiences and developments that mankind achieved, anything else, including philosophical systems, governments, and religion, was of only incidental significance. Science, not morality or religion, was the determinant of human evolution.

Liebig spelled out the real meaning of his lecture to Wöhler (29 July 1866):

> Even if the powers of governments and of the churches had been in union with science, it would not have led science to make a single further advance or caused any discoveries to happen earlier. As adversaries, they have not in the least hindered its progress. That's my meaning. Without the discoveries of scientists, Luther would have been burned at the stake like Huss; with the discovery of the shape of the earth [sic] the "heaven" of the church collapsed, as did "hell" with the clarification of the nature of fire, and a belief in witches and spells fell with atmospheric pressure, and Nature's "will" and "abhorrence" became like an empty dream.[52]

51. J. Liebig, "Die Entwickelung," *Reden* (1874), p. 323; *Development of Science* (Edinburgh, 1867), p. 23.

52. Hofmann, *Liebig's und Wöhler's Briefwechsel,* vol. 2, p. 217; compare Liebig to Schönbein 1 August 1866, in G.W.A. Kahlbaum and E. Thon, eds., *Justus von Liebig und Christian Friedrich Schönbein Briefwechsel* (Leipzig, 1900), p. 221.

Wöhler's comment in reply, that the lecture would stir up a hornet's nest, appears to have been misplaced; but the lecture did seem to confirm Bavarian rumours that Liebig was a rank materialist and therefore to play into the hands of the ultramontane party, which believed that all foreigners surrounding the king were undermining religious orthodoxy.[53] One of the best anecdotes told of Liebig concerns the Catholic bishop of Mainz, who had an audience with King Ludwig of Hessen-Darmstadt after Liebig had gone to Munich. The zealous bishop, urging the king to sweep materialist thought from his kingdom, misunderstood the king's reference to the fact that he had known Liebig's father, a materialist (i.e., hardware merchant); he nodded his head vehemently, saying that his son was one too.[54] But was Liebig a philosophical materialist?

Historians such as Pat Munday have certainly thought so; others, such as Frederick Gregory, have denied it. Gregory points out that materialism as a philosophical system is the atheistic belief that everything in the world, including human beings, are material entities, that there is no separation or distinction between the human mind and body, and that categorically there is no immaterial transcendental being.[55] Such anti-idealism was not Liebig's philosophical position. The common theme of Liebig's scattered and limited references to religion and on the relationship between science and religion show that a teleological outlook ran through his work on agricultural and animal chemistry.[56] Liebig's essentially liberal Protestant outlook probably helps to explain his *rapport* with the English mind and its enthusiasm for natural theology.

There is no doubt at all that as a student and young chemist Liebig thought of himself as a materialist. He told the poet Platen in 1823 that it would be difficult to abandon materialism now that he was tied to it.[57] And the tenor of his work with Wöhler on uric acid derivatives and with organic chemistry generally was reductionist and materialistic in the sense that organic matter was claimed to obey the same laws of physics and chemistry as mineral substances. However, greater familiarity with physiological chemistry in the 1840s must have given him pause, for like William Prout before him, Liebig used the concept of vital force to explain the development and direction of chemical actions in metabolism. Moreover, as we have seen, his commitment to explaining putrefaction by molecular motions also forced him to appeal to a vital force that promoted the stability of animal and vegetable systems.

53. Liebig to Wöhler 15 June 1859.
54. Volhard, ii, 351.
55. F. Gregory, *Scientific Materialism in Nineteenth-Century Germany* (Dordrecht, 1977).
56. O. Sonntag, "Religion and science in the thought of Liebig," *Ambix* 24(1977), 159–69.
57. Volhard, i, 37.

Liebig's apparent inconsistency is easily understood when it is realized that he made a distinction between organic chemistry and organized chemistry. Whereas organic chemicals outside a living vegetable or animal environment obeyed the same rules as their inorganic counterparts, as did molecules undergoing fermentation or putrefaction, inside living systems they were under the control of a vital force.[58] This distinction accounts for a good deal of the apparent discrepancies between the three sections of *Animal Chemistry*. On the other hand, there is no doubt that Liebig was inconsistent and confusing whenever he discussed physiological chemistry. Why, for example, did he not argue that physiological transformations were purely chemical? In one confusing sense he did, for all forces, whatever their kind – mechanical, electrical, heat, or vital – possessed the twin aspects of motion and resistance. Just as chemical force depended upon molecular arrangements and the nature of the atoms in the array, so the vital force could be considered as,

> a peculiar property, which is possessed by certain material bodies, and [which] becomes sensible when their elementary particles are combined in a certain arrangement or form.[59]

But if vital force is merely another latent property of chemical atoms that is manifested only in certain molecular situations in living systems, it is difficult to see why any distinction is necessary.

The most likely answer, as many historians have struggled to explain,[60] is that Liebig saw no *purely* chemical way to explain growth, the resistance that living systems possessed to external destructive forces, and the inability of the chemist or the physiologist to create a living cell. Only through a vital force, he believed, could one explain the forms characteristic of a living system. Although he made no explicit acknowledgement, it seems highly likely that Liebig had been led to this position from reading Johannes Müller's *Handbuch der Physiologie des Menschen,* which had appeared between 1835 and 1837.[61] Liebig was always cavalier with sources, but in this instance, no doubt, having attacked *Naturphilosophie* so vehemently, he would not have dared cite a physiologist whose work contained many echoes of that speculative philosophy.

58. Timothy O. Lipman, "Vitalism and reductionism in Liebig's physiological thought," *Isis* 58(1967), 167–85.
59. J. Liebig, *Animal Chemistry,* 1842, p. 198.
60. Besides Lipman, "Vitalism and reductionism," see more recently, Kenneth L. Caneva, *Robert Mayer and the Conservation of Energy* (Princeton, 1993); note also T. O. Lipman, "The response to Liebig's vitalism," *Bulletin History of Medicine* 40(1966), 511–24; and Vance M. D. Hall, "The role of force or power in Liebig's physiological thought," *Medical History* 24(1980), 20–59.
61. Hall, "Role of force or power," 28–36.

The result was not experimental science but a position as metaphysical as the reductionist aspirations of his critics. Moreover, as Lipman has pointed out, unlike the *chemical* speculations of his *Animal Chemistry*, Liebig's vitalistic explanations had little heuristic power to aid research:

> It could not tell where to go in the study of the organism but only served as an explanation of where one had been. While it did not hinder methodology, Liebig's vitalistic explanation did not promote it.[62]

The way forward, as Claude Bernard was to show in his experiments on the glycogenic function of the liver, was to limit questions and answers to observed functions and connections between different organs in a living system.[63]

Liebig's "establishment" antireductionist viewpoint was not to the liking of Moleschott, Büchner, Vogt, and other German and Dutch materialists, who, for political and social reasons, were using the rhetoric of science as a weapon against religion and idealism.[64] In one of his lectures given in the presence of King Max in 1856, Liebig dismissed materialists like Moleschott as dilettantes "who stroll at the edge of natural science [and who were like] children in knowledge and natural law."[65] Ironically, as students of chemical physiology, Karl Vogt and Ludwig Büchner (who were both students of Liebig) and Jacob Moleschott (a student of Mulder's) may have taken their inspiration to become materialists from Liebig. Moleschott answered the teleology and vitalism of Liebig's *Chemische Briefe* in 1852 with his own set of twenty physiological letters on revelation, the indestructibility and circulation of matter, the soul, and the purpose of life! Cheekily dedicating the book to Liebig, ironically these materialistic letters became as famous as Liebig's own among free-thinking circles.[66] As a committed teleologist, it was also impossible for Liebig to embrace wholeheartedly the scientific reductionist programme of Müller's disciples and pupils: Ludwig, Helmholtz, Virchow, and Du Bois Reymond. Consequently, he has been seen as having the difficult task of steering "a middle course between two sorts of dynamical physiology, between the Scylla of vitalism (and of Naturphilosophie) and the Charybdis of outright reductionism."[67] To his enemies, critics, most devout pupils, and later historians alike, Liebig's blunt refusal to espouse the

62. Lipman, "Vitalism and reductionism," p. 184.
63. F. L. Holmes, *Claude Bernard and Animal Chemistry* (1974).
64. Gregory, *Scientific Materialism.*
65. Volhard, ii, 351; the lecture became Letter 23 in *Chemical Letters.*
66. J. Moleschott, *Der Kreislauf des Lebens. Physiologische Antworten auf Liebigs Chemische Briefe* (Mainz, 1852); further eds. 1855, 1857, and 1862.
67. Hall, "Role of force or power," p. 57.

"winning side" of physiological reductionism has been an embarrassment and black mark. Coupled with his penchant for hasty judgements, arrogance, misrepresentation of his critics' views, and his dilatoriness in acknowledging his own mistakes, we may have an explanation for the relative Anglo-German neglect of Liebig's physiology after his death. On the other hand, it must be admitted that, although Liebig "may not have wanted to make the total substitution of science for religion that some of his contemporaries carried through, yet with his more extravagant claims for science [in his popular writings] he unwittingly tended in that direction."[68]

Finally, it is worth emphasising that even though Liebig was one of the few serious scientists of the middle of the nineteenth century not to oppose a vital force on methodological grounds, even here his attitude proved stimulating. In 1837 Liebig had published in *Annalen* a paper by his friend Mohr, which argued that heat, light, electricity, magnetism, cohesion, chemical affinity, and motion were all the consequences of mechanical forces. Implicitly this raised the question of their mutual convertibility and the origins and sustainability of force. In *Animal Chemistry* Liebig made it clear that he found the idea of force without matter as inconceivable.

As *Animal Chemistry* was published, Julius Robert Mayer (1814–78), a Tübingen-trained doctor, was working as a surgeon on a ship bound for Java. The consequence of his famous observation there that there was little difference between venous and arterial blood in the tropics is well known. The fact that less oxygen was needed to maintain animal heat in a warm climate (as Liebig, contrariwise, had said of more oxygen for Eskimos in the Arctic) led Mayer to deduce that mechanical forces, chemical forces, and heat were interconvertible. From this assumption and from calculations based upon previously published values of specific heat capacities, he was able to arrive at a value for what was later called the mechanical equivalent of heat. Aware of the connection with the theory of animal heat he had developed in Part I of *Animal Chemistry,* Liebig published Mayer's work in *Annalen* in 1842.

Initially, Mayer seems to have been content with a vitalistic explanation of life, but a detailed critical reading of *Animal Chemistry* caused a change of heart. In 1845, in response to Liebig's confusing treatment of vital force, Mayer published *Organic Motion in Its Connection with the Exchange of Matter.* He was now clear that heat was motion, and that if motion arose from a chemical (or in Liebig's case, vital) force, this force could be neither created nor destroyed. "Different forces," he urged, "can

68. Sonntag, "Religion and science," p. 167.

be transformed into one another. There is in truth only a single force."[69] Mayer had clearly grasped that the vital force could not be inexhaustible and that "in all physical and chemical processes the given force remains a constant quantity." To identify the cause of mechanical work in an animal with the expenditure of vital force was absurd, he concluded:

> Since we perceive a chemical process, in the exchange of matter, an adequate basis for the continued existence of living organisms, we must therefore protest against the proposal of a vital force in the sense of Liebig. . . . But as regards Liebig's hypothesis of the expenditure of a vital force to produce mechanical effects, that appears to be still more venturesome than the positing of such a *vis occulta* in and of itself.[70]

Intriguingly, Liebig did not reply to Mayer's attack – unless he simply lumped him together with the materialists he had dismissed in 1856. Meanwhile, in 1847 Helmholtz had placed the law of conservation of *force* upon a mathematical footing, and by the mid-1850s, conservation of *energy* had become the foundation stone of thermodynamics and basic to the further exploration of physiological chemical dynamics. It was not, however, until 1858 that Liebig firmly grasped the idea of energy in a popular lecture on "the transformation of forces."[71] The lecture was immediately incorporated into the fourth edition of *Chemical Letters* (1859) but restricted discussion to Mayer's demonstration that motion was converted into heat. Elsewhere, in the same volume, he forcefully reiterated his commitment to the vital force.

Liebig's lifestyle in Munich was quite different from that in Giessen. No longer tied to a laboratory bench, his position at court and in the academy encouraged him to indulge in both scientific statesmanship and meta-science. Opportunities for reflection on the philosophy of science came in particular from what he saw as an incorrect obsession with practical results in both English and German farming practice, from his entry into physiological chemistry, and from the disturbing political and social implications of materialist thought.

Liebig was not the only nineteenth-century scientist who was critical of Bacon. In Scotland, David Brewster had been equally contemptuous in the 1850s. Such cumulative criticisms had the required effect of leading historians and philosophers to abandon the view of Bacon as the founder of

69. Caneva, *Robert Mayer*, p. 260.
70. *Ibid.*, p. 265.
71. *Ibid.*, p. 219. Liebig, "Ueber die Verwandlung der Kräfte" in *Vorträge, wissenschaftliche, gehalten zu München* (Braunschweig, 1858), pp. 581–96; the lecture was incorporated into Letter 13 of *Chemical Letters*, 4th ed., 1859.

experimental science and of scientific method, while retaining him as a visionary who had seen the need for the coordination of research. On the other hand, Liebig undoubtedly misunderstood Bacon's original intentions – mainly because he concentrated his attack on the posthumous ragbag of *Sylvia* and on Bacon's rational (and implausible) reconstruction of the induction of heat. He certainly misunderstood the utilitarianism of Bacon's *New Atlantis,* which was really a benevolent form of philanthropy whereby science was to have benefitted all mankind, not merely the individual inventor. In any case, if invention was the mother of science, as Liebig claimed in 1866, then Bacon's intentions ought to have been helpful to science. Above all, Liebig failed to grasp the incomplete nature of Bacon's enterprise and to understand that what had survived emphasized induction, whereas what Bacon had probably intended to emphasise was that theory and hypothesis would form part of the whole enterprise.

But Liebig's misinterpretations are really beside the point, for it was the fact that British landowners appealed to utilitarianism, to praxis over "truth," that had annoyed Liebig, and he had put this down to their amateur Baconianism. To Liebig, this was to put the cart before the horse. There could be no progress without ideas or questions in view:

> To resolve an enigma, we must have a perfectly clear conception of the problem. There are many ways to the highest pinnacle of a mountain; but those only can hope to reach it who keep the summit constantly in view. All our labour and all our efforts, if we strive to attain it through a morass, only serve to cover us more completely with mud; our progress is impeded by difficulties of our own creation, and at last even the greatest strength must give way when so absolutely wasted.[72]

Yet, when it came to the investigation of vegetable and animal physiology, Liebig's original materialistic notions proved, in his eyes, inadequate. The enigma of "life" could never be solved because by definition it was the domain of a vital force whose ultimate nature could never be determined, even though its manifestations were plain to see.

72. J. Liebig, *Animal Chemistry,* 1842, p. 125.

12

Death and Assessment

> We were the first pioneers in unknown regions, and the difficulties in the way of keeping on the right path were sometimes insuperable. Now, when the paths of research are beaten roads, it is a much easier matter; but all the wonderful discoveries which recent times have brought forth were then our own dreams, whose realization we surely and without doubt anticipated.[1]

In 1869 the flamboyant and unconventional Münster sculptress Elisabet Ney (1833–1907) successfully ingratiated herself into the confidence of the Wagner-obsessed and mad King Ludwig II of Bavaria. An immediate result of this relationship was the commission of the imposing statue of the king that stands today in his extravagant folly, the Palace of Herrenchiemsee. At the end of the year, a few months before she exiled herself to America, Ney took it upon herself to offer Ludwig some political advice:

> Last evening your Majesty asked: "Is there a man in whose generosity and high-mindedness I dare to believe?" Your Majesty asked me whether I knew one. As then, also today, I can answer: "Indeed I know of one;" . . . Once again let me repeat the name of the man of whom I spoke last night; a man who knows no purpose in life other than to stand always openly for truth, who is recognized as supreme not only in his own field, but in the whole realm of thought. Justus von Liebig is his name.[2]

There was, of course, a hidden agenda in these remarks. As a friend of Liebig's (who had purchased one of her sculptures, the *Sursum*), she would have been aware of Liebig's unhappiness in losing the position of honorary scientific advisor to the Bavarian Court which he had held under Ludwig's father, Maximilian II, and which had brought him to Munich from Giessen in 1852. Nevertheless, it underlines the considerable esteem that the public held the great organic chemist during his lifetime.

1. J. Liebig, "Autobiography," in *Ann. Rep. Smithsonian Institution* (1891), p. 267.
2. J. Fortune and J. Burton, *Elizabet Ney* (London, 1943), pp. 127–8.

In January 1853 Liebig and his wife celebrated twenty-five years of happy marriage. As happens to this day in German academic communities, the event was celebrated publicly with pomp and ceremony. The great room of the Bavarian Palace was filled with flowers, shrubs, and garlands, the focus of the room being a garlanded bust of Liebig that had recently been executed by the classical sculptor Johann von Halbig (1814–82). Commissioned on behalf of Liebig's grandchildren, it was to be placed in the entrance hall of the chemical laboratories of the university. At the great feast, which must have reminded Liebig of the similar event in Glasgow in 1844, toasts were drunk to the king, science, art, and industry, as well as to Liebig, the "father of organic chemistry and founder of agricultural and physiological chemistry." Following Liebig's thanks, Pettenkofer toasted the whole of Liebig's school, which he metaphorically described as a great chemical family whose banner was "science and truth." A final toast by their family friend, the historian von Kobell, was made to Liebig's wife and their children and grandchildren before the festivities culminated in music and dancing.[3]

The only one of the Liebigs' children to miss this ceremony was their eldest son Georg (1827–1903), who was then practising as an army doctor in India, where in 1855 he was elected professor of geology and mineralogy at a college in Calcutta. He owed this position, as well as his previous medical appointment, to the patronage of English friends he and his father had made in Britain when Georg stayed in London and Oxford with Hofmann and Buckland in 1846. To help Georg with his Indian lectures, Liebig bound up copies of his course of Giessen chemistry lectures and sent them to Georg in November 1856. The spotless condition of this beautifully illustrated volume today suggests that it remained unused.[4] Within a few years of his Indian preferment, however, Georg decided to return to Germany, where, as we have seen, his father engineered an opening for him as chief medical officer at Reichenhall. Georg, who might have made an independent career as a physiologist – he was the first investigator to notice the alteration in the acidity of blood after muscle contraction – was always overshadowed by his father's fame.[5] He proved to be a competent director of health facilities at Reichenhall, married in 1863, produced several grandchildren for Liebig, and published a couple of monographs on environmental medicine.[6] One of the grand-

3. Volhard, vol. ii, pp. 343–7.
4. O. P. Krätz and C. Priesner, *Liebigs Experimentalvorlesung* (Weinheim, 1983).
5. Georg Liebig, "Ueber die Respiration der Muskeln," *Arch. Anat. Physiologie* (1850), 393–416; see J. S. Fruton, *Molecules and Life* (New York, 1972), pp. 281–3.
6. Krätz and Priesner, *Liebig's Experimentalvorlesung*, pp. 25–46; and W. H. Brock, ed., *Liebig und Hofmann . . . Briefen* (Weinheim, 1984), pp. 18–19.

children, Heinrich (1877–1962), continued the family tradition of chemistry, and in 1957 he became the first recipient of the Liebig Medal conferred by the Justus Liebig University (as the University of Giessen became known after World War II). Heinrich's marriage was childless, so that the direct line of Liebigs ended with the death of his wife Katherina Freifrau von Liebig in 1983.

If Georg was overshadowed by his father, whom he closely resembled in face and physique, it was the fate of Hermann to be overshadowed by both father and brother. Unlike Georg, Hermann (1831–94) is rarely mentioned by Liebig in his correspondence, and one is left with the impression that he scarcely lived up to his father's no doubt inflated expectations.[7] Following education at the Giessen Gymnasium, in 1848 Hermann was sent to live with his uncle Georg Liebig at Darmstadt so that he could study science at the Gewerbeschule there before matriculating at the University of Giessen in 1851. Under his father's wing, Hermann followed chemical courses in Giessen as well as agricultural courses in Munich before being sent to England in 1855 to learn about British agricultural practice.[8] In 1856, after spending some months in Edinburgh with William Gregory, he went to Hungary to manage the estates of a Count Hadik.[9] These experiences led to disagreements between father and son over agriculture. On returning to Bavaria, and helped financially by his father, Hermann became a gentleman farmer at Schorn, near Starnberg. After divorcing his first wife Anna Frank in 1871, he married Auguste Linder in 1873. His son by the first marriage, Eugen (1868–1925), became a civil servant; the other, Hans (1874–1931), born after his grandfather's death, became a chemist and actually taught at the University of Giessen before becoming a professor at Halle, where he became notorious for his extreme political views.

The Liebigs' eldest daughter, Agnes (1828–62), a blond beauty, was recognized as extremely clever but weak in health. There were some misgivings when she married the philosopher Moritz Carrière (1817–95) in 1853. After producing two grandchildren for Liebig, Agnes died of tuberculosis on 29 December 1862. Like Darwin and Huxley, who similarly lost a beloved daughter and son, Liebig was devastated and unable to overcome his grief for more than a year. On receiving condolences from his Swiss friend Schönbein, he confessed that nothing could ever heal the sorrow: "I still think of how profound and penetrating her mind was. One

7. *Hessische Biographien,* vol. 3 (Darmstadt, 1934; repr. Wiesbaden, 1973), pp. 377–80; includes a list of Hermann's agricultural writings.
8. Liebig to Mohr, 14 December 1855, in G.W.A. Kahlbaum, ed., *Justus von Liebig und Friedrich Mohr in ihren Briefen* (Leipzig, 1904), p. 145.
9. Gregory to Liebig, 4 January 1856, Liebigiana, Staatsbiblithek, München.

can never adjust to such a loss, however many years pass. We must, however, bow and accept it as a higher decree which has to be accepted with humility."[10] This is one of Liebig's few hints of a firm belief in God. Carrière's loss, which brought him closer to his in-laws, was further compounded when one of the children of the marriage died at the age of ten.[11]

Liebig took some consolation in the even greater beauty of his second daughter, the dark-haired Johanna (Nanny, 1836–1925), who had bewitched Wöhler when she was only seventeen.[12] She also captivated Hofmann in 1854 after he had tragically lost his English wife to scarlet fever. There were serious negotiations for a marriage, but in the end she rejected Hofmann and chose the son of the Liebigs' next-door neighbour in Munich, a young surgeon named Karl Thiersch (1822–95). Their marriage took place in August 1855, and with Thiersch's promotion to a chair in Erlangen, and then Leipzig, Liebig saw little of Nanny again and the eight grandchildren of the happy marriage. Two of the grandchildren, Amalie and Lina, married eminent men, the theologian Adolf von Harnack and the historian Hans Delbrück, respectively. The only present-day descendants of Liebig come from the marriage of another daughter, Agnes Thiersch, to a dentist Friedrich Hesse. As was mentioned in the first chapter, the Liebigs' youngest daughter Marie (1845–1920) never married after tragically losing her fiancé. She stayed single to look after her elderly parents and to keep her mother company in widowhood.

The poet Platen had described Liebig at Erlangen as of slim build, of friendly seriousness, regular features, and large brown eyes with dark eyebrows.[13] The dashingly handsome man of the earliest surviving sketch is still recognisable in Wilhelm Trautschold's famous portrait of 1840 and in the first photographs of Liebig as a family man.[14] Always of slim build, Liebig remained in relatively good health until 1835 when, worn out by chemical investigations, financial worries, and negotiations with his government, he had a nervous breakdown. A period of recuperation at Bad

10. G. W. A. Kahlbaum and E. Thon, eds., *Justus von Liebig und Christian Friedrich Schönbein Briefwechsel* (Leipzig, 1900), p. 150.
11. See W. Diehl, "Moritz Carrières Lebenserinnerungen (1817–1847)," in *Archiv für hessische Geschichte* 10 (1914), 133–301, where Carrière describes his love for Agnes. Twenty years after her death he wrote a sonnet, "Agnes, Liebeslieder und Gedankendichtungen" (Leipzig, 1883).
12. Wöhler to Liebig, 20 September 1853, in A. W. Hofmann, ed., *Aus J. Liebig's und F. Wöhler's Briefwechsel,* 2 vols. (Braunschweig, 1888), vol. 2, p. 8.
13. Quoted by R. Winderlich, "Liebig," in G. Bugge, ed., *Das Buch der grossen Chemiker,* vol. 2 (Berlin, 1930), p. 2.
14. On Trautschold (1815–76), see Christian Rausch, "Wilhelm Trautschold," in *Ludoviciana Festzeitung zu dritten Jahrhundertfeier der Universität Giessen* (Giessen, 1907), pp. 91–2.

Salzhausen and the news that his salary was to be increased and his laboratory facilities expanded restored him to his energetic self. Apart from minor ailments, he remained in good health for the remainder of the Giessen period.

Liebig was forty-nine years of age when he settled down to a very different lifestyle in Munich. Giessen had obviously taken its toll; his letters to Wöhler rapidly succumbed to increasing references to middle-aged ailments of aching joints, headaches, dyspepsia, insomnia, and depression. In a famous remark to Kekulé, Liebig said that a chemist must work without regard for his health, and Liebig supremely exemplified this maxim.[15]

As a liberal-minded latitudinarian Protestant in a deeply Catholic country, Liebig was despised by ultramontane sympathisers as a threat to their values. Liebig and the other north German intellectuals who had been brought to Munich inevitably tended to sympathise with Prussia in its continual wrangles with the Austrians, with whom the Catholic Bavarians sided. While Maximilian II was alive, surrounded by intellectuals who shared liberal attitudes towards biblical criticism and the apparently increasing tendency of science towards materialism, all was well. Liebig had enjoyed a close personal friendship with his king, and he was genuinely shocked by his death in March 1864.[16] The year 1864 also witnessed the promulgation by Pope Pius IX of the Bull *Quanta Cura,* with its challenging and notorious appendix *Syllabus Errorum.* To European Protestants and liberal Catholics generally, this attack on modern science and critical theology was an assault on the rational eucumenical attitudes that had prevailed in Maximilian's court.

Liebig had no high regard for the new, eccentric King Ludwig, whom he had dismally failed to interest in chemistry when he instructed him as a teenager.[17] Ludwig's passions were music and architecture, not science and industrialization. Embroiled in continuing controversies over the reform of Bavarian agriculture, Liebig quickly found himself attacked in the ultramontane press as an unpatriotic materialist. Coming as these strident attacks did in the midst of Liebig's campaign to dispose of London's sewage for agricultural purposes, Liebig (as we have seen) thought seriously of quitting Munich for London. His depressed spirits were lifted in

15. "If you wish to become a chemist, you must be prepared to sacrifice your health. Whoever does not ruin his health by studying will not amount to much in chemistry these days." Kekulé, quoted by R. E. Oesper, *The Human Side of Scientists* (Cincinnati, 1975), p. 108.
16. See Liebig's memorial address to the Munich Academy, *Reden* (Leipzig, 1874), pp. 330–31.
17. F. Gerard, *The Romance of Ludwig II of Bavaria* (London, 1899), p. 31.

November 1864, however, when he received a petition from 838 manufacturers, industrialists, and intelligentsia who included Catholics as well as Protestants, pleading with him to ignore the "silly" attacks on him in the press and to stay in their beloved city.[18] Liebig, who did not know any of the memorialists personally, was gladdened by this sign of support from a group of independent-minded men who were determined to see Bavaria industrialised and modernised.[19] Despite a flattering, but perhaps only formal, call to Berlin to succeed Mitscherlich, who had died in August 1863, Liebig decided to stay in Munich.[20]

The *Syllabus Errorum* seriously affected Liebig's Catholic friend Ignaz von Döllinger (1799–1890). The son of a professor of anatomy at Würzburg, Döllinger had entered the Catholic priesthood in 1822 and, as professor of church history at Munich, had become Bavaria's greatest theologian and liberal Catholic. Döllinger, who saw parts of the *Syllabus* as being directed against him personally, spearheaded a campaign to head off the pope's intended announcement of papal infallibility. When this dogma was duly promulgated in 1871, it split the Catholic community in Munich, with Protestants forming a sympathetic alliance with Döllinger's group of liberal Catholics. Liebig and his son-in-law, Moritz Carrière (a philosopher who vigorously defended freedom of thought), were both approached in June 1871 by leading Protestants who asked them to support Döllinger, who had been excommunicated in April. It is clear that Liebig gave Döllinger every support through the Munich Academy, which honoured Döllinger in several ways. For their part, civic authorities were also able to overrule local ultramontane officials and to shower him with civic honours as well.

Having survived a severe dose of dysentery in 1853, Liebig remained in good health for most of his first ten years in Munich. When the novelist George Eliot and her consort, the critic and physiologist George Henry Lewes, visited Liebig in May 1858 (Eliot was completing *Adam Bede*), they were both struck by his begrimed skin and nails that were black to the roots. "He looks best in his laboratory with his velvet cap on," they recorded, "holding little phials in his hand and talking of kreatin and kreatinine in the same way that well-bred ladies talk of scandal."[21]

18. Volhard, vol. ii, pp. 389–91, includes Liebig's fulsome reply. A smaller-scale attempt had been made to keep him at Giessen; see Chapter 11.
19. Liebig to Wöhler, 22 November 1864, Hofmann, *Liebig's und Wöhler's Briefwechsel*, vol. 2, p. 170; Kahlbaum, *Liebig und Mohr in ihren Briefen*, pp. 215–16.
20. Liebig to Wöhler, 27 February 1865, Hofmann, *Liebig's und Wöhler's Briefwechsel*, vol. 2, p. 179.
21. David Williams, *Mr George Eliot. A Biography of George Henry Lewes* (London, 1983), p. 180.

On holiday with Wöhler and Kopp in Passau in the autumn of 1859, Liebig slipped in the snow and severely injured his knee. Thiersch, who by then had moved with Nanny to Erlangen, was sent for, and he diagnosed a broken patella. Liebig was forced to abandon his public lectures (Pettenkofer took over) and to lie prone or sitting with his leg supported by a stool, for eight weeks. A local artist Franz Pocci sketched a delightful caricature of Liebig in bed holding a gigantic retort (piss pot?), with his leg suspended from the ceiling.[22] The knee injury, which left him permanently lame, was compounded in November 1862 when he slipped again while attending the theatre in Munich with Carrière. This injury again incapacitated him for a month. Liebig was clearly prone to falls, for it was only his wife's presence of mind and her umbrella that saved him from another serious accident a few years later.[23] It says much for Liebig's international fame (as well as the transformation of the Victorian press) that the knee injury of October 1859 became front page news, as did his serious illness in 1870. The lameness confined him much more to the house during the remainder of his days. He very much regretted being unable to visit England to see his god-daughter Ethel Harley, the daughter of Emma Muspratt and Robert Harley.[24] Although his astonishing literary output was little diminished by these accidents, he was decreasingly active in the laboratory. He took longer holidays in the company of male friends (his wife rarely accompanied him) but returned increasingly weary.

He told an old friend, Dalwigk, in 1866:

> I don't know whether others who grow old have the same feelings, but since I've reached my sixties memories and pictures of Darmstadt and my earliest times there come back, whereas thirty years ago they were totally effaced from memory. It was truly the time for youth when one looked forward to the challenge and strife of life. How different it is in old age when life is only interesting to look back on![25]

It was probably at this time that he drafted the set of reminiscences that were published posthumously by Georg Liebig.

Liebig had been made perpetual president of the Bavarian Academy of Sciences in 1858, a position that gave him ample local prestige and opportunities for patronage. His final performance on the world's stage occurred during the spring of 1867, when, on behalf of the academy, he was

22. See reproduction in Krätz and Priessner, *Liebig's Experimentalvorlesung*, p. 56.
23. Liebig to Wöhler, 1 November 1866, in Hofmann, *Liebig's und Wöhler's Briefwechsel*, vol. 2, pp. 220–2. See Harley's obituary of Liebig, *British Medical Journal* 26 April 1873.
24. A. Tweedie, *George Harley* (1899) , p. 168.
25. Liebig to Dalwigk, 8 March 1866, in Anon., *Aus dem Briefwechsel von Justus Liebig mit dem Minister Reinhard Freiherrn von Dalwigk* (Darmstadt, 1903).

chosen to be a member of the international jury that was to judge the exhibits in Group X of the Exposition Universelle in Paris. The international jury subsequently decided to elect Liebig as its president. According to Volhard,[26] Liebig undertook this task with some anxiety, though in the event he enjoyed himself enormously despite some displeasure at finding that his recent work on meat extract and baby's milk was little known in France and that his *Natural Laws of Husbandry* was unknown even to Dumas.[27]

Although the Great Exhibition in Hyde Park in 1851 had become the model for later world's fairs, the Paris exhibition of 1867 was in some ways far more significant. For the first time, countries showed their wares in the Champ de Mars in different buildings instead of under one great glass panopoly. Its chief organiser, the socialist Frederic Le Play, hit upon the idea of family and work as running themes and of using the exposition not merely as a shopwindow for international trade but as an educational resource.[28] Liebig was much impressed, he told his Munich friend Charles Boner, with

> the "histoire du travail," under which name all is brought together which, from the primeval age to the seventeenth century, is known by positive examples of labour in France, Switzerland and England; the implements and the weapons of the dwellers in the Lake cities of the middle ages, &c. That in the same park the huts of Swedish peasants, Egyptian temples, Moorish palaces, Chinese coffee-houses, may be seen, you already know from the newspapers.[29]

Indeed, Group X, which Liebig was to judge, consisted of "dwellings for the poor, constructed on sanitary principles, and at small cost, and articles exhibited with the special object of improving the physical and moral condition of the people."[30]

Among chemical products, German steel, glass, and paper won prizes against English and French competition. German machinery also did well, and the country overcame all opposition in prizes for teaching and instructional equipment.[31] Liebig evidently found judgement irksome, as his fellow jurors were not prepared to accept his views as chairman (as he thought they would have done in England!); instead the committee argued

26. Volhard, vol. ii, p. 404.
27. Liebig to Schönbein, 1 June 1867, in Kahlbaum, *Liebig und Schönbein Briefwechsel*, pp. 253–5.
28. Paul Greenhalgh, *Ephemeral Vistas. The Expositions Universelles, Great Exhibitions and World's Fairs, 1851–1939* (Manchester, 1988).
29. R. M. Kettle, *Memoirs and Letters of Charles Boner* (London, 1871), vol. 2, p. 330.
30. Quoted in Greenhalgh, *Ephemeral Vistas*, p. 146.
31. Volhard, vol. ii, p. 405.

informally amongst itself as if in a restaurant.[32] The exhibition clearly revealed Napoleon III's overwhelming desire to demonstrate that France, not Great Britain, was the leading progressive industrial state. In this he was only too successful. Visiting English jurists, such as Playfair and Frankland, were quite dismayed by Britain's poor showing in the prize lists and returned to London with warnings of an industrial crisis unless Britain's system of technical education was transformed.[33]

Among the 150 of the world's chemists assembled in Paris in April 1867 were many of Liebig's British, French, and German pupils and friends, including Frankland, Kane, Playfair, and Wheatstone from the British Isles; Fehling, Hofmann, Kekulé, Schröter, and Varrentrapp from Germany; and from France, Balard, Becquerel, Berthelot, Boussingault, Cahours, Chevreul, Deville, Dumas, Pelouze (who was dying), and Wurtz. It was a time for final reconciliation with Dumas as well as with Hofmann, his most distinguished pupil, with whom relations had become fairly frosty since Hofmann's assumption of the Berlin chair in 1865. At a grand banquet of chemists on 22 April, at which Liebig's extract was served, Dumas praised the British chemical community and the learned academies of Europe; Playfair extolled Dumas and the French; Balard toasted Liebig and the "flaming intellectual hearth" of Giessen; Liebig movingly commemorated his teachers, Gay-Lussac and Thenard. Hofmann (always a brilliant speechmaker) toasted the relations between, and an alliance of, science and industry and coupled this with praise for the absent ninety-one-year-old Chevreul.[34]

At another state banquet, Liebig found himself guest of honour sitting next to Louis-Napoleon himself. Napoleon engaged him in conversation on agricultural reform and the recycling of sewage. "Napoleon is a remarkable man," Liebig told Wöhler, "not to talk to, but to listen to, and he is quick on the uptake" (Liebig to Wöhler, 26 May 1867). But to Boner he confided:

> He [Napoleon] has formerly lived freely, and must bear the consequences now; but he may live to be as old as the rocks: and as to what concerns his political position, I have spoken to men of high standing and of all classes of opinion, and no one wishes for any other dynasty, but they look forward anxiously to the future, as the Prince is not likely to have strong health. The men of science forgive him for having

32. Liebig to Wöhler, 26 Mary 1867, in Hofmann, *Liebig's und Wöhler's Briefwechsel*, vol. 2, pp. 233–4.
33. See D. S. L. Cardwell, *The Organisation of Science in England*, 2nd ed. (London, 1972), Chap. 5.
34. C. A. Browne, "The banquet des chimistes, Paris, April 22, 1867," *J. Chem. Educ.* 15(1938), 253–9.

> laid such fetters on freedom of speech , but it is by no means certain that freedom of speech in France would be of the same use that it is in England, for the French are an irritable, capricious people. If God were to send an angel to rule over them, in ten years they would be tired of him, and be longing to have a devil instead, just by way of change.[35]

Two years later Prussia was at war with France in an effort to preempt the historical French influence over the southern states of Germany and to ensure the unification of the German empire in 1871. France's Second Empire of Napoleon III collapsed, and Louis-Napoleon was exiled to England.

Here something should be said about Liebig's political views. Born and bred during the Napoleonic Wars, Liebig died soon after the Franco-Prussian War and the unification of Germany, having lived through the turmoil of 1848 and earlier revolutionary disturbances in Hessen-Darmstadt. Any indifference that he had shown towards politics as a young man was clearly dispelled by the events of Vormärz – the continued agricultural unrest in Hessen that was fueled by a series of bad harvests and emigration, the absence of free trade and political rights, and the revolutionary activities of the *Sonderbund* group of students at Giessen who were led by Liebig's future son-in-law Moritz Carrière, and which included favoured pupils such as Friedrich Bopp (1825–49) and Carl Vogt, as well as his own son, Georg.[36]

The news of King Louis-Philippe's exile after the Parisian mob rose up against him in February 1848 had immediately caused unrest in Giessen, and Liebig had found his usual laboratory class of forty reduced to ten (Liebig to Wöhler, 21 October 1848). It was during this period that Liebig, along with many other professors, formed a civic militia to help guard the town.[37] Liebig revealed his sympathy for the republican cause to his friend Vieweg (3 March 1848):

> If the government of this state were not so blind, there would be no revolution. So it was for the unfortunate Louis-Philippe and his lamentable ministers who, after exchanges never previously heard of in the history of mankind, had to be forced into exile in his 8th year of rule. Thus it comes to us as well. Everyone sees this except the highest in the land who ought to be able to see the farthest.[38]

35. Kettle, *Memoirs and Letters*, pp. 330–31.
36. See Krätz and Priessner, *Liebig's Experimentalvorksung*. The *Sonderbund* took their name from the separatist war between Catholic and Protestant cantons of Switzerland in 1847.
37. *Ibid.*, p. 29; Volhard, vol. i, p. 176.
38. M. and W. Schneider, eds., *Justus von Liebig Briefe an Vieweg* (Braunschweig, 1986).

Liebig seems to have warmly supported moderate constitutional reforms and to have believed that men had to be free for their well-being and intellectual development. Disillusion set in when Bopp died in prison from brain fever (meningitis) following his participation in the Baden revolt. The peace of May 1848 under Prussian martial law and the election of ineffectual deputies to the "pre-Parliament" National Assembly at Frankfurt, including his son Georg as well as Vogt and Dieffenbach, were depressing to him. "Freedom is like a fine wine which is bought at a price," he told Mohr. "Only the wise, who drink from it in moderation find a genuine delight in it, while the proletariat only thinks that being free entitles them to get drunk."[39] Liebig fully supported the people's demands for freedom of the press, assize courts, and what would follow from them, but he had doubts about the creation of a national state of northern Germany because he felt that "nationality" tended to mean different things in different parts of Germany. Equally, the prospect of a greater German-Austrian nation filled him with disquiet. He had mixed feelings, therefore, when Frederick William of Prussia refused the offer made by the German National Assembly in June 1849 to make him Emperor of a Germany of united northern states. The inevitable consequence of Frederick's refusal was counterrevolution as "shaken rulers gradually regained their confidence and reasserted control, sometimes with the aid of Prussian troops restoring order elsewhere."[40] Perhaps this was why Liebig was a sympathetic observer at Richard Cobden's Frankfurt Peace Conference in 1850. Liebig shared the Englishman's belief that negotiation was a more civilised way than war.[41]

In 1866, when Prussia went to war with Austria over the Schleswig-Holstein issue, Bavaria and the other southwest German states were caught in a trap. With the defeat of its traditional Catholic Austrian ally in July of that year, Bavaria had to look either to France or to union with the "blood and iron" of Prussia. But, as Liebig commented to Mohr on 12 August 1866, union with a reactionary France would do little to alleviate the hardships and impoverished lives of the common people and would perpetuate the power of princes to rule. Liebig, therefore, concluded, after attending a mass rally in Munich on 12 August 1866, that it would be best for Bavarians to open diplomatic negotiations with Prussia and that independent sovereignty for Bavaria, Württemberg, and Baden would only cause Germany harm in the long term.[42] In fact, the full union of the

39. Kahlbaum, *Liebig und Mohr in ihren Briefen,* p. 99.
40. Mary Fulbrook, *A Concise History of Germany* (Cambridge, 1993), p. 122.
41. E. K. Muspratt, *My Life* (London, 1917), p. 24.
42. Liebig to Schönbein, in Kahlbaum, *Liebig und Schönbein Briefwechsel,* pp. 220–1.

southern German states with Bismarck's North German Confederation came about only as a result of the Franco-Prussian War of 1870–71.

Although Liebig had once viewed the rise of Prussia with considerable distaste and alarm, despite the anti-French feelings inevitably whipped up by war,[43] he was magnanimous in praise of the defeated French. His brave speech of reconciliation given to the Munich Academy of Sciences on 28 March 1871 was widely reported. After noting that, since the foundation of their Bavarian Academy 112 years before in 1759, Germany had become a nation, he went on:

> It is characteristic of the German peoples, with their knowledge of languages, their understanding of foreign nations, and their cultural-historical viewpoint, to be fair to other nations even to the extent of being less than generous to themselves. Hence we should never forget to thank the great French philosophers, mathematicians and scientists who, in so many fields, were our teachers and master-builders.[44]

And he concluded by recalling again his own personal indebtedness to the education he had received in Paris forty-eight years before. His peroration looked forward to forgiveness and a renewed spirit of internationalism through scientific cooperation. The Franco-Prussian War also revealed the generous side of Liebig's nature. When he had heard that the wife of Charles Barrewil (1817–70) had fled penniless to Boulogne during the siege of Paris, he quietly arranged for 500 francs to be made available to her; and he arranged for Henri St.-Claire Deville to be sent news of his wife's safety in Geneva.

These political developments were the accompaniment of aging and its inevitable bereavements. In 1868 Liebig lost three of his close correspondents: Pelouze in France, Daubeny in England, and Schönbein in Switzerland. He remarked wearily to Frau Schönbein in his letter of condolence:

> At my age death becomes familiar, though that does not make it any less a tragedy to their families. Lately I have lost a great number of my best and oldest friends, and the consequence of such bereavements is the reflection that we shall be next in the series. So it shall be![45]

His spirits were raised in the summer of the same year, however, when a group of Bavarian farmers commissioned two bronze busts of himself and Wöhler to stand in the newly erected Munich Polytechnic. Designed by the sculptress Elizabeth Ney, the matrices were also used to mould two

43. In correspondence with Kolbe, Liebig had harsh things to say about the French conduct of the war. See A. J. Rocke, *The Quiet Revolution. Hermann Kolbe and the Science of Organic Chemistry* (Berkeley, 1993), pp. 344, 347.

44. J. Liebig, "Nach dem Friedensschluss," in *Reden* (1874), pp. 331–34.

45. Liebig to Frau Schönbein, in Kahlbaum, *Liebig und Schönbein Briefwechsel,* p. 275.

marble versions that Hofmann commissioned to be placed in his magnificent new Chemical Institute in Berlin that year (Hofmann to Liebig, 24 September 1868). Liebig was proud, as Ney's busts of Bismarck, Garibaldi, and Jacob Grimm had achieved considerable acclaim.[46] Hofmann was also able to give Liebig the good news that the Royal Society of Arts in London had decided to award Liebig their Albert Medal.

Despite such gratifying acknowledgements of his past contributions to chemistry and his homely dabbling at research on breadmaking, he remained in a general state of ennui. When in 1870 he was invited by the London Chemical Society to deliver a Faraday Memorial Lecture in 1871, he declined, telling Wöhler that he no longer had the energy to travel so far or to prepare an important lecture in a foreign language.[47] He had been seriously ill that year with cerebral meningitis and painful carbuncles on his neck. Family and friends, as well as Liebig himself, had not expected him to recover. He made his will, ordered his coffin, and directed that his body should be packed in charcoal and buried in Darmstadt.

It was at this time that he wrote to his surviving sister Elizabeth Knapp:

> I have as many religious doubts as other men, but for different reasons, and what the churches teach I fully understand and believe that their dogmas are not entirely useless, especially for mankind. But my familiarity with Nature and her laws has convinced me that we should not worry ourselves about death and our future life since everything is so infinitely wisely arranged that anxiety about what happens after death has no place in the soul of the scientist. Everything has been cared for, and whatever becomes of us is surely for the best.[48]

As Sonntag has noted in his essay on Liebig's religious views,[49] these remarks were not just words of reassurance to a devout sister, for he shared similar sentiments with Bischoff as well as Reuning. He told the latter in 1870, "I find everything so wisely ordained that the very question, what will become of me after the end of my life, little concerns me. What will become of me is surely for the best, and of that I am quite certain."[50]

In 1871 Reuning was the first recipient of the Liebig gold medal that the peripatetic Versammlung deutscher Land- und Forstwirte had decided to

46. Volhard, vol. ii, p. 407.
47. Liebig to Wöhler, 3 December 1870, in Hofmann, *Liebig's und Wöhler's Briefwechsel*, vol. 2, p. 303.
48. Cited by Hans Linser, "Liebig und der Chemismus des Lebens," *Giessener Universitätsblätter* 6(April 1973), p. 16.
49. O. Sonntag, "Religion and science in the thought of Liebig," *Ambix* 24(1977), 164.
50. Liebig to Reuning, 29 November 1870, in R. Echtermeyer and G. von Liebig, eds., *Briefwechsel zwischen J. von Liebig und Theodor Reuning* (Dresden, 1884), p. 190.

strike in gratitude for Liebig's contributions to the modernisation of German agriculture. In 1869, the Versammlung commissioned a portrait by the artist Brehmer for the medal's face, and an allegorical group was designed for the obverse by Ludwig Thiersch, the artist brother of Liebig's son-in-law. Although Brehmer was chosen because he had shown skill in designing a Gauss medal, his initial portrait did not meet with Wöhler's approval and a second version had to be made.[51]

Astonishingly, Liebig pulled through his illness, and in June 1870 his coffin (with a companion one for Henriette) was packed away in the university laboratory's lumber room. However, though he took up his old activities and he did a little writing, his old energy and drive had completely disappeared. His last letters to Wöhler in the Spring of 1873 goodhumouredly referred to his enervating insomnia.

> I planned to write to you yesterday, but I had a bad night, no sleep at all, and I lay the whole day on the sofa, tired and exhausted. I thought of you, your sound sleep, your good appetite, the normal operation of all your functions. Can the aged succomb to insomnia alone without having a real illness? It's vegetative life with recuperation at night, and if this fails, the lamp gradually flickers out.[52]

The end came in April 1873, when he caught a severe cold after falling asleep in his garden. The cold turned into pneumonia and he spent much of the early part of April in a coma, fed only with cognac, wine, and his own meat extract. He died surrounded by his family at 5.30 P.M. on Friday, 18 April 1873, just a month short of his seventieth birthday. As he had wished, there was an autopsy conducted by his friend Theodor Bischoff, who found fatty degeneration of the heart muscles, which explained his langour since 1870. Brain sections showed his headaches to have been caused by hypertrophy of the cerebral membrane caused by the earlier meningitis. Bischoff was fascinated by the hypothesis that brain size and weight reflected intelligence. In a monograph published in 1880 he concluded, however, that although "the brains of Gauss, Hausmann, Dupuytren, Liebig, Tiedemann, etc., were not as heavy as those of Byron and Cuvier, it does not follow that their talents and accomplishments were not just as good as those whose brains were materially well-endowed."[53]

It is unclear why his earlier wish to be buried in Darmstadt was ignored. As was customary in Bavaria, the body was laid out for inspection by his

51. Volhard, vol. ii, pp. 409–10; *Lancet,* 14 May 1870, p. 710.

52. Liebig to Wöhler, 3 April 1873, in Hofmann, *Liebig's und Wöhler's Briefwechsel,* vol. 2, p. 360.

53. M. Kutzer and E. Heuser, "Gehirn und Wissenschaft," *Medizinhistorisches Journal* 23(1988), 325–41, at p. 328.

Munich friends. Volhard records that "although the flashing eyes which lent such a fascinating expression to the face were closed, the head nevertheless was wondrously handsome." Two death masks were made by Kreitmeyer, a plaster worker in the Bavarian National Museum, and later presented to the Deutsches Museum.[54]

The funeral took place on 20 April, when Liebig's body was interred in the Südfriedshof Cemetery. The simple gravestone was embellished by a bust of Liebig by Michel Wagmüller, which Volhard had persuaded Liebig to sit for only a few months before his death. A space was left in the grave for his widow Henriette, who was to survive her husband by a further eight years, until 1881.

Liebig's many pupils naturally wanted to see a more public memorial to the great chemist; but given the three competing venues – his birthplace, Darmstadt; Giessen, the town he had made famous; and Munich, the city where he had spent twenty-one years of his life – it soon became clear that each would want its own memorial. Munich's large marble statue, also designed by Michael Wagmüller and executed by Wilhelm von Ruemann, was finally unveiled by Adolf von Baeyer in the Maximiliansanlagen in August 1883. It showed Liebig seated and surrounded by allegorical figures.[55] Succeeded temporarily by the faithful Volhard, Baeyer had been called to Liebig's chair in 1875. He was quickly dissatisfied by the facilities Liebig had enjoyed. By demolishing Liebig's house (the former laboratory of the Munich Academy), he was able to erect a large set of teaching and research laboratories for 200 students, which began operations in the spring of 1878. In 1887 a German society for professional chemists who were working in industry was founded (the equivalent of the British Society for Chemical Industry). In 1896 the eminent industrial chemist Carl Druisberg suggested that this society should be renamed the *Verein* (Union) *deutscher Chemiker.* It was this union that struck another Liebig medal in 1903, using the same Brehmer design as the Liebig agricultural medal. It was first awarded to Baeyer for his synthesis of indigo. Later, during World War I, Druisberg, together with Emil Fischer, established a Liebig Scholarship Foundation (Liebig Stipendium-Verien) to help fund deserving research students. The foundation raised over 1 million Marks, only to see their efforts dissipated during post-war inflation.[56]

Meanwhile, the faithful Hofmann had campaigned for funds in Prussia and in Great Britain to erect a large memorial in Giessen by the Berlin

54. Robert Sommer, "The death mask of Justus von Liebig," *J. Chem. Educ.* 11(1934), 503–5. See Volhard, vol. ii, p. 425.
55. *Berichte* 16(1883), 3103–20.
56. Druisberg to Fischer, 9 and 26 February 1917, Fischer papers, Bancroft Library, University of California, Berkeley.

sculptor Schaper. This was unveiled by Hofmann in 1890.[57] After surviving the heavy bombing suffered by Giessen during World War II, the Denkmal was vandalised in May 1945; only Liebig's head survives. Last but not least, a much cheaper and therefore less baroque bronze statue of Liebig by the Darmstadt-born sculptor Bersch was erected in Darmstadt in 1903. Like the one at Giessen, this too was destroyed during the last war. Walhalla, near Regensburg, one of the romantic extravagancies of King Ludwig I in 1840, also commemorated Liebig with a copy of the bust by Elizabeth Ney.

Liebig had been unable to accept the British Chemical Society's invitation to deliver the Faraday Lecture in 1871. It fell to Hofmann to use a similar invitation in 1875 to memorialise his teacher. Parts of this long oration had already formed the substance of Liebig's German obituary, which was read to the Deutsche Chemische Gesellschaft in 1873 and printed in the *Berichte*.[58] Another Anglo-German pupil, Ludwig Thudichum, similarly commemorated Liebig's achievements before the Royal Society of Arts in 1875. William Shenstone, who had studied chemistry and pharmacy at the Pharmaceutical Society before becoming a science teacher at Clifton College, was then able to use both Hofmann's and Thudichum's writings as the basis of a short English biography in 1895. With the centenary of Liebig's birth in May 1903, there were elaborate commemorations in Darmstadt, Giessen, and Munich. It was these that prompted the historian Adolph Kohut , as well as Volhard, to publish their biographies of Liebig in 1906 and 1908, respectively.[59] Having been seen by his contemporaries as part of the trunk (*Stammbaum*) of nineteenth-century chemistry, in death Liebig would now be viewed as one of the great figures (*Grossmacht*) of the newly unified German nation.[60]

Liebig's laboratory in Giessen remained the University's Chemical Institute until the retirement of his successor Heinrich Will in 1888, when a new and larger building was erected on the Ludwigstrasse. The old barracks building then reverted to clinical use until March 1920 when, through the efforts of Robert Somner and with the financial help of Merck, it was converted into the Liebig Museum. Although severely damaged by bombing in 1944, the museum was restored and reopened in 1952. Liebig's birthplace in Darmstadt, which had been similarly transformed into a museum in 1928, was not so fortunate; it was a permanent casualty of the war and of postwar urban renewal. As for Munich, not to

57. *Berichte* 23 (1890), Part 2, 3331–6, and Part 3, 792–816.
58. *Berichte* 6 (1873), 465–73.
59. Commemorative lectures and publications for Liebigian anniversaries of 1898, 1903, 1923, 1928, and 1953 are listed in Paoloni's *Bibliographie* (1968).
60. I thank Pat Munday for this observation.

be outdone, with the help of Liebig's family and descendants, the city became the repository for Liebig's *Nachlass,* to the great benefit of historians of nineteenth-century chemistry.

Liebig was a complex human being, full of contradictions and inner conflicts. At one moment genial, charming, pleasant, and affectionate, in another he was difficult, emotional, easily provoked, and on the lookout for quarrels. Always overworked and overworking, because of his obsessive determination to make chemistry the fundamental science for modern societies, it thrived. Liebig was wilful but never arrogant. When Mohr portrayed Liebig as "a conqueror" in an anonymous essay, Liebig gently rebuked his friend:

> It seems to me that you have used too strong a set of colours. When I look back over my life it seems the effectiveness of one person is trifling, and that one must give way to the next generation; the old are not to be converted. I see this with farmers. First the old must die, at least their spokesmen, and thus it was that I experienced success when I was young.[61]

61. Liebig to Mohr, 1 December 1867, in Kahlbaum, *Liebig und Mohr in ihren Briefen*, p. 238.

Appendix 1
Carl Wilhelm Bergemann's Report to the Prussian Minister on the Chemical Laboratory at Giessen in 1840[1]

The chemical laboratory of the University of Giessen, presently under the direction of Professor Liebig, is to be found in a building located approximately 200 paces from the town boundary, and devoted exclusively to this subject. It has stood in its present enlarged state for about two years. It forms the continuation of another grand-ducal building with which it is now joined by its new extension. This new part is single-storeyed, while the older, projecting part has two storeys, of which the upper part is the official residence of Professor Liebig. The enclosed sketch of the ground plan [page 334] from memory will, perhaps, help to make the short description of the furnishings and fittings more comprehensible.

The entire building is divided in two parts. The first, ABCDE, is one-storeyed and was built only a couple of years ago. It is wholly built in stone, covered by a flat roof in the form of a Dornsches roof [a form of plaster roof that used a single covering extending over the whole surface rather than the individual slates customary until then]. The second, older projecting part of the building DEFG consists, by contrast, of two floors, the lower of which was previously at the disposal of Professor Liebig and his students alone.

The first room forms the Lecture Theatre (1). This stands adjacent to the large laboratory. It is very practically lit by four facing windows, and the door (2) leading in from the garden is used as an entrance by those students who are simply attending lectures. The internal furnishings of the lecture room are very simple and might almost be called sparse. In this auditorium I found some 65 to 70 auditors crowded together on four or five rows of benches (3) arranged in a semicircle.

A large table (4) equipped with built-in compartments and racks is

1. Translated from Regine Zott and Emil Heuser, *Die streitbaren Gelehrten. Justus Liebig und die preussischen Universitäten* (Berlin, 1992), pp. 173–81. I am grateful to Peter Graves, Department of German, University of Leicester, for correcting and improving my initial translation. C. W. Bergemann (1804–84) was professor of pharmacy at the University of Bonn.

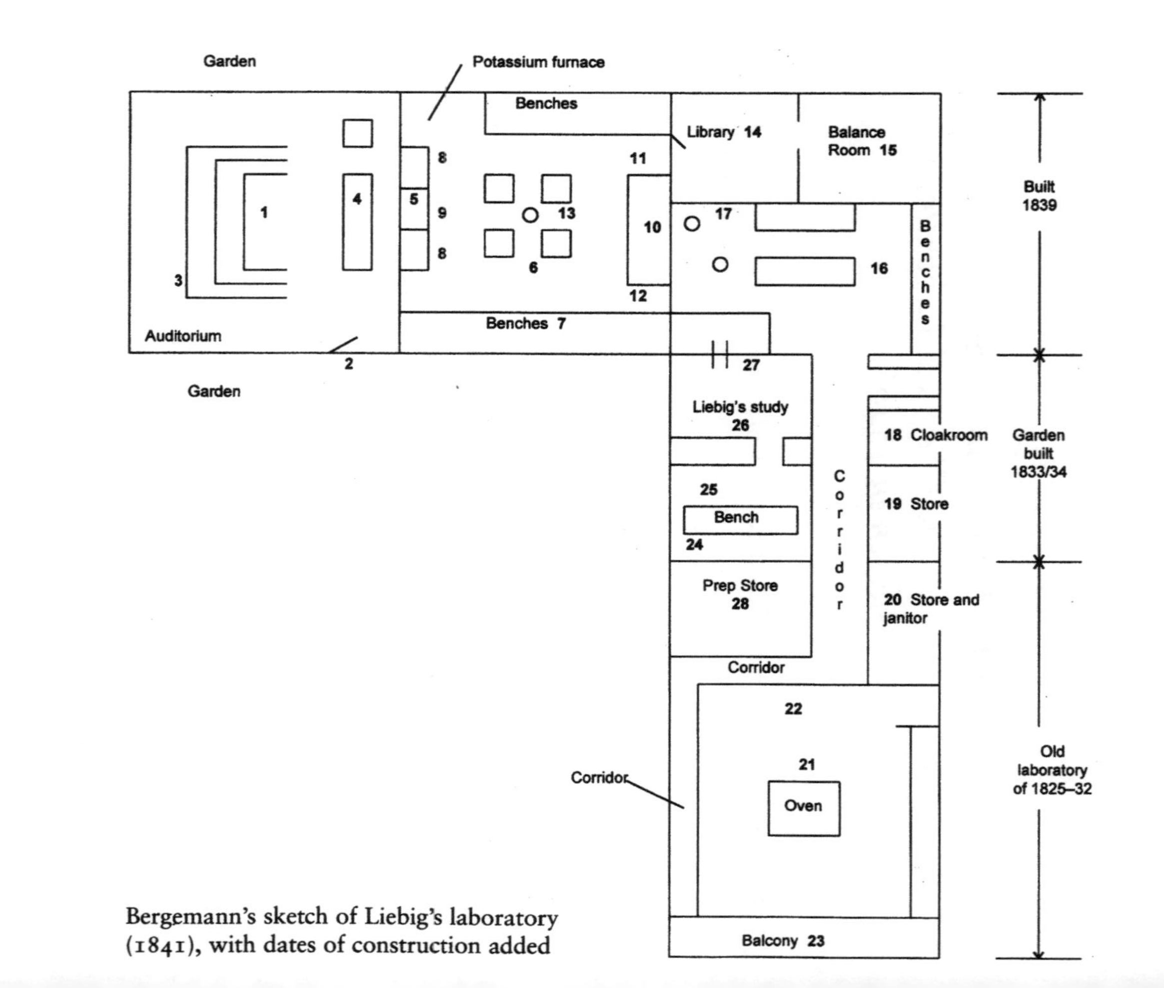

Bergemann's sketch of Liebig's laboratory (1841), with dates of construction added

covered with the reagents, glass apparatus, lamps, and so on, that are needed at the time. This is the place from which Liebig delivers his lectures. The auditorium is in no way special or superior to the chemical theatres of other universities. The central section of the wall (5) facing the seated audience is arranged so that part of it is removable; in this way a connection can be made with the laboratory and specifically with the furnaces erected there. There are no cupboards in this lecture theatre for the storage of reagents, models, and the like.

Next to this room is the large laboratory (6); it too is covered by a flat roof, and like the auditorium, it derives its light from four opposite-facing windows. In addition, against its wall there are workbenches that run almost the whole length of the laboratory (7) and so arranged that at each one, five or six individuals can undertake chemical work. The tables are equipped with compartments and pigeonholes of varying sizes in which the workers store the objects being used in their investigations, and on the same are shelves of reagents which are in such good supply that usually they can be shared in the experiments of every two students. Under these tables are to be found recesses, and in these containers from which water can be drawn and drained through a tube. The tables set up against both walls have a similar arrangement. On the wall dividing the laboratory from the lecture theatre stand six smelting furnaces (8). These are made entirely of iron plates and are so constructed that under each furnace are to be found two containers, one of which stores coal, while the other stores various instruments. Each furnace has its own chimney, which emerges on the flat roof through an earthernware flue. These furnaces are separated by a very practical device. On the part that projects as a smoke hood there are sliding windows that completely shut off the furnaces from the laboratory and that can be operated very easily by means of a simple device. The side walls above the furnaces are likewise closed off with glass windows. This arrangement has the advantage that the most varied kinds of work can be carried out simultaneously, as if in separate rooms; also acids and other materials dangerous to health can be evaporated without harming the investigators in the slightest. The objects in the room are also protected from all risk, and also against dust.[2] The central part of this row of furnaces is taken up by two or three that are not completely closed off (9). These can be used in lectures by means of the removal of the partition that separates the laboratory and lecture theatre at this point. The mercury apparatus is also set up here. Finally, on the same wall I noticed yet another furnace of particular construction, where experiments were being

2. Fume cupboards had been introduced into some earlier laboratories but never to the extent of Giessen.

conducted to prepare large quantities of potassium by means of intense heat.[3] The fourth wall of the laboratory contains a large drying kiln (10). It is so constructed that a large sandbath can be first raised to a high temperature, and from this the warmth can be conducted through several flues and pipes into particular rooms, where it can be regulated for several purposes. The whole iron-plated furnace is similarly provided with sliding windows so that the items to be dried are protected from dust and contamination. In this same wall there are two further doors (11 & 12), one of which leads into another smaller laboratory, the other into a study [library]. In the middle of this [large] laboratory, four smaller work places for the use of students are found with compartments, pigeonholes, drawers, and so on. The laboratory is heated in winter by means of a large, iron brown-coal-burning stove in the centre of the room (13). Its construction is similarly remarkable. It provides a partial underfloor heating by means of the way that the smoke-extraction pipe in the middle of the furnace is linked to a brick conduit through which the smoke passes on its way to the chimney. As I have already said, one passes from this laboratory through the door into a small study, where apart from the necessary writing materials there is a small student library; leading from this room is another in which balances are kept.

The second door of the large laboratory (12) leads into a second smaller laboratory (16), where the less-advanced students work. This is lit by one window, but in addition it gets some light from above. The arrangements and contents of this room are exactly the same as those in the larger laboratory. Work benches with repositories for reagents, glass, funnels, and so on occupy the walls as well as the middle of the laboratory; in addition, there is a drying oven and a smelting furnace with the same arrangement as previously described above. In addition, this area contains an apparatus for distilling water continuously in a tin-still (17), to which a drying oven is likewise attached.

From this laboratory one enters the older, right-angled wing of the building DEFG, whose upper floor constitutes Professor Liebig's living quarters. The rooms here, apart from the old laboratory, are divided by a corridor into two sections. Those to the left are small and are for secondary purposes. The first room here (18) is used as the students' cloakroom, the second (19) serves as a storeroom in which there is also a table used for glassblowing; the third room (20) at the end of the corridor serves both as

3. This was the slightly dangerous project of the laboratory's janitor Aubel, who reduced caustic potash (potassium hydroxide) over red-hot iron filings. Vaporised potassium was then condensed in a neighbouring retort. The process was sufficiently profitable for Aubel to buy several houses in the neighbouring town of Wiesbeck, where he eventually became mayor.

a storeroom and as the janitor's sleeping quarters. Immediately adjoining this is the old, long-established laboratory (21), adjacent to the street and also the main entrance. Here too are work benches of the usual form and construction, and further apparatus for executing such work as needs to be done with large quantities of materials that would be difficult to handle in the other laboratories. There are also cupels and sandbaths for large retorts, flasks (22), and the like; in addition, in the centre of the room there is a large four-sided stove equipped with a smoke hood and likewise constructed of iron plates in which smelting furnaces of various dimensions, cupels, and so on are to be found. The lower section of this furnace is used for storing coals. A glass door leading to the street forms the main entrance to this laboratory. In front of this door there is another area protected from rain and shaped like a balcony whose projecting roof is supported on columns (23). Here large preparations can also be undertaken in which considerable amounts of steam or strong fumes are evolved.[4]

On the eastern, or right-hand, side of the previously mentioned corridor is the main entrance to the laboratory and to the living quarters of Professor Liebig (24). Next to this is a small room (25) used as a laboratory, where Professor Liebig himself can work. This contains apparatus and arrangements of reagents, balances, and so forth, further smelting and drying ovens, as well as writing desks. From this area one reaches Professor Liebig's study (26), from which he can overlook the advanced student laboratory through a small closeable window (27). In front of this lies a further room (28), which is filled with glass cabinets containing stored chemical preparations.

A storeroom for models, ovens, and other apparatus is not connected with the laboratory and similarly there is an absence of many objects, even very familiar ones, that are customarily found in other laboratories. Everything here is directed towards the practical work of students; in this respect the main effort is directed to the arousing and maintaining of their interest in science. All the equipment found in the building is geared to the study of general and analytical chemistry, which also forms the main thrust of the lectures. The application of chemical doctrines to other sciences, arts, and commerce is touched on only briefly and incidentally. No concessions are made to nonchemical students specialising in these other areas.

The Grossherzgl. Hessen government provides 1,500 florins annually from its funds for the chemical laboratory; in addition, the first laboratory assistant receives 300 florins, and the janitor an annual 150 florins. How-

4. The colonnaded terrace of the former guardhouse.

ever, as the most superficial observation shows, these sums are insufficient to defray all the expenses that occur here. From information that Professor Liebig gave to me, it is clear that he himself adds the sum of about 1,500 florins derived from students' fees to cover the maintenance of the laboratory in its current extent of use.

Students who wish to use the laboratory in their study of chemistry pay, for each day of the week they spend in the laboratory, 1 carolin per semester.[5] Thus, the ones who work daily in the laboratory pay 6 carolins, those who work two or three days have to pay 3 carolins. Work starts at 9 o'clock in the morning and continues without a break until 5 o'clock. For their fees, the students are provided with everything that they need for smaller investigations, as well as sufficient reagents and apparatus. All the equipment is handed to them in sufficient amount and in the best condition when they first enter the laboratory, and they return it when they leave. Any student who wants to undertake quantitative analysis must provided himself with an accurate balance, together with the necessary weights, and all these weighings are made in a special room. Experience has shown, however, that beginners often consume disproportionately large quantities of many of the materials provided, so that the laboratory thereby incurs too great a loss. This drawback has been overcome by requiring students to pay for consumables. For example, there used to be a wasteful use of methylated spirits for heating. Accordingly, this is now purchased in bulk by the laboratory technician and distributed to the students at cost price, so that only a third part of the cost is really paid for and the other two thirds are obtainable free.[6] This arrangement has a double advantage insofar as the students become accustomed to an economical use of materials, and a modest subvention is provided to the laboratory fund. However, many investigations require reagents and general resources in such quantities that they cannot be supplied [afforded] by the laboratory. The students obtain such items at their own expense because the consumption of materials for a single piece of research can often exceed the total of the fees. While I was in Giessen, for example, I was aware of a research worker who was obliged to buy a hundredweight of sulphuric ether [ethyl ether] in a succession of researches. Those students who work all day from morning to evening in the laboratory and who are quite skilled have to reckon on finding on average an additional 300 florins annually for such expenses.

It stands to reason that great things can be achieved, given such excel-

5. A carolin, or karlin, was a South German currency worth 3 gulden.
6. This is not clearly expressed. Presumably Bergemann meant that by buying in bulk the laboratory obtained methylated spirits at two-thirds the price.

lent means and equipment. Chemists cannot be trained in the lecture theatre but have to be created in laboratories, where students have the personal opportunity to undertake all manner of research from the simplest to the most complex. This was the guiding principle of Professor Liebig's arrangements, and it is what he practices. Students must first acquire a general knowledge of chemistry and then proceed from the simple to the more complex by concerning themselves only with the manipulation of individual objects that are relevant to their future occupations. For everybody who wishes to derive profit through a special application of science must first acquire a general view of it, for only then will he be able to perceive the coherence of the phenomena that present themselves in nature or in experimental situations, and thereby be able to recognise the relevance of a particular instance that presents itself. He will then be in a position to see the whole scope of the field to which he is devoting his activities and be able to introduce improvements in the general interest or to his own particular one. The degree of precision with which the chemical investigations are carried out in Professor Liebig's laboratory is well known because most of them are published in the *Annalen der Chemie und Pharmacie;* likewise, many distinguished chemists have been trained there. At the present time I found about forty people in the laboratory who were studying practical chemistry under the direction of Professor Liebig. Not all of them were Germans; some came from the most distant parts of Europe. Almost eighteen of them worked daily from morning until evening, and among them were teachers from foreign universities and other people whose names are well known in science. That the study of chemistry and pharmacy is significant for students of medicine is shown by investigations of areas of pharmaceutical, physiological, and pathological chemistry undertaken by them, yet the pharmaceutical knowledge of our medical students [in Prussia] is usually passed over as the teaching of pharmacy is unfortunately confined to lectures. The laboratory work at Giessen is not entirely confined to mineral analysis, but research into the composition of organic compounds is presently the chief topic of interest. An effort is also made to answer questions whose solutions are of a more general interest, but such investigations are much less frequently undertaken here than is the case at other laboratories. With so large a corps of skilled workers, in cases where there is room for argument, several can undertake research independently of each other, and the results be compared.

If it be asked, in what ways are the laboratory facilities at Giessen distinguished from those at other universities, the answer will be simply that Professor Liebig commands large resources which he uses exclusively for the practical instruction of students. In other respects, little happens and the applications of chemical doctrines to arts and commerce are seldom touched upon. The technologist will not find any models here by which the construction of simple machines is made plain, the miner and metallurgist will find nothing that will help him in his education. Yet, the purpose of a university education should be to allow students the opportunity to get to know about the construction of objects that are intimately concerned with these subjects. In the absence of such machinery, how on earth are they to derive such knowledge? To be sure, if part of his task were to provide machines and models, as in other universities, it is true that Professor Liebig would then be forced to cut back on the important practical work. For these things are expensive even though they keep their value for a long time and are of use for as long as the educational institution exists. In Giessen it is said, indeed, that the laboratory's resources are used solely on experimental work.

Quite apart from his well-recognized and undoubted talent for teaching, this policy of Professor Liebig's certainly brings him large numbers of young people to do practical work in science. The economic security of the Giessen laboratory even allows him to accept students from other countries. The arrangement and support of a suitable laboratory is so very expensive that only a few will be in a position to afford such a facility. At Giessen they will find everything they need, and students have only to provide for themselves what they consume in large quantities. Given the sum of 300 florins per annum, as the advanced students pay here on average for their materials, they could never afford such research in their own laboratories; moreover, they have access to a teacher. They derive instruction mutually through scientific conversation and by observing the experience of other individuals. Although each retains personal possession of his research, the publication of the most interesting results is encouraged by Professor Liebig. Even if it is true that students are offered similar benefits at other laboratories, there is no denying that this laboratory (through the generosity of the Grossherzgl. Hessian government and through the efforts of Professor Liebig) has achieved special recognition for its education of chemists, whereas the general education of chemists at our other universities by means of lectures and chemical apparatus is attained by inferior means. It must be acknowledged finally that the uncommon talent of Professor Liebig, by which he is able to awaken in beginners a lively interest in science, has brought about the favourable reputation of the laboratory there and attracted its many visitors. Only through a breadth of vision, such as Professor Liebig possesses, is it possible with the help of just two assistants, always to be able to supervise the varied work of so many individuals simultaneously and to a degree through which teaching and advice can be given to each person immediately.

If this "Report" on the study of chemistry as it is promoted at Giessen demonstrates that the arrangements there contain much that is good and worthy of imitation, nevertheless, by the same token, its flaws (as I have set them out) should not be ignored. Nor can we accept the presumptuous statements of opinion on Prussian administration and teaching institutions that are made in Professor Liebig's publication "On the State of Science and on the Study of Chemistry in Prussia."

Appendix 2
The British and American Network of Liebig's Pupils, Disciples, and Giessen Graduates

Sources: Unless otherwise indicated, information has been abstracted from Conrad (1985), Fruton (1990), van Klooster (1956), Morrell (1972), and Wankmüller (1967).

Allan, James (1825–66), b. Edinburgh; Giessen mat. 1844; PhD 1846 on zinc salts of uric acid. Analytical chemist in Manchester (1849–54) and in Sheffield from 1854. See *Chemical News* 13(1866), 167, and *J. Chem. Soc.* 20 (1867), 386.

Anderson, Thomas (1819–74), b. Leith, Scotland where trained in medicine in Edinburgh 1841; Giessen mat. 1843/4. Regius Prof. Chemistry in Glasgow (1852–74). Opposed Liebig's teachings on agricultural chemistry. See T. B. Anderson, "The forgotten chemist," *Chemistry in Britain* May 1992, 442–44.

Bailey, Henry, b. Wolverhampton; Giessen mat. 1848. Later career not known.

Bastick, William (1818–1903), b. London; Giessen mat. 1842/3. Became pharmaceutical chemist in Buckingham and published in this field. His attacks on Pereira and Bell in the *Chemical Record* (which he edited) caused a crisis in the Pharmaceutical Society 1851–52. See C. P. Cloughly, J. G. L. Burnby, and M. P. Earles, *My Dear Mr Bell: Letters from Dr Jonathan Pereira to Mr Jacob Bell, London, 1844 to 1853* (American Institute for the History of Pharmacy, 1987).

Bernays, Albert James (1823–92), b. Derby; Giessen mat. 1841/2 and in 1853 granted PhD *in absentia* for his book *Household Chemistry* (1852). Private laboratory in Derby (1845–55); lecturer in chemistry at St. Mary's Hospital, London (1855–60) and St Thomas's Hospital (1860–92). See *Dictionary of National Biography.*

Blake, James Elliott (1814–93), b. Liverpool; medical degree 1842; Giessen mat. 1843. Married Julia Sophie Muspratt (1828–90), a daughter of James Muspratt. Emigrated to Australia, where he helped create the Australian wine industry with Liebig's correspondent James King. See D. W. F. Hardie, "The Muspratts and the chemical industry in Britain," *Endeavour* 14(1955), 29–33.

Blyth, John (1814–71), b. Jamaica of Scottish parents. Trained in medicine at Edinburgh 1839; studied with Graham; Giessen mat. 1843. Worked with Hofmann at RCC (1845–47); Royal Agricultural College, Cirencester

(1847–49); professor of chemistry Queen's College Cork (1849–72). Translated Liebig's *Letters on Modern Agriculture* (1859), the fourth edition of *Chemical Letters* (1859), and *Natural Laws of Husbandry* (1863). See *J. Chem. Soc.*, 25 (1871), 343.

Böttinger, Wilhelm Heinrich (1820–1874), b. Heilbronn; Giessen mat. 1843; PhD on analysis of wood 1844. Joined Hofmann at RCC 1845 and then chemist to Bass at Burton on Trent. See R. G. Anderson, *J. Institute of Brewing* 95(1989), 337–45.

Boyd, Christopher M., from Belfast; Giessen mat. 1842/3. Later career unknown.

Boyd, John, from Ireland (Belfast?); Giessen mat. 1841/2. Later career unknown.

Breed, Daniel, b. New York. Not in Giessen matriculation list but definitely there in 1850. PhD for work on analysis of brain. Translated Will's *Outline of Chemical Analysis*, 3rd ed. (Boston, 1855).

Brewer, William Henry (1828–1910), b. Poughkeepsie, N.Y., the son of a farmer. After training at the Sheffield Scientific School, Yale, he toured Germany, spending a year with Bunsen at Heidelberg, and 1856 with Liebig in Munich. Professor of agriculture at Yale from 1864 to 1903. Created Connecticut Agricultural Station with S. W. Johnson. See *American Dictionary of Biography.*

Brodie, Benjamin Collins (1817–80), b. London; Giessen mat. 1844/5 with PhD on beeswax 1850. Private laboratory London (1847–55); Waynefleete Professor of Chemistry Oxford (1855–73). Developed graphite process that Liebig helped to patent in Germany. See W. H. Brock, ed., *The Atomic Debates. Brodie and the Rejection of the Atomic Theory* (Leicester, 1967).

Brown, George, b. Glasgow; Giessen mat. 1852. Later career unknown.

Brown, John, b. Glasgow; Giessen mat. 1848/9. Later career unknown.

Buckland, Francis (Frank) Trevelyan (1826–80), b. Oxford, son of geologist William Buckland; Giessen mat. 1845; hosted Liebig's son, Georg, in London in 1845. Abandoned chemistry for natural history. See G. H. O. Burgess, *The Curious World of Frank Buckland* (London, 1967).

Bull, Buckland, b. Hartford, Connecticut. Giessen mat. 1848; worked on emulsin. Later career unknown.

Bullock, John Lloyd (1812–1905), b. London; Giessen mat. 1839. Druggist, pharmacist, and chemical operator in London. Supported foundation of RCC with Gardner and involved in quinidine affair with Liebig. See G. K. Roberts, *Hist. Stud. Physical Sciences*, 7(1976), 437–85.

Cameron, William, b. Dingwall; Giessen mat. 1838/9. Later career unknown.

Campbell, Robert Corbett (1817–c. 1840), b. Glasgow; student with Graham in Glasgow; Giessen mat. 1838. Worked on ferrocyanides. Later career unknown.

Cane, Lyons, b. Dublin; Giessen mat. 1850. Later career unknown.

Carnegie, John Linde, b. Arbroath; Giessen mat. 1851. Later career unknown.

Clark, J. Edward, an American who was briefly with Liebig at Munich before studying with Wöhler at Göttingen in 1857.

Clunie, Thomas, b. Glasgow; Giessen mat. 1845/6. Later career unknown.

Crighton, James, b. Glasgow; Giessen mat. 1846/7. Later career unknown.

Crum, Alexander (1828–93), b. Thornliebank, Glasgow, Scotland, son of calico manufacturer Walter Crum. Giessen mat. 1844/5. Joined family firm; MP 1880–85. Sister Margaret married William Thomson, Baron Kelvin.

Darby, Stephen (?–1911), b. Cookham. Giessen mat. 1847. One later publication on diastase. Also wrote local history.

Daubeny, Charles B. (1795–1867), b. Stratton, Gloucestershire. Professor of chemistry at Oxford from 1822. Not a pupil but staunch supporter of Liebig's agricultural endeavours. See *Dictionary Scientific Biography.*

Detmar, Moritz, b. Manchester; Giessen mat. 1839/40. Later career unknown.

Dieffenbach, Ernst Johann (1811–55), b. Giessen, where mat. in medicine 1828. Political refugee to Switzerland where was medical doctor in Zurich 1835. Emigrated to London, where involved in expedition to New Zealand. Returned to Giessen 1849 and was extraordinary professor mineralogy 1850–55. No chemical publications, but advised Liebig to publish *Chemical Letters* in 1843. See *Hessische Biographien* (Darmstadt, 1918; reprint Wiesbaden, 1973), vol. 2, pp. 146–50.

Dunlop, Charles, b. Glasgow; Giessen mat. 1842. Worked in bleaching industry in Scotland and held patents for recovery of manganese. See J. R. Partington, *History of Chemistry,* vol. 4 (London, 1964), p. 903.

Eatwell, William (1819–99), b. Byculla, Bombay; Giessen mat. 1837. Student of medicine at Glasgow (MD 1840). Practised medicine in India 1841–57; principal of Calcutta Medical College 1857–61.

Faber, William Leonard, b. New York; Giessen mat. 1850. Became metallurgist and mining engineer in America.

Ferguson, Alex, b. Glasgow; Giessen mat. 1852. Later career unknown.

Fownes, George (1815–49), b. London; Giessen mat. 1838 and PhD on the atomic weight of carbon 1841. Professor of chemistry at University College, London, until his death. Author of *Manual of Elementary Chemistry* (London, 1844), whose later editions were edited by Hofmann. See *Dictionary Scientific Biography.*

Francis, William (1817–1904), b. London, the illegitimate son of printer and science publisher Richard Taylor; Giessen mat. 1841/2; PhD on *cocculus indicus* 1842. Co-founder and editor *Chemical Gazette.* See W. H. Brock and A. J. Meadows, *The Lamp of Learning. Taylor & Francis and the Devlopment of Science Publishing* (London, 1984).

Frankland, Edward (1825–1899), b. Churchtown, Lancashire. Worked with Playfair in London (1845–47), where collaborated with Kolbe. Taught at Queenwood College, Hampshire (1847); mat. Marburg 1847; PhD under Bunsen 1849 and spent autumn of 1849 with Liebig. Professor of chemistry, Owens College, Manchester (1851–57); lecturer in chemistry at St Bartholomew's Hospital, London (1857–64); professor of chemistry, Royal Institution (1863–69); professor of chemistry, Royal College of Mines (1865–85). Formulated concept of valence. See C. A. Russell, *Edward Frankland* (Cambridge, 1996).

Gardner, John (1804–80), b. Great Coggeshall; trained as apothecary (LSA) and obtained honorary PhD in medicine at Giessen 1843. Translated Liebig's

Familiar Letters and other writings. Involved in creation of RCC 1845 but sacked as secretary over quinidine affair. Worked at Apothecaries Hall, but later career unknown. See G. K. Roberts, *Hist. Stud. Phys. Sciences* 7(1976), 437–85.

Genth, Frederick Augustus (1820–93), b. Wächtersbach. After studies at Heidelberg, 1839–41, he matriculated at Giessen in 1841, but took his doctorate at Marburg in 1845. After assisting Bunsen he emigrated to Philadelphia in 1848, where he established a private analytical laboratory and acted as professor of chemistry at the University of Pennsylvania 1872–88. See *Dictionary Scientific Biography.*

Gibbs, Oliver Wolcott (1822–1908). b. New York; Giessen mat. 1846 following studies in Berlin. Not impressed by Giessen. Professor of chemistry at New York's Free Academy 1849–63, and at Harvard 1871–87. A notable inorganic chemist. See *Dictionary Scientific Biography.*

Gilbert, Joseph Henry (1817–1901), b. Hull; trained with Graham at University College; Giessen mat. 1840. Partnered Sir John Bennet Lawes at Rothamsted and opposed Liebig's views on nitrogen and other agricultural issues. See *Dictionary Scientific Biography* under "J. B. Lawes."

Gladstone, John Hall (1827–1902), b. London. Studied chemistry with Graham; Giessen mat. 1847/8; PhD on formation of urea from fulminic acid 1848. Lecturer in chemistry at St. Thomas's Hospital, London (1848–50); Fullerian Professor Chemistry at Royal Institution (1874–77). Worked on chemical kinetics and closely involved in science education. See *Dictionary Scientific Biography.*

Glassford, Karl F.O., b. Glasgow; Giessen mat. 1844. Later career not known.

Gregory, William (1803–58), b. Edinburgh, the last of the "academic Gregories." Trained in medicine Edinburgh 1828. Worked at Giessen in 1835 and 1841. Professor of chemistry at Andersonian Institute Glasgow (1837–38); Aberdeen (1839–44); Edinburgh (1844–58). Translator of Liebig's work and wrote several important chemistry texts. Later involved in mesmerism and spiritualism and lost his reputation. Named a son James Liebig Gregory. See W. H. Brock and S. Stark, "Liebig, Gregory and the British Association," *Ambix* 37(1990), 134–46.

Guest, August, from London; Giessen doctorate 1846. Later career not known.

Halcrow, Benjamin, b. in Lancashire; Giessen mat. 1848. Later career not known.

Harley, Robert (1829–96), b. Haddington, Scotland. Graduated in medicine at Edinburgh in 1850. Further studies in Paris and Würzburg before matriculating at Giessen in 1854, where he lodged with Hermann von Liebig. Professor of practical physiology at University College 1855 until he went blind in 1869. Not a pupil of Liebig's but through his marriage to Emma Muspratt, became close friend of Liebig's and supporter of his medical and nutritional ideas. Liebig became godfather of his daughter Ethel. See memoir, Mrs Alec Tweedie, *George Harley, FRS. The Life of a London Physician* (London, 1899).

Henry, William Charles (1804–92), b. Manchester and member of the Henry dynasty of chemists and manufacturers of magnesia. Medical degree Edinburgh 1827; student of chemistry in Berlin; Liebig's first English pupil in 1836. Published biography of John Dalton, but otherwise acted the country gentleman. See W. V. Farrar, "William Charles Henry," *Ambix* 24(1977), 1–26.

Hodges, John Frederick (1815–99), b. Downpatrick, Ireland. Trained in medicine at Dublin and practised medicine at Newcastle and Downpatrick before Giessen PhD on Peruvian matico 1843. Further work at Giessen 1845. Professor of chemistry Belfast (1845–99). Several books and articles on agricultural chemistry. See *Chemical News* 79(1899), 315. A son, christened George Liebig Hodges, became a consul official in Japan and offered Liebig information on Japan. Letter Hodges to Liebig, 27 October 1871, Staatsbibliothek München.

Hofmann, August Wilhelm (1818–92), b. Giessen where mat. in law 1836 but changed to chemistry; PhD on coal tar 1841. Liebig's chief assistant 1843–45. Docent at Bonn and Extraordinary Professor 1845 before became director of RCC (1845–65). Professor of chemistry Berlin 1865 until death. Liebig's principal London contact. See *Dictionary Scientific Biography.*

Horsford, Eben Norton (1818–93). b. Moscow, New York. Giessen mat. 1844, where worked on nitrogen content of foodstuffs. Professor of chemistry at Harvard 1847–63. Founded Rumford Chemical Company for manufacture of his and Liebig's baking powder and other products. See *Dictionary Scientific Biography.*

James, Hugo, b. Edinburgh; Giessen mat. 1843/4. Later career not known.

Jamieson, Alexander John, b. Arbroath, Scotland. Giessen mat. 1845/6. Later career not known.

Johnson, Carl, b. New York; Giessen mat. 1848 and took PhD in 1851 on the analysis of cheese. Later career unknown.

Johnson, Samuel William (1830–1909). b. Kingsboro, New York. A student of Norton's at Yale, he studied with Erdmann at Leipzig before working in Liebig's private laboratory in Munich in 1852. Translated Liebig's polemic against Lawes, *The Relation of Chemistry to Agriculture* (Albany, 1855). Distinguished American agricultural chemist. See *American Dictionary Biography.*

Jones, Henry Bence (1813–1873), b.Yoxford. Student of Graham's and Fownes's while reading medicine University College London 1839–41; Giessen mat. 1841; PhD on plant proteins 1843. Physician St George's Hospital from 1842 and Secretary Royal Institution 1860–73. Specialised in clinical medicine of urinary diseases and was much influenced by Liebig's *Animal Chemistry.* See N. G. Coley, *Notes & Records Royal Society* 28(1973–4), 31–56.

Kane, Robert John (1809–90), b. Dublin. Professor at Apothecaries Hall in Dublin 1831; BA (Dublin) 1835. Lecturer (then professor) in natural philosophy, Royal Dublin Society 1834–7. Worked with Liebig 1836 on methyl alcohol. President of Queen's College, Cork 1847–90. Knighted 1846. Important work in radical theory, including the identification of what Liebig called "ethyl." Important Irish educationist. See *Dictionary Scientific Biography.*

Kennedy, Edward Shirley, Giessen doctorate 1851. Later career unknown.

Krüger, J., b. London; Giessen mat. 1851. Later career not known.

Knighton, William, from Cheltenham; Giessen doctorate 1851. Later career unknown.

Kyd, John, b. Arbroath, Scotland; Giessen mat. 1848. Later career unknown.

Lindsay, Thomas, b. Glasgow; Giessen mat. 1847. Popular science writer.

Macadam, Stevenson (1829–1901), b. Glasgow. Gained Giessen PhD 1853 for work on iodine in plants. Acted as chemical consultant in Edinburgh and lectured in medical school of Edinburgh University. Active in civic chemistry of water supply.

Macfarlane, William, b. Glasgow; Giessen mat. 1845/6. Later career unknown.

Mackenzie, Kenneth Smith (1832–1900), b. London; Giessen mat. 1850; landed gentleman in Scotland. According to Edmund Muspratt, Mackenzie's private tutor would not allow him to mix with drinking, non-church-going students. See E. K. Muspratt, *My Life* (London, 1917), p. 23.

Maddrell, Robert, b. Isle of Man; Giessen mat. 1845/6. Later career unknown.

Mark, M. Frederick, from London; Giessen doctorate 1851. Later career unknown.

Magnay, C. J., b. London; Giessen mat. 1843/4. Later career unknown.

Matthiessen, Augustus (1831–70), b. London; Giessen mat. 1852; PhD 1853; worked with Bunsen at Heidelberg 1853–57; private laboratory in London 1857–61; lecturer in Chemistry at St. Mary's Hospital 1862–8 and St. Bartholomew's Hospital 1869–70. Worked on alkaloids and electrical conductivity. Committed suicide. See *J. Chem. Soc.*, 24 (1871), 615.

Mertz, Philipp, b. Manchester; Giessen mat. 1843/4. Later career unknown.

Miller, William Allen (1817–70), b. Ipswich; Giessen mat. 1840; studied medicine King's College, London (MD 1842), where he was professor of chemistry 1845 until suicide in 1870. Pioneer of spectroscopy. See *Dictionary Scientific Biography.*

Mitchell, Alexander, b. Glasgow; Giessen mat. 1839. Later career unknown.

Murray, Francis (1826–?), b. Glasgow; matriculated Giessen 1845. Later career unknown.

Muspratt, Edmund Knowles (1833–1923), b. Liverpool; Giessen mat. 1850; studied medicine Munich, where attended Liebig's public lectures. Joined family firm. A close friend of Liebig's. See Muspratt's *My Life* (1917).

Muspratt, Frederick (1825–72), b. Liverpool; Giessen mat. 1843; joined father's alkali works as manager; involved with Sheridan Muspratt in Liebig's manure and quinoidine ventures 1845–47. See *J. Chem. Soc.*, 26 (1873), 780.

Muspratt, James Sheridan (1821–71), b. Dublin; Giessen mat. 1843; PhD on indigo 1844. Worked with Hofmann at RCC 1845–48 before opening Royal College of Chemistry in Liverpool. Compiled *Chemistry, Theoretical, Practical and Analytical* (1854–60). See *J. Chem. Soc.*. 24 (1871), 620.

Muspratt, Richard (1822–85), b. Liverpool; student with Graham in Glasgow 1837–39; studied privately with Liebig at Giessen in 1840. Joined family firm, and managed Flint works. Also involved in local politics. Named his son James Liebig Muspratt (1844–1907) and had him educated at the RCC and UCL. See *Chemical Trade Journal* 40(1907), 469.

Noad, Henry Minchin (1815–77), b. Shawford. Member of Electrical Society; student of Hofmann's at RCC 1845–47. Professor of chemistry St George's Hospital 1847–77. Awarded honorary PhD by Liebig 1851. Published text on electricity. See *J. Chem. Soc.*, 33(1878), 233.

Ogden, Jacob James, from Manchester; Giessen doctorate 1846.

Parkinson, Robert (1831–1913), b. London; Giessen PhD on valeraldehyde 1853. Partner of chemical works in Bradford.

Paul Benjamin Horatio (1827–1917), b. Grettisham; Giessen mat. 1847/8; PhD on alkaloids 1848. Science journalist and editor *Pharmaceutical Journal* 1870–1912. Translated several papers of Liebig's into English. Was critical of Liebig on agricultural and manuring matters. See *Pharmaceutical Journal* 1917.

Penny, Frederick (1816–69), b. Glasgow. Awarded honorary PhD by Liebig 1842 for a work on the salts of nitric acid. Professor of chemistry at Anderson's Institution, Glasgow. Did valuable work on atomic weights and analysis. See *Dictionary Scientific Biography.*

Pinto, E. A., from Calcutta; Giessen mat. 1839/40. Later career unknown.

Playfair, Lyon (1818–98), b. Chunar, Bengal, of Scottish parents; Giessen mat. 1839/40; PhD on myristic acid 1840. Translated *Agricultural Chemistry.* Advised Prince Albert on Great Exhibition 1851. Professor of chemistry, Edinburgh 1858–68. Became MP and scientific statesman. Knighted 1883 and made a peer in 1892. See *Memoirs* by W. Reid (London, 1899).

Porter, John Addison (1822–66). b. Catskill, New York; Giessen mat. 1847 and worked on analysis of dung. Professor of analytical chemistry at Yale 1852–56, professor of organic chemistry 1856–64, and dean of the Sheffield Scientific School at Yale 1861–64.

Price, David Simpson (1823–88), b. Margate; Giessen mat. 1847; Liebig thought he lacked the drive necessary for a research chemist. Became superintendent of a technical museum at Crystal Palace, Sydenham, and published on metallurgical chemistry. See *J. Chem. Soc.*, 55(1889), 294.

Radcliff, William, b. Liverpool; Giessen mat. 1841. Became a doctor and published nothing on chemistry.

Redwood, Theophilus (1806–92), pharmacist; at Hofmann's instigation, Liebig awarded him an honorary PhD in 1852. See Peter H. Thomas, "Medical Men of Glamorgan. No.3," *Glamorgan Historian* 3(1966), 101.

Richardson, Thomas (1816–67), b. Newcastle. The first British student to register formally at Giessen, in 1836. Industrial chemist in Newcastle where manufactured superphosphates. Edited technical dictionary *Chemical Technology,* 2 vols. (1848–51; 2nd ed. 1855) which was loosely based on Friedrich Knapp's *Lehrbuch der chemischen Technologie* (Braunschweig, 1844). He also translated Liebig's *Elements of Chemistry* (1836; 2nd ed. 1839). This book is exceedingly rare. See *Dictionary National Biography.*

Rogers, John Robinson, b. Honiton, Devon; Giessen mat. 1846/7; PhD on composition of human faeces 1848. Courted Agnes Liebig, but his poor prospects as an apothecary in Devon prevented the match.

Ronalds, Edmund (1819–89), b. London; nephew of the electrician Sir Francis Ronalds. Giessen mat. 1842; PhD on oxidation of wax 1842; further continental studies at Jena, Berlin, Heidelberg, Zurich, and Paris before becoming professor of chemistry, Queen's College, Galway (1849–56). Ended career as consultant for Bonnington Chemical Works in Edinburgh. Edited technical dictionary with Richardson. See *J. Chem. Soc.*, 57 (1890), 456.

Rosengarten, Samuel George (1827–1908), b. Philadelphia; Giessen mat. 1847 and worked on oxidation of benzene. Rejoined family chemical business in Philadelphia.

Rowney, Thomas Henry (1817–94), b. London; student with Graham; with Hofmann at RCC 1845–46; awarded PhD at Giessen for thesis on sebacic acid 1852. Succeeded Ronalds as professor of chemistry Queen's University, Galway 1856. Several papers on fats and oils.

Rue, Warren De La (1815–89). Not a pupil. Joined his father's printing house and invented the first envelope-making machine. Founding member of Chemical Society and co-editor with Hofmann of English translation of Liebig and Kopp's *Jahresbericht*. Honorary PhD during Liebig's tenure as dean, 1851. See *Dictionary Scientific Biography* and S. Lorna Houseman, *The House that Thomas Built. The Story of De La Rue* (London, 1968).

Schedel, Henry Edward (1804–56), b. London; Giessen mat. 1847/8. Active author, but not on chemical subjects.

Schunck, Edward (1820–1903), b. Manchester; Giessen mat. 1840; PhD on oxidation of aloe and lichens 1841; further studies in Berlin; in Manchester did pioneering work on the chemistry of dyestuffs. See *Dictionary Scientific Biography.*

Smith, Robert Angus (1817–84), b. Glasgow; pupil of Graham's; Giessen mat. 1840; PhD 1841. Assisted Playfair in Manchester 1842–45; private consultant from 1845 and became country's leading authority on air pollution and administration of alkali acts. A civic chemist. See *Dictionary Scientific Biography.*

Squire, William Stevens, b. London; Giessen mat. 1851/2 on recommendation of Sir James Clark. Also worked with Hofmann before becoming involved in sulphuric acid production. Helped Liebig with mirror patents. See S. Miall, *A History of the British Chemical Industry* (London, 1931).

Smith, John Lawrence (1818–83). b. Charleston, South Carolina. Giessen mat. 1841 after medical studies in South Carolina. Further studies in Paris before becoming professor of chemistry at Louisiana State University 1850–52 and University of Virginia 1852–54 and professor of medical chemistry at Louisville University 1854–66. Named the mineral Liebigit in 1848.

Stein, James. b. Glasgow; Giessen mat. 1848/9. Later career unknown.

Stenhouse, John (1809–80), b. Glasgow; pupil with Graham; Giessen mat. 1839; PhD on hippuric acid 1840. Lecturer St Bartholomew's Hospital, 1851–57; opened own private laboratory 1860. Crippled with polio. See *J. Chem. Soc.*, 39(1881), 185.

Sullivan, William Kirby (1821–93), b. Cork; Giessen mat. 1842/3. Worked at

Museum of Irish Industry 1847–73 and was president of Queen's College, Cork 1873–93. Interested in beet sugar.

Summer, Thomas Jefferson. b. Columbia, South Carolina; mat. Giessen 1846, where analysed cotton seeds. Died prematurely soon after and results published by his countryman, C. M. Wetherill (below).

Thomson, Robert Dundas (1810–64), b. Eccles, Scotland; nephew of Thomas Thomson, professor of chemistry at Glasgow. Trained in medicine at Glasgow (MD, 1831). Made trip to India and began medical practice in London; Giessen mat. 1842. Assisted uncle in Glasgow 1842–52. Lecturer St Thomas's Hospital 1852–56. Very active in sanitary movement. See *J. Chem. Soc.*, 18 (1865), 345.

Thomson, Thomas, Jr. (1817–78), b. Glasgow, son of professor of chemistry at Glasgow; Giessen mat. 1837. Read medicine at Glasgow (MD, 1839); worked under Liebig at Giessen in 1839 on pectic acid in carrots. Emigrated to India as physician; professor of botany Calcutta Medical School 1854–61. Friend of Georg Liebig's. See *Dictionary National Biography.*

Thudichum, Ludwig (1829–1901), b. Büdingen; Giessen mat. in medicine 1847 and doctorate in medicine 1851; attended Liebig's lectures. Private practice of medicine in London 1853 onwards; professor of chemistry St George's Place Medical School 1858–65; director of pathological laboratory at St Thomas's Hospital 1865–71. Several important investigations for Chief Medical Officer John Simon. Pioneer of brain chemistry using spectroscopy. Wrote an interesting memoir of Liebig. See D. L. Drabkin, *Thudichum Chemist of the Brain* (Philadelphia, 1958).

Tilley, Thomas George (?–1849), b. Brentwood; Giessen mat. 1840/1; PhD on oxidation of castor oil 1841. Further Continental studies before becoming professor of chemistry, Queen's College, Birmingham 1845–49. Publications on plant materials. See *J. Chem. Soc.*, 2 (1849), 352.

Tindal, C. G. Worked in meat trade in Queensland and became involved in meat extract business. He implied he had studied with Liebig. See J. T. Critchell and J. Raymond, *A History of the Frozen Meat Trade* (London, 1912; rep. 1969).

Turner, Wilton (1811–55), brother of Edward Turner. Evidence that Wilton worked at Giessen. Taught chemistry at Sydenham College, London. Became partner of Christian Allhusen at Newcastle. Together with Liebig and Gregory, revised the sixth edition of Edward's *Elements of Chemistry* (1842). According to Vogt, Edward Turner also spent some time at Giessen. See J. F. Allen, *Some Founders of Chemical Industry* (London, 1907), p. 234, and J. C. Poggendorff, *Biographisch-literarisches Handwörterbuch.*

Wallace, William (1832–88), b. Edinburgh; Giessen mat. 1849; PhD under Will 1857. Analytical and consulting chemist in Glasgow where also city analyst. A civic chemist. See *J. Chem. Soc.*, 55 (1889), 297.

Wetherill, Charles Mayer (1825–71). b. Philadelphia; following graduation at the University of Pennsylvania, he practised as an analytical chemist before matriculating at Giessen in 1847. PhD thesis on sulphur compounds. After further studies in Paris, he set up private analytical laboratory in Philadelphia before becoming the first chemist in the Department of Agriculture at the

Smithsonian Institution. From 1864 he was professor of chemistry at Lehigh University. See E. F. Smith, *J. Chem. Educ.*, 6(1929), 1076–89.

Wheatstone, Charles (1802–75), b. Gloucester; not a pupil, electrician and inventor; professor of experimental physics at King's College, London. Keen for Liebig to take chair of chemistry there in 1845 following the death of J. F. Daniell. Awarded honorary PhD by Liebig in 1852. See *Dictionary Scientific Biography.*

Whitney, Josiah Dwight (1819–96). b. Northampton, Massachusetts. Giessen mat. 1846. Became chemical geologist on state surveys and professor of geology at Harvard 1875–96. See *Dictionary Scientific Biography.*

Williamson, Alexander William (1824–1904), b. Wandsworth; Giessen mat. 1844; PhD on bleaching salt and ozone 1845. Private laboratory in Paris where became disciple of Comte. Professor of chemistry at University College, London 1819–87. Developed the theory of etherification in 1850. See *Dictionary Scientific Biography.*

Wornum, Conrad, b. London; Giessen mat. 1845/6. Later career unknown.

Wrightson, Francis Trippe, b. Birmingham; Giessen mat. 1844; studied for doctorate at Marburg with Bunsen and Kolbe (1854). Involved in agricultural chemistry in Warwickshire. Is portrayed in Hirst's *Journals* as an ineffective and weak individual. See W. H. Brock and R. M. MacLeod, *Natural Knowledge in Social Context. The Journals of T. A. Hirst* (London, 1980).

Bibliography

Only Liebig's principal works are cited; for full listings of his works, see Paoloni (1968).

Acton, Elizabeth. *Modern Cookery* (London, 1845; 3rd ed. 1855). Revised by Mrs Beeton 1865 and reprinted by Elek Books (London, 1966), edited by Penelope Farmer.

Anderson, T. B. "The forgotten chemist [Thomas Anderson]," *Chemistry in Britain* May 1992, 442–44.

Anon., "Dinner at Glasgow in honour of Professor Liebig," *Lancet,* ii (2 November 1844), 170–77.

Anon., *Aus dem Briefwechsel von Justus von Liebig mit dem Minister Reinhard Frhrn von Dalwigk* (Darmstadt, 1903).

Apple, Rima D. *Mothers and Medicine. A Social History of Infant Feeding 1890–1950* (Madison, Wis., 1987).

Aulie, R. P. "The mineral theory," *Agricultural History* 48(1974), 369–82.

Bancroft-Cooke, Reginald, trans. *The Sonnets of Karl August Georg Max Graf von Platten-Hallermünde* (Boston, 1923).

Bell, Gerda Elizabeth. *Ernest Dieffenbach. Rebel and Humanist* (Palmerston North, New Zealand, 1976).

Benfey, O. Theodor. *From Vital Force to Structural Formulas* (New York, 1964; reprint, Philadelphia, 1993).

Benfey, O. Theodor, ed. *Classics in the Theory of Chemical Combination* (New York, 1963).

Berichte der Justus Liebig-Gesellschaft zu Giessen. Band 1. Symposium, *150 Jahre Agrikultur Chemie* (Giessen, 1990).

Berl, E. "Justus Liebig, May 14 [sic], 1803–April 18, 1873," *J. Chem. Educ.* 15(1938), 553–62.

"The Liebig House and the Kekulé Room at Darmstadt," *J. Chem. Educ.,* 6(1929), 1869–81.

Berl, Ernst, ed. *Briefe von Justus Liebig nach neuen Funden* (Giessen, 1928).

Liebig und die Bittersalz-und Salzsäurefabrik zu Salzhausen (Berlin, 1931).

Billig, Christine. *Das Pharmaziestudium an der Universität Gießen von den Anfängen im 17. Jahrhundert bis 1938* (PhD, Liebig Universität Gießen, 1994).

Blunck, Richard. *Justus von Liebig* (Berlin, 1938).

Bornstein, P., ed. *August Graf von Platen. Der Briefwechsel,* 4 vols. (Munich, 1921; reprint Hildesheim, 1973).

Borscheid, Peter. *Naturwissenschaft, Staat und Industrie in Baden, 1848–1914* (Stuttgart, 1976).

Braun, J., and O. Wallach, eds. *Briefwechsel zwischen J. Berzelius und F. Wöhler*, 2 vols. (Leipzig, 1901).

Brock, W. H. "Found in the Othmer Library V: Justus von Liebig" [on the *Complete Works*], *Chemical Heritage* 11 (Summer 1994), p. 12.

"Liebig's laboratory accounts," *Ambix* 19(1972), 47–58.

"Liebig buys platinum from Janety the younger," *Platinum Metals Review* 17(1973), 102–4.

"Liebigiana: Old and new perspectives," *History of Science* 19(1981), 201–18.

Brock, W. H., ed. *Justus von Liebig und August Wilhelm Hofmann in ihren Briefen (1841–1873)* (Weinheim, 1984). For supplement, see Heuser & Zott (1988).

Brock, W. H., and Susanne Stark. "Liebig, Gregory and the British Association, 1837–1842," *Ambix* 37(1990), 137–47.

Browne, C. A. "The 'Banquet des chimistes,' Paris, April 22, 1867," *J. Chem. Educ.* 15(1938), 253–59.

Büchler, Anne, and Rolf Schumacher, eds. *Die Nachlässe von Martius, Liebig und den Brüdern Schlagintweit in der Bayerische Staatsbibliothek* (Wiesbaden, 1990).

Bumm, Peter. *August Graf von Platen. Eine Biographie* (Paderborn, 1990).

Burgess, G. H. O. *The Curious World of Frank Buckland* (London, 1967).

Caneva, Kenneth L. *Robert Mayer and the Conservation of Energy* (Princeton, N.J., 1993).

Carrière, Justus, ed. *Berzelius und Liebig ihre Briefe von 1831–1845* (Munich, 1892; 2nd ed., 1898; repr. Göttingen, 1991).

Cazden, Robert E. *A Social History of the German Book Trade in America to the Civil War* (Columbia, S.C., 1984).

Coley, N. G. "Henry Bence Jones, MD, FRS (1813–1873)," *Notes and Records Royal Society* 28(1973), 31–56.

Conrad, Willi. *Justus von Liebig und sein Einfluss auf die Entwicklung des Chemiestudiums und des Chemieunterrichts an Hochschules und Schulen* (PhD, Technischen Hochschule Darmstadt, Darmstadt, 1985).

Cottrell, P. L. *Investment Banking in England 1856–1881. A Case Study of International Financial Society*, 2 vols. (New York and London, 1985).

Crellin, John K. *The Development of Chemistry in Britain through Medicine and Pharmacy, 1700–1850* (PhD, University of London, 1969).

Critchell, James T., and Joseph Raymond. *A History of the Frozen Meat Trade* (London, 1912; reprint 1969).

Crosland, M. P. *Gay-Lussac, Scientist and Bourgeois* (London, 1978).

Cross, E. "Chemistry of meat products" in M. D. Curwen, ed., *Chemistry in Commerce* (London, 1936), vol. 1, pp. 103–12, 211–17.

Crossley, J. C. *The Location and Development of the Agricultural and Industrial Enterprises of Liebig's Extract of Meat Company in the River Plate Countries 1865–1932*, 3 vols. (PhD, University of Leicester, 1973).

Crossley, J. C., and R. Greenhill. "The River Plate beef trade" in D. C. M. Platt, ed., *Business Imperialism 1840–1930* (Oxford, 1977).

Davidis, Henriette. *Kraftküche von Liebig's Fleischextrakt* (Braunschweig, 1870).

Davy, H. *Elements of Agricultural* Chemistry (London, 1813).

von Dechend, H., ed. *Justus von Liebig in eigenen Zeugnissen und solchen seiner Zeitgenossen,* 2nd ed. (Weinheim, 1963).

Dictionary of Scientific Biography, ed. C. C. Gillispie, 16 vols. (New York, 1970–80).

Diehl, Wilhelm. "Moriz Carrières Lebenserinnerungen (1817–1847)," *Archiv für hessische Geschichte und Altertumskunde* 10(1914), 133–301.

Drabkin, David L. *Thudichum, Chemist of the Brain* (Philadelphia, 1958).

Driscoll, W. P. *The Beginnings of the Wine Industry in the Hunter Valley* (Newcastle History Monographs No. 5) (The Council of the City of Newcastle, NSW, 1969).

Droßmar, Fred. *Das publizistische Wirken von Justus von Liebig* (PhD, Freie Universität Berlin, 1964).

Drummond, J. C., and A. Wilbraham. *The Englishman's Food,* 2nd ed. (London, 1957; reprint 1991).

Dumas, J. L. "Liebig et son empreinte sur l'agronomie moderne," *Revue d'historie des sciences* 18(1965), 73–108.

Echtermeyer, R., and G. von Liebig, eds. *Briefwechsel zwischen Justus v. Liebig und Theodor Reuning über landwirtschaftliche Fragen aus den Jahren 1854 bis 1873* (Dresden, 1884).

Eyler, John M. *Victorian Social Medicine. The Ideas and Methods of William Farr* (Baltimore, 1979).

Farrar, W. V. "Science and the German university system," in M. P. Crosland, ed., *The Emergence of Science in Modern Europe* (London and Basingstoke, 1975), pp. 161–79.

Farrar, W. V. and K. R., and E. L Scott. "The Henrys of Manchester. Part 6, William Charles Henry," *Ambix* 24(1977), 1–26.

Felschow, Eva-Marie, and E. Heuser, eds. *Universität und Ministerium im Vormärz. Justus Liebigs Briefwechsel mit Justin von Linde* (Giessen, 1992).

Finegold, Harold. "The Liebig–Pasteur controversy," *J. Chem. Educ.* 31(1954), 403–6.

Finlay, Mark R. "Early marketing of the theory of nutrition; the science and culture of Liebig's extract of meat," in H. Kamminga and A. Cunningham, eds., *The Science and Culture of Nutrition, 1840–1940* (Cambridge, 1995).

"The German agricultural experiment stations and the beginnings of American agricultural research," *Agricultural History* 62(1988), 41–50.

"Quackery and cookery. Justus von Liebig's extract of meat and the theory of nutrition in the Victorian age," *Bulletin History of Medicine* 66(1992), 404–18.

"The rehabilitation of an agricultural chemist: Justus von Liebig and the seventh edition," *Ambix* 38(1991), 155–66.

"Science and practice in German agriculture. Justus von Liebig, Hermann von

Liebig, and the agriculture experimental stations," in W. R. Woodward and R. S. Cohen, eds., *World Views and Scientific Discipline Formation* (Dordrecht, 1991), pp. 309–20.

Freeman, Sarah. *Muttons and Oysters. The Victorians and Their Foods* (London, 1989).

Fruton, Joseph S. *Contrasts in Scientific Style. Research Groups in the Chemical and Biochemical Sciences* (Philadelphia, 1990).

"The Liebig research group – a reappraisal," *Proceedings American Philosophical Society,* 132(1988), 1–66.

Molecules and Life (New York, 1972).

Fulbrook, Mary. *A Concise History of Germany* (Cambridge, 1990).

Geison, G. L. "Scientific change, emerging specialities and research schools," *History of Science* 19(1981), 20–40.

Gibson, A., and W. V. Farrar. "Robert Angus Smith, FRS, and 'sanitary science,' *Notes & Records Royal Society* 28(1974), 24–62.

Glas, E. *Chemistry and Physiology in Their Historical Relations* (Delft, 1979).

"The Liebig–Mulder controversy. On the methodology of physiological chemistry," *Janus* 63(1976), 27–46.

"Methodology in the emergence of physiological chemistry," *Studies in the History & Philosophy of Science* 9 (1978), 291–312.

Goddard, N. P. W. *The Royal Agricultural Society of England and Agricultural Progress 1838–1880* (PhD, University of Kent, Canterbury, 1981).

Goddard, Nicholas. *Harvests of Change. The Royal Agricultural Society of England 1838–1988* (London, 1988).

Good, H. G. "On the early history of Liebig's laboratory," *J. Chem. Educ.* 13 (1936), 557–62. [Contains translations from Vogt (1896).]

Gregory, F. *Scientific Materialism in Nineteenth-Century Germany* (Dordrecht, 1977).

Gregory, William. "The cerebral development of Dr Justus Liebig; with remarks," *Phrenological Journal* 18(1845),54–60.

Gundel, H. G. "Liebig als Dekan der philosophischen Fakultät der Universität Giessen 1846 und 1851," *Giessener Universitätsblätter* 6(1973), 58–80.

Güssefeld, O. E., ed. *Justus von Liebig und Emil Louis Güssefeld Briefwechsel 1862–1866* (Leipzig, 1907).

Gustin, Bernard Henry. *The Emergence of the German Chemical Profession 1790–1867* (PhD, University of Chicago, 1975).

Habacher, Marie. "Auf der Suche nach dem 'Od.' Karl Ludwig Freiherr von Reichenbach und Karl Wilhelm Mayrhof, zwei Verbündte contra Justus von Liebig," *Clio Medica* 14 (1980), 105–18.

"Der Plan zur Berufung Justus von Liebigs nach Wien 1840/41," *Österreichische Akademie der Wissenschaften. Philisophische-Historische Klasse Sitzungsberichte* 242, Band 3, 1964, pp. 1–41.

Hall, A. D. *The Book of Rothamsted Experiments* (London, 1905).

Hall, Vance. *A History of the Yorkshire Agriculture Society 1837–1987* (London, 1987).

Hall, V. M. D. "The contribution of the physiologist William Benjamin Carpenter

(1813–1885) to the development of the principles of the correlation of forces and the conservation of energy," *Medical History* 23(1979), 129–55.

"Justus Liebig and agricultural chemistry in Britain, c. 1840–1880," unpublished paper given to the British Society for the History of Science, 24 May 1980.

"The role of force or power in Liebig's physiological chemistry," *Medical History* 24(1980), 20–59.

Hamlin, Christopher. *What Becomes of Pollution: Adversary Science and the Controversy on the Self-Purification of Rivers in Britain 1850–1900* (PhD, University of Wisconsin, 1982). Published in facsimile, New York and London, 1987.

Hardie, D. W. F. "The Muspratts and the chemical industry in Britain," *Endeavour* 14 (1955), 29–33.

Harrison, Tony. *Square Rounds* (London, 1992).

Hayford, Margaret Trautschold. *C. F. Wilhelm Trautschold 1815–1877. A Preliminary Catalogue Illustrated,* priv. print. Ringwood, N.J., 1980 (copy Wellcome Institute History of Medicine).

Heilbron, J. L. "The affair of the Countess Görlitz," *Proceedings American Philosophical Society* 138 (1994), 284–316.

Heilenz, Siegried. *Eine Führung durch das Leibig-Museum in Gießen* (Verlag Leibig-Gesellschaft, Gießen, 1994).

Hein, Wolfgang-Hagen. *Die Deutsche Apotheker* (Stuttgart, 1960).

Hessische Biographien, 3 vols. (Darmstadt, 1918–34; reprint Wiesbaden, 1973).

Heuser, E., ed. *Justus von Liebig und Emil Erlenmeyer in ihren Briefen von 1861–1872* (Mannheim, 1988).

Justus von Liebig und der Pharmazeut Friedrich Julius Otto in ihren Briefen von 1838–1840 und 1856–1867 (Mannheim, 1989).

Heuser, E., and R. Zott, eds. *Justus von Liebig und August Wilhelm Hofmann in ihren Briefen. Nachträge 1845–1869* (Mannheim, 1988), supplements Brock (1984) edition. Bound with Heuser (1988).

Heuss, Theodor. *Justus von Liebig vom Genius der Forschung* (Hamburg, 1942).

Hofmann, A. W. *The Lifework of Liebig in Experimental and Philosophic Chemistry* (London, 1875), offprinted from *Journal Chemical Society,* 28(1875), 1065–1140, and reprinted in Chemical Society's *Faraday Lectures 1869–1928* (London, 1928).

Hofmann, A. W., ed. *Aus Justus Liebig's und Friedrich Wöhler's Briefwechsel in den Jahren 1829–1873,* 2 vols. (Braunschweig, 1888; repr. one vol., Göttingen, 1982).

Holmes, F. L. "Introduction" to J. Liebig, *Animal Chemistry* (facsimile ed. New York, 1964).

"Liebig," in *Dictionary of Scientific Biography,* vol. 8 (New York, 1973), pp. 329–50.

Claude Bernard and Animal Chemistry (Cambridge, Mass., 1974).

"The complementarity of teaching and research in Liebig's laboratory," *Osiris* 5 (1989), 121–64.

"Justus Liebig and the Construction of Organic Chemistry," in Seymour H.

Mauskopf, ed., *Chemical Sciences in the Modern World* (Philadelphia, 1993), pp. 119–34.
Homburg, Ernst. *Van beroep 'Chemiker.' De opkomst van de industriele chemicus en het polytechnische onderwijs in Deutschland (1790–1850)* (Delft, 1993). Has English summary.
Hubensteiner, Benno. *Bayerische Geschichte* (München, 1981).
Hudson, Derek, and Kenneth W. Luckhurst. *The Royal Society of Arts 1754–1954* (London, 1954).
Hufbauer, Karl G. *The Formation of the German Chemical Community (1700–1795)* (PhD, University of California, Berkeley, 1970).
Jones, Paul R. *Bibliographie der Dissertationen amerikanischer und britischer Chemiker an deutschen Universitäten 1840–1914* (Deutsches Museum, München, 1983).
"Justus von Liebig, Eben Horsford and the development of the baking powder industry," *Ambix* 40(1993), 65–74.
Kahlbaum, George W. A., ed. *Justus von Liebig und Friedrich Mohr in ihren Briefen von 1834–1870* (Leipzig, 1904; reprint Leipzig, 1970).
Kahlbaum, George W. A., and E. Thon, eds. *Justus von Liebig und Christian Friedrich Schönbein Briefwechsel 1833–1868* (Leipzig, 1900).
Kargon, Robert H. *Science in Victorian Manchester. Enterprise and Expertise* (Manchester, 1977).
Keen, Roger. *The Life and Work of Friedrich Wöhler (1800–1882)* (PhD, University of London, 1976).
Kettle, R. M. *Memoirs and Letters of Charles Boner,* 2 vols. (London, 1871). Contains three Liebig letters.
King, James. "On the growth of wine in New South Wales," *Journal Royal Society of Arts* 4(1855–56), 575–78.
Kleinert, Andreas, ed. *Justus von Liebig "Hochwohlgebornen Freyherr," Die Briefe an Georg von Cotta und die anonymen Beiträge zur Ausburger Allgemeinen Zeitung* (Mannheim, 1979).
Klickstein, Herbert S. "Charles Caldwell and the controversy over Liebig's *Animal Chemistry,*" *Chymia* 4 (1953), 129–57.
van Klooster, H. S. "Liebig and his American pupils," *J. Chem. Educ.*, 33 (1956), 493–97.
"The story of Liebig's Annalen der Chemie." *J. Chem. Educ.*, 34 (1957), 27–39.
von Kobell, Luise. *Unter den vier ersten Königen Bayers nach Briefen und eigenen Erinnerungen,* 2 vols. (München, 1894).
Kohut, Adolph. *Justus von Liebig. Sein Leben un Wirken* (Giessen, 1904).
Krätz, Otto P. "Historische Experimente (1850). Justus von Liebig, Bereitung von moussirendem Maiwein," *Chemie Experiment und Didaktik* 1 (1975), 247–50.
"Historische Experimente (um 1856). Justus von Liebig, Experimental-Vorlesungen über Selen, Phosphor, Kohlenstoff und Kohlenwasserstoffe," *Chemie Experiment und Didaktik* 2 (1976), 71–76.
"Historische Experimente (1858). Justus von Liebig und Dr Beeg, Die Farbe der

Spiegel und der teint der französinversuche zur Versilberung von Glas," *Chemie, Experiment und Didaktik* 2 (1976), 401–6.

Krätz, Otto P., and Claus Priessner. *Liebigs Experimentalvorlesungen* (Weinheim,1983).

Krohn, Wolfgang, and Wolf Schäfer. "The origins and structure of agricultural chemistry," in Gerard Lemaine et al., eds., *Perspectives on the Emergence of Scientific Disciplines* (The Hague, 1976), pp. 27–52; repr. (with variations) in W. Schäfer, ed., *Finalization in Science* (Dordrecht, 1982), pp. 17–52.

Kutzer, Michael, and E. Heuser. "Gehirn und Wissenschaft: Theodor Ludwig Wilhelm von Bischoffs Sektionsbefund am Gehirn von Justus von Liebig," *Medizinhistorisches Journal* 23 (1988), 325–41.

von Laubmann, G., and L. von Schleffler, eds. *Die Tagebücher der Grafen August von Platen,* 2 vols. (Stuttgart, 1896; reprint Hildesheim, 1969).

Lawes, J. B., and J. H. Gilbert. "Reply to Baron Liebig's *Principles of Agricultural Chemistry,*" *J. Roy. Agricultural Soc.* 16 (1855), 411–98.

Liebig, J. "Über einen neuen Apparat zur Analyse organischen Körper," *Pogg. Ann. Physik* 21 (1831), 1–43.

Anleitung zur Analyse organischer Körper (Braunschweig, 1837).

Instructions for the Chemical Analysis of Organic Bodies (Glasgow, 1839). Trans. W. Gregory.

Die organische Chemie in ihrer Anwendung auf Agricultur und Physiologie (Braunschweig, 1840).

Organic Chemistry in its Applications to Agriculture and Physiology (London, 1840). Trans. L. Playfair.

Die organische Chemie in ihrer Anwendung auf Physiologie und Pathologie (Braunschweig, 1842).

Animal Chemistry or Organic Chemistry in Its Applications to Physiology and Pathology (London, 1842). Trans. W. Gregory.

Familiar Letters on Chemistry (London, 1843).

Chemische Briefe (Heidelberg, 1844).

Chemistry and Physics in Relation to Physiology and Pathology (London, 1846).

Chemische Untersuchungen über das Fleisch und seine Zubereitung zum Nahrungsmittel (Heidelberg, 1847).

Research on the Chemistry of Food (London, 1847). Trans. W. Gregory.

Untersuchungen über einige Ursachen der Säftebewegung im thierischen Organismus (Braunschweig, 1848).

Researches on the Motion of the Juices in the Animal Body (London, 1848). Trans. W. Gregory.

Die Grundsätze der Agricultur-Chemie (Braunschweig, 1855).

Principles of Agricultural Chemistry, with especial reference to the late researches made in England (London, 1855). Trans. W. Gregory.

Complete Works on Chemistry (Philadelphia, 1856).

Naturwissenschaftliche Briefe über die moderne Landwirtschaft (Leipzig and Heidelberg, 1859).

Letters on Modern Agriculture (London, 1859). Trans. J. Blyth.

Die Chemie in ihrer Anwendung auf Agricultur und Physiologie, 2 vols (Braunschweig, 1862). Vol. 1, *Der chemische Process der Ernährung der Vegetabilien;* Vol. 2, *Die Naturgesetze des Feldbaues.*

The Natural Laws of Husbandry (London, 1863). Trans. J. Blyth of Liebig (1862), vol. 2.

"Lord Bacon as natural philosopher," *Macmillan's Magazine* 8 (1863), 237–49, 257–67.

Letters on the Subject of the Utilisation of the Metropolitan Sewage (London, 1865).

"Was Lord Bacon an Impostor?" *Fraser's Magazine* 76 (April 1867), 482–95.

Reden und Abhandlungen, ed. by G. von Liebig and M. Carrière (Leipzig and Heidelberg, 1874); reprint Wiesbaden, 1965).

"Autobiography," in *Ann. Rep. Smithsonian Institution* (1891), 257–68; translated by J. Campbell Brown from *Berichte Chem. Ges.* 23 (1890), 785–816 and reprinted from *Chemical News* 63 (1891), 265–7, 276–8.

Lipman, Timothy, O. "The response to Liebig's vitalism," *Bulletin of the History of Medicine* 40 (1966), 511–24.

"Vitalism and reductionism in Liebig's physiological thought," *Isis* 58 (1967), 167–85.

Löw, R. *Pflanzenchemie zwischen Lavoisier und* Liebig (München, 1977).

McCosh, F. W. J. *Boussingault* (Dordrecht, 1984).

McGee, Harold. *The Curious Cook* (New York, 1990).

Maddison, R. E. W. "An unpublished letter of Justus Liebig [to John Stenhouse]," *Annals of Science* 45 (1988), 191–92.

Mayne, Xavier. [*pseud.* Edward Irenaeus Prime Stevenson]. "The life and diary of an Uranian poet: August von Platen (1796–1835)" in his *The Intersexes. A History of Similisexualism as a Problem in Social Life* (priv. print., 1908; reprint Arno Press, New York, 1975), chapter 13, separately paginated at end of volume.

Mepham, T. B. "Humanising milk: The formulation of artificial feeds for infants (1850–1910)," *Medical History* 37 (1993), 225–49.

Moraw, Peter. *Kleine Geschichte der Universität Giessen* (Giessen, 1982).

Morrell, J. B. "The chemist breeders. The research schools of Liebig and Thomas Thomson," *Ambix* 19 (1972), 1–46.

Moulton, F. R., ed. *Liebig and after Liebig. A Century of Progress in Agricultural Chemistry* (Washington, D.C., 1942).

Munday, E. Patrick III. *Sturm und Dung: Justus von Liebig and the Chemistry of Agriculture* (PhD, Cornell University, 1990).

Munday, Pat. "Social climbing through chemistry: Justus Liebig's rise from the *niederer Mittelstand* to the *Bildungsbürgertum,*" *Ambix* 37 (1990), 1–19.

"Sturm und Dung. Liebig's metamorphosis from organic chemistry to the chemistry of agriculture," *Ambix* 38 (1991), 135–54.

Muspratt, Edmund K. *My Life and Work* (London, 1917).

Oesper, Ralph E. "Justus von Liebig – Student and Teacher," *J. Chem. Educ.* 4 (1927), 1461–76.

Orend, Friedrich. "Henriette Davidis und Liebig" in *Beruf der Jungfrau. Henriette Davidis und Burgerliches Frauen verstandnis im 19. Jahrhundert* (Oberhausen, 1990).

Ostwald, W. "Liebig," in R. Zott, ed., *Wilhelm Ostwald zur Geschichte der Wissenschaft* (Leipzig, 1985).

Paoloni, Carlo. *Justus von Liebig. Die chemischen und naturwissenschaftlichen Abhandlungen (1822–1873). Bibliographisch nach dem Sachinhalt geordnet* (unpublished holograph, Milano, 1995).

Justus von Liebig. Eine Bibliographie sämtlicher Veröffentlichungen (Heidelberg, 1968).

Partington, J. R. *A History of Chemistry,* vol. 4 (London, 1964).

Paul, Benjamin H. "Manure," in H. Watts, ed., *A Dictionary of Chemistry,* vol. 3 (London, 1882).

Pelling, Margaret. *Cholera, Fever and English Medicine 1825–1865* (Oxford, 1978).

Perren, Richard. *The Meat Trade in Britain 1840–1914* (London, 1978).

von Pettenkofer, Max. "Liebig's scientific achievements," *Contemporary Review* 29 (April 1877), 865–87.

Phillips, J. P. "Liebig and Kolbe, critical editors," *Chymia* 11 (1966), 89–97.

Praetorius, Otfried. "Liebigs Nachkommen," *Familie und Volk,* offprinted c. 1954 (Liebig Museum).

Priessner, C. "Liebig," in *Neue Deutsche Biographie,* Band 4 (Berlin, 1984), 497–501.

Reid, T. Wemyss. *Memoirs and Correspondence of Lyon Playfair* (London, 1899; reprint Jemimaville, Scotland, 1976).

Rezneck, Samuel. "The European education of an American chemist and its influence in nineteenth-century America: Eben Norton Horsford," *Technology & Culture* 11(1970), 366–86.

Roberts, Gerrylynn K. "The establishment of the Royal College of Chemistry: an investigation of the social context," *Historical Studies Physical Sciences* 7 (1976), 437–85.

The Royal College of Chemistry (1845–1853): A Social History of Chemistry in Early Victorian Britain (PhD, Johns Hopkins University, 1973).

Rocke, Alan J. *Chemical Atomism in the Nineteenth Century. From Dalton to Cannizzaro* (Columbus, Ohio, 1984).

The Quiet Revolution. Hermann Kolbe and the Science of Organic Chemistry (Berkeley, 1993).

Rocke, Alan J., and E. Heuser, eds. *Justus von Liebig und Hermann Kolbe in ihren Briefen, 1846–1873* (Mannheim, 1994).

Rossiter, Margaret. *The Emergence of Agricultural Science. Justus Liebig and the Americans, 1840–1880* (New Haven, 1975).

Justus von Liebig and the Americans. A Study in the Transit of Science, 1840–1880 (PhD, University of Yale, 1971).

Russell, E. John. *A History of Agricultural Science in Great Britain 1620–1954* (London, 1966), Chapters 3–4.

"Rothamsted and its experimental station," *Agricultural History* 16 (1942), 161–83.
Sachtleben, R. "Nobel prize winners descended from Liebig," *J. Chem. Educ.* 35 (1958), 73–5.
Sauer, B., and H. Haupt, eds. *Ludoviciana Festzeitung zu dritten Jahrhundertfeier den Universitäten Giessen* (Giessen, 1907), No. 3. [On the Giessen laboratory, probably by Volhard.]
Scharrer, K. "Justus von Liebig and today's agricultural chemistry," *J. Chem. Educ.* 26 (1949), 515–18.
Schierz, Ernest R. "Justus von Liebig: The sponsor of a prize contest in chemistry," *J. Chem. Educ.* 6 (1929), 973–6.
"Liebig's student days," *J. Chem. Educ.* 8 (1931), 223–31.
Schling-Brodersen, Ursula. *Entwicklung und Institutionalisierung der Agrikulturchemie im 19.Jahrhundert: Liebig und die Landwirtschaftlichen Versuchsstationen* (Braunschweig, 1989).
"Liebig's role in the establishment of agricultural chemistry," *Ambix* 39 (1992), 21–31.
Schlödler, Friedrich. "Das chemische Laboratorium unserer Zeit" [on Giessen], *Westermann's Jahrbuch der Illustrierten Deutschen Monatshefte* 38 (April–September 1875), 21–47.
Schneider, M. W. "Das *Handwörterbuch* in Liebigs Biographie," in W. Dressendörfer and W. D. Müller-Jahncke, eds., *Orbis Pictus. Kultur- und pharmaziehistorische Studien* (Frankfurt, 1985), pp. 247–54.
Schneider, Margarete and Wolfgang, eds. *Justus von Liebig Briefe an Vieweg* (Braunschweig, 1986).
Schütt, Hans-Werner. *Eilhard Mitscherlich. Baumeister am Fundament der Chemie* (München, 1991).
Scott, John M. "Karl Friedrich Mohr, 1806–1879. Father of volumetric analysis," *Chymia* 3 (1950), 191–203.
Servos, John W. "Research schools and their histories," in G. L. Geison and F. L. Holmes, eds., *Research Schools Historical Reappraisals,* a special issue of *Osiris* 8 (1993), 3–15.
Shenstone, W. A. *Justus von Liebig. His Life and Work (1803–73)* (London, 1901).
Smith, E. F. "Charles Mayer Wetherill, 1825–1871," *J. Chem. Educ.,* 6 (1929), 1076–89.
Smith, J. Graham. "Fréderic Kuhlmann, pioneer of platinum as an industrial catalyst," *Platinum Metals Review* 32 (1988), 84–90.
Snelders, H. A. M. "The Mulder–Liebig controversy elucidated by their correspondence," *Janus* 69 (1982), 199–221.
Snelders, H. A. M., ed. *The Letters from Gerrit Jan Mulder to Justus Liebig (1838–1846)* (Amsterdam, 1986).
Sommer, Robert. "The Liebig Laboratory and Liebig Museum in Giessen," *J. Chem. Educ.* 8 (1931), 211–22.
Sonntag, Otto. "Liebig on Francis Bacon and the utility of science," *Annals of Science* 31 (1974), 373–86.

"Religion and science in the thought of Liebig," *Ambix,* 24 (1977), 159–69.

Spalt, Georg. *Das Geschlecht Liebig* (priv. print., Reinheim, 1974).

Steil, Hans. "Katalog des Archivbestandes des Liebig-Museums in Giessen," *Giessener Universitätsblatter* 6 (April 1873), 90–108.

Stephens, M. D., and G. W. Roderick. The Muspratts of Liverpool," *Annals of Science* 29 (1972), 287–311.

Sterling, John. "The state of society in England (1828)," in *Essays and Tales,* 2 vols. (London, 1848), vol. 2,

Strube, Irene. *Justus von Liebig* (Leipzig, 1973).

Szökefalvi-Nagy, Z. "Justus von Liebigs Einfluss auf der Chemie Ungarns," *NTM: Schriftenreihe für Geschichte der Naturwissenschaften* 16 (1979), 118–20.

Teuteberg, Hans-Jürgen. *Die Rolle des Fleischextrakts für die Ernährungswissenschaften und den Aufstieg der Suppenindustrie* (Stuttgart, 1990).

Thomas, Ulrike. "Philipp Lorenz Geiger and Justus Liebig," *Ambix* 35 (1988), 77–90.

Thudichum, J. L. W. "On the discoveries and philosophy of Liebig, with especial reference to their influence upon the advancement of arts, manufactures and commerce," Cantor Lectures, *J. Soc. Arts* 24 (1875–76), 80–6, 95–100, 111–16, 125–28, 141–45; also separately printed as *The Discoveries and Philosophy of Liebig* (London, 1876).

Tower, Donald B. *Hensing, 1719. An Account of the First Chemical Examination of the Brain and the Discovery of Phosphorus Therein* (New York, 1983).

Turner, R. S. "Justus Liebig versus Prussian chemistry. Reflections on early institute building in Germany," *Hist. Stud. Physical Sciences* 13 (1982), 129–62.

Tutzke, Dietrich. "Liebigs Beitrag zur Spiegel Industrie und der Kampf der Arzte gegen Hydrargyrose in den Speigelfabriken," *NTM. Schriftenreihe für Geschichte der Naturwissenschaften* 14 (1977), 92–8.

Tweedie, Mrs A. *George Harley, FRS, The Life and Work of a London Physician* (London, 1899).

Vogt, Carl. *Aus meinem Leben: Errinnerungen und Rückblick* (Stuttgart. 1896). [See also Good (1936).]

Volhard, Jacob. *Justus von Liebig,* 2 vols. (Leipzig, 1909).

Walker, Frederic. "Early history of acetaldehyde and formaldehyde," *J. Chem. Educ.* 10 (1933), 546–51.

Wankmüller, A. "Ausländische Studierende der Pharmacie und Chemie bei Liebig in Giessen," *Deutsche Apotheker Zeitung* 107 (1967), 463–6; offprinted as *Tübinger Apothekengeschichtliche Abhandlungen,* Heft 15 (Stuttgart, 1967).

"Die Verlagsverträge von Liebig und Geiger aus den Jahren 1832 und 1833 über die *Annalen der Pharmazie,*" *Giessener Universitätsblatter* 6 (1973), 81–9.

Wilmot, Sarah. *"The Business of Improvement": Agriculture and Scientific Culture in Britain c. 1770–c.1870* (Reading, 1990).

Winderlich, R. "Justus Liebig" in G. Bugge, ed., *Das Buch der Grossen Chemiker,* 2 vols. (Berlin, 1930), pp. 1–30.

Yeo, Richard. "An idol in the market place. Baconianism in nineteenth-century Britain," *History of Science* 23 (1985), 251–98.

Young, Mrs H. M. *Liebig Company's Practical Cookery Book. A Collection of New and Useful Recipes in Every Branch of Cookery* (London, 1893).

Zott, Regina. "The development of science and scientific communication: Justus Liebig's two famous publications of 1840," *Ambix* 40 (1993), 1–10.

Zott. Regina, and E. Heuser. *Die streitbaren Gelehrten. Justus Liebig und die preußischen Universitäten* (Berlin, 1992).

Index

Liebig's pupils are listed in Appendix 2 and are not indexed here unless they are also mentioned in the text. For L. *read* Liebig.

www.ingramcontent.com/pod-product-compliance
Ingram Content Group UK Ltd.
Pitfield, Milton Keynes, MK11 3LW, UK
UKHW040703180125
453697UK00010B/354